# INFINITE DIMENSIONAL LINEAR CONTROL SYSTEMS
## THE TIME OPTIMAL AND NORM OPTIMAL PROBLEMS

NORTH-HOLLAND MATHEMATICS STUDIES 201

(Continuation of the Notas de Matemática)

Editor: Jan van Mill
*Faculteit der Exacte Wetenschappen*
*Amsterdam, The Netherlands*

ELSEVIER

2005

Amsterdam – Boston – Heidelberg – London – New York – Oxford
Paris – San Diego – San Francisco – Singapore – Sydney – Tokyo

# INFINITE DIMENSIONAL LINEAR CONTROL SYSTEMS
## The Time Optimal and Norm Optimal Problems

H.O. FATTORINI
*University of California*
*Department of Mathematics*
*Los Angeles, California 90095*
USA

ELSEVIER

2005

Amsterdam – Boston – Heidelberg – London – New York – Oxford
Paris – San Diego – San Francisco – Singapore – Sydney – Tokyo

**ELSEVIER B.V.**
**Radarweg 29**
**P.O. Box 211, 1000 AE Amsterdam**
**The Netherlands**

ELSEVIER Inc.
525 B Street, Suite 1900
San Diego, CA 92101-4495
USA

ELSEVIER Ltd
The Boulevard, Langford Lane
Kidlington, Oxford OX5 1GB
UK

ELSEVIER Ltd
84 Theobalds Road
London WC1X 8RR
UK



First edition 2005

Library of Congress Cataloging in Publication Data
A catalog record is available from the Library of Congress.

British Library Cataloguing in Publication Data
A catalogue record is available from the British Library.

ISBN: 0-444-51632-8
ISSN: 0304-0208

♾ The paper used in this publication meets the requirements of ANSI/NISO Z39.48-1992 (Permanence of Paper).
Printed in The Netherlands.

To
Natalia
María Elena
Sonia, José María, Cristina and Carlos
Susana and David

# PREFACE

One of the first infinite dimensional control systems to come under scrutiny was

$$y'(t) = Ay(t) + u(t)\,, \quad y(0) = \zeta \tag{1}$$

where $y(t)$ takes values in a Banach space $E$ and $A$ is the infinitesimal generator of a strongly continuous semigroup $S(t)$. Research on the time optimal problem started in the early sixties and branched into other optimal control problems. However, many basic questions were unsolved at the end of the last century. Since then, some new results have clarified the panorama but the subject is still in need of proofs of, or counterexamples to many natural conjectures and it remains a live research area with many actual and potential applications.

In spite of being, in a sense, the simplest infinite dimensional control system, the equation (1) models some important control processes such as those described by the parabolic equation

$$\frac{\partial y(t,x)}{\partial t} = Ay(t,x) + u(t,x) \tag{2}$$

where $A$ is an elliptic operator in the space variables $x = (x_1, x_2, \ldots, x_m)$ in a domain $\Omega$ of $m$-dimensional Euclidean space $\mathbb{R}^m$; the domain of $A$ is restricted by boundary conditions. The *control* $u(t,x)$ satisfies bounds of the type

$$\int_\Omega |u(t,x)|^p dx \le C \tag{3}$$

for some $p$, $1 \le p < \infty$, or

$$|u(t,x)| \le C\,. \tag{4}$$

These bounds determine the *state space* $E$ in which (2) is modeled. For the bound (3) the space is $E = L^p(\Omega)$. For the uniform bound (4) we take $E = L^\infty(\Omega)$, (or, rather $E = C(\overline{\Omega})$). The most physically significant cases are (4) for heat processes and (3) with $p = 1$ for diffusions. On the other hand, $p = 2$ leads to the simplest mathematics since the state space $L^2(\Omega)$ is a Hilbert space.

We consider two optimal control problems for the equation (1). Both of them include a *target condition*

$$y(T) = \bar{y}\,. \tag{5}$$

In the *norm optimal problem* we minimize

$$\operatorname*{ess.\,sup}_{0\le t\le T} \|u(t)\|$$

among all solutions of (1) satisfying the initial condition $y(0) = \zeta$ and the target condition (5); the control interval $0 \le t \le T$ is fixed. In the *time optimal* problem the controls satisfy a fixed bound such as

$$\operatorname*{ess.\,sup}_{0\le t\le T} \|u(t)\| \le 1\,,$$

and we minimize $T$ subject to the initial and target condition.

The time optimal problem received privileged attention from the very start of control theory, but this has been less the case for the norm optimal problem. It was known for a long time that time optimality implies norm optimality, but that the two problems are far from equivalent in the infinite dimensional setting seems to have been realized much more recently. However, there are many situations (determined by conditions on the semigroup $S(t)$ or on the target $\bar{y}$) where time and norm optimality are essentially equivalent.

Most of this book deals with the relation among time and norm optimality and *Pontryagin's maximum principle*

$$\langle S(T-t)^* z, \bar{u}(t)\rangle = \max_{\|u\|\le 1} \langle S(T-t)^* z, u\rangle \tag{6}$$

($\|u\| \le$ minimum norm for the norm optimal problem). The maximum principle with $z \in E^* =$ dual of $E$ is a necessary and (almost) sufficient condition for time and norm optimality in finite dimension.[1] The finite dimensional theory extends to the equation (1) when $S(t)E = E$ for $t > 0$ (in particular when $S(t)$ is a group) but the similarities with the finite dimensional case end here. In general, special assumptions on the target $\bar{y}$ are needed to make (6) a necessary condition for optimality (with $z$ in a space larger than $E^*$) and, conversely, special assumptions on $z$ are needed to make (6) a sufficient condition for optimality of $\bar{u}(t)$. In fact, *singular* optimal controls (those that do *not* satisfy Pontryagin's maximum principle, or satisfy it only in a weak form) are the main actors in various places of this monograph.

[1] Even in finite dimension, sufficiency of the maximum principle for time optimality requires additional conditions on the initial condition $\zeta$ and/or the target $\bar{y}$.

Much of the material is independent of the maximum principle. Under suitable conditions on $z$, (6) implies the *bang-bang principle*

$$\|u(t)\| = 1 \quad \text{a. e.} \tag{7}$$

but (7) can be also be proved without intercession of the maximum principle for time optimal controls. Other results (some depending on the maximum principle, some not) include various well posedness properties of control problems, that is, continuous dependence of optimal controls on parameters of the system such as the initial and target conditions.

This monograph is organized as follows. The first two sections of Chapter 1 contain a survey of some finite dimensional results with an outline of infinite dimensional systems in the third. The aim is reveal that some infinite dimensional results are descendants of finite dimensional theorems. In some cases, however the "family resemblance" is slight, and many other results have no counterpart in finite dimension. We have included some references to the early history of infinite dimensional control theory in **1.3**.

Chapter 2 and Chapter 3 deal with the system (1) in an arbitrary Banach space $E$ with a view towards the modeling of partial differential equations such as (2) in $L^p(\Omega)$ for $1 < p < \infty$. However, we do also other equations; for instance, some of the most interesting examples in **2.6** and **2.7** use the "proto-hyperbolic" equation $y_t(t,x) = -y_x(t,x)$. These results suggest that a systematic study of the maximum principle (or, rather, of its interpretation) for equations of hyperbolic type would be worth undertaking, but this is not attempted here.

Chapter 4 is on the modeling of the equation (2) in $C(\overline{\Omega})$. Due to existence requirements, the control space must be expanded to $L^\infty(\Omega)$ and we can request only weak measurability of the controls with respect to $t$; this corresponds to driving (2) with controls in $L^\infty((0,T)\times\Omega)$.

Chapter 5 is on the modeling of the equation (2) in $L^1(\Omega)$. The control space $L^1((0,T)\times\Omega)$ places us in a adverse existence situation, thus we must replace it by a space of (weakly measurable) controls taking values in the space $\Sigma(\overline{\Omega})$ of Borel measures in $\overline{\Omega}$.

There is an obvious parallelism between the two models in Chapters 4 and Chapter 5, so much so that one can translate results from one case to the other using a "replacement chart" where $C(\overline{\Omega})$ is replaced by $L^1(\Omega)$, $L^\infty(\Omega)$ is replaced by $\Sigma(\overline{\Omega})\ldots$ and so on. However, the similarity does not go all the way. Although both the geometries of $L^\infty(\Omega)$ and $\Sigma(\Omega)$ are devoid of smoothness, the control space $L^\infty(\Omega)$ allows uniqueness results for optimal controls not unlike those in smooth spaces, while in the space $\Sigma(\overline{\Omega})$ uniqueness breaks down completely both in the time optimal and the norm optimal problems. In a sort of compensation, in the $\Sigma(\overline{\Omega})$ setting the bang-bang principle (7) is, in certain situations, essentially sufficient for both time optimality and norm optimality, a result that is not known to hold in any other space.

The treatments in Chapters 4 and 5 could have been unified by means of the theory of Phillips adjoints, but the gain in brevity and conciseness would not outweigh the additional insight that each of the parallel theories afford.

The first two sections of Chapter 6 deal with some results that are known to hold for a very restricted class, self adjoint semigroups in Hilbert spaces, and examine the possibility of generalizations to other semigroups and Banach spaces. In the last section we include some recent references on the time and norm optimal problems as well as on problems not treated in this book but related to the material in one way or another, among them minimization of functionals other than time or norm (in particular, the linear-quadratic problem) and a sketch of the methods used in the control theory of semilinear equations. Some of these methods lean on the linear theory.

Infinite dimensional control theory is increasingly becoming the calculus of variations of the new century. We believe this monograph will have something to say to specialists (even if in a restricted area). On the other hand, the book is also accessible to beginners; the only advanced prerequisite is a course in basic linear functional analysis including semigroup theory.

ACKNOWLEDGEMENTS. In the spring of 2002 I was invited by the Istituto Nazionale di Alta Matematica Francesco Severi, sponsored by Dr. Piermarco Cannarsa, to lecture and do research on the time and norm optimal problem at the Dipartamento di Matematica, Università di Roma II Tor Vergata. This motivated the writing of a set of notes that, after some refinements and additions, became this monograph. In particular, most of the material in **2.6**, **2.7**, **6.1** and **6.2** (not in print at the time) owes much to conversations with Dr. Cannarsa and interaction with colleagues and students at Tor Vergata. My thanks go also to David Mora, who did the final printing and assisted in many other ways.

# CONTENTS


PREFACE vii

CHAPTER 1: INTRODUCTION

1.1. Finite dimensional systems: the maximum principle 1
1.2. Finite dimensional systems: existence and uniqueness 9
1.3. Infinite dimensional systems 16

CHAPTER 2: SYSTEMS WITH STRONGLY MEASURABLE CONTROLS, I

2.1. The reachable space and the bang-bang property 27
2.2. Reversible systems 36
2.3. The reachable space and its dual, I 47
2.4. The reachable space and its dual, II 56
2.5. The maximum principle 65
2.6. Vanishing of the costate and nonuniqueness in norm optimality 77
2.7. Vanishing of the costate for time optimal controls 87
2.8. Singular norm optimal controls 96
2.9. Singular norm optimal controls and singular functionals 108

CHAPTER 3: SYSTEMS WITH STRONGLY MEASURABLE CONTROLS, II

3.1. Existence and uniqueness of optimal controls 117
3.2. The weak maximum principle and the time optimal problem 125
3.3. Modeling of parabolic equations 134
3.4. Weakly singular extremals 143
3.5. More on the weak maximum principle 152
3.6. Convergence of minimizing sequences to optimal controls 163

CHAPTER 4: OPTIMAL CONTROL OF HEAT PROPAGATION

4.1. Modeling of parabolic equations 173
4.2. Adjoints 180
4.3. Adjoint semigroups 187
4.4. The reachable space 191
4.5. The reachable space and its dual, I 197
4.6. The reachable space and its dual, II 207
4.7. The maximum principle 215
4.8. Existence, uniqueness and stability of optimal controls 225
4.9. Examples and applications 231

CHAPTER 5: OPTIMAL CONTROL OF DIFFUSIONS
5.1. Modeling of parabolic equations 243
5.2. The reachable space and its dual, I 252
5.3. The reachable space and its dual, II 258
5.4. The maximum principle 266
5.5. Existence of optimal controls; uniqueness and stability of supports 273
5.6. Examples and applications 285
CHAPTER 6: APPENDIX
6.1. Self adjoint operators, I 295
6.2. Self adjoint operators, II 301
6.3. Related research 305
REFERENCES 309
NOTATION AND SUBJECT INDEX 319

CHAPTER 1

# INTRODUCTION

**1.1. Finite dimensional systems: the maximum principle.** Half a century ago, Bellman, Glicksberg and Gross [1956] figured out the time optimal controls $\bar{u}(t)$ for the linear, time invariant ordinary differential system

$$y'(t) = Ay(t) + Bu(t)\,, \quad y(0) = \zeta \tag{1.1.1}$$

where $A$ (resp. $B$) is a $n \times n$ matrix (resp. a $m \times n$ matrix). *Controls* $u(t)$ are measurable $\mathbb{R}^m$-valued vector functions $u(t) = (u_1(t), \dots, u_m(t))$ and a control is *admissible* if

$$\|u(t)\| \le 1 \quad \text{a. e. in} \quad 0 \le t \le T \tag{1.1.2}$$

where $\|\cdot\|$ is the *maximum norm*

$$\|u\| = \max(|u_1|, \dots, |u_m|)\,. \tag{1.1.3}$$

In the *time optimal problem* we drive the initial condition $\zeta \in \mathbb{R}^n$ to a target $\bar{y} \in \mathbb{R}^n$ in minimum time $T$ by means of an admissible control $u(t)$; "driving to a target $\bar{y}$ in time $T$" means the *target condition*

$$y(T, \zeta, u) = \bar{y} \tag{1.1.4}$$

must be satisfied, where $y(t, \zeta, u)$ is the solution $y(t)$ of (1.1.1). In the *norm optimal* problem, the controls are $m$-vector functions with measurable, essentially bounded components, and one drives $\zeta$ to the target $\bar{y}$ in a fixed time interval $0 \le t \le T$, so that (1.1.4) is satisfied; the objective is to minimize the norm

$$\|u(\cdot)\|_{L^\infty(0,T;\mathbb{R}^m)} = \operatorname*{ess.\,sup}_{0 \le t \le T} \|u(t)\|\,. \tag{1.1.5}$$

It was proved in Bellman *et al.* [1956] that every time optimal control $\bar{u}(t)$ satisfies

$$\langle B^* S(T-t)^* z, \bar{u}(t) \rangle = \max_{\|u\| \le 1} \langle B^* S(T-t)^* z, u \rangle \tag{1.1.6}$$

in the control interval $0 \le t \le T =$ optimal time, where $S(t) = e^{tA}$ and the *multiplier* $z \in \mathbb{R}^n$ is not zero. This result was immediately submerged by *Pontryagin's maximum principle*, announced by Boltyanski, Gamkrelidze and Pontryagin in [1956]. The maximum principle could deal with nonlinear equations and general cost functionals and included the linear time-invariant result as a very particular case. However, the proof in Bellman *et al.* [1956] was so elementary that invited generalizations to infinite dimensional spaces, these generalizations motivated by the modeling of optimal problems for partial differential equations (for more on this see **1.3**). Here is the proof of (1.1.6), which holds as well for the norm optimal problem.

**Theorem 1.1.1.** *(a) Let $\bar{u}(t)$ be time optimal in the interval $0 \leq t \leq T$. Then the maximum principle* (1.1.6) *holds for a multiplier $z \neq 0$. (b) Let $\bar{u}(t)$ be norm optimal in the interval $0 \leq t \leq T$. Then $\bar{u}(t)$ satisfies the maximum principle*

$$\langle B^*S(T-t)^*z, \bar{u}(t)\rangle = \max_{\|u\|\leq\rho} \langle B^*S(T-t)^*z, u\rangle \tag{1.1.7}$$

*for a multiplier $z \neq 0$, where*

$$\rho = \|\bar{u}(\cdot)\|_{L^\infty(0,T;\mathbb{R}^m)} \,. \tag{1.1.8}$$

*Proof.* Solutions of (1.1.1) are given by the variation-of-constants formula[1]

$$y(t) = y(t,\zeta,u) = S(t)\zeta + \int_0^t S(t-\sigma)Bu(\sigma)d\sigma \,. \tag{1.1.9}$$

The *reachable space* $R^\infty(T)$ of (1.1.1) is the subspace of $\mathbb{R}^n$ of all elements of the form

$$y = \int_0^T S(T-\sigma)Bu(\sigma)d\sigma \tag{1.1.10}$$

with $u(\cdot) \in L^\infty(0,T;\mathbb{R}^m)$ (that is, the space of all $y \in \mathbb{R}^n$ to which we can drive from $\zeta = 0$ with controls in $L^\infty(0,T;\mathbb{R}^m)$) and $B^\infty_\rho(T)$ is the subset of $R^\infty(T)$ of all elements of the form (1.1.10) with[2]

$$\|u(t)\| \leq \rho \quad \text{a. e. in } 0 \leq t \leq T \,. \tag{1.1.11}$$

Clearly, $B^\infty_\rho(T)$ is a convex subset of $\mathbb{R}^n$ containing the origin. To show $(b)$, assume that $\bar{u}(t)$ is norm optimal, and let $\rho$ be given by (1.1.8) (that is, let $\rho$ be the optimal norm, which means the norm of the optimal control). The target condition is

$$S(T)\zeta + \int_0^T S(T-\sigma)B\bar{u}(\sigma)d\sigma = \bar{y} \,,$$

[1] Due to the controls being merely measurable, solutions are no more than absolutely continuous; the derivative $y'(t)$ exists a. e. and satisfies (1.1.1) a. e.

[2] Note that (1.1.10) will hold for many $u(\cdot)$; only one has to satisfy (1.1.11).

so that

$$\bar{y} - S(T)\zeta = \int_0^T S(T-\sigma)B\bar{u}(\sigma)d\sigma \in B_\rho^\infty(T).$$

Moreover, $\bar{y} - S(T)\zeta$ is a *boundary point* of $B_\rho^\infty(T)$. In fact, if it were an interior point, since $0 \in B_\rho^\infty(T)$ we would have

$$r(\bar{y} - S(T)\zeta) \in B_\rho^\infty(T)$$

for some $r > 1$. This means

$$r(\bar{y} - S(T)\zeta) = \int_0^T S(T-\sigma)Bu(\sigma)d\sigma$$

for some control $u(t)$ satisfying the constraint (1.1.11) or, equivalently,

$$\bar{y} = S(T)\zeta + \int_0^T S(T-\sigma)B\frac{u(\sigma)}{r}d\sigma\,,$$

hence we can drive from $\zeta$ to $\bar{y}$ in time $T$ with a control $u(\sigma)/r$ whose norm is

$$\left\|\frac{u(\cdot)}{r}\right\|_{L^\infty(0,T;I\!R^m)} \le \frac{\rho}{r} < \rho\,.$$

This contradicts the norm optimality of $\bar{u}(t)$. Having proved that $\bar{y} - S(T)\zeta$ is a boundary point of $B_\rho^\infty(T)$, we can separate it from $B_\rho^\infty(T)$ by means of a nonzero $z \in I\!R^n$; this means

$$\langle z, y\rangle \le \langle z, \bar{y} - S(T)\zeta\rangle \tag{1.1.12}$$

for every $y \in B_\rho^\infty(T)$, or

$$\left\langle z, \int_0^T S(T-\sigma)Bu(\sigma)d\sigma\right\rangle \le \left\langle z, \int_0^T S(T-\sigma)B\bar{u}(\sigma)d\sigma\right\rangle \tag{1.1.13}$$

for every $u(\cdot)$ satisfying (1.1.11). This inequality can be written

$$\int_0^T \langle z, S(T-\sigma)Bu(\sigma)\rangle d\sigma \le \int_0^T \langle z, S(T-\sigma)B\bar{u}(\sigma)\rangle d\sigma$$

or

$$\int_0^T \langle B^*S(T-\sigma)^*z, u(\sigma)\rangle d\sigma \le \int_0^T \langle B^*S(T-\sigma)^*z, \bar{u}(\sigma)\rangle d\sigma \tag{1.1.14}$$

which is equivalent to (1.1.7). This ends the proof of $(b)$.

The proof of $(a)$ is similar. Again, $\bar{y} - S(T)\zeta \in B_1^\infty(T)$ and to do the separation argument we need to show that $\bar{y} - S(T)\zeta$ is a boundary point of

$B_1^\infty(T)$. If $B_1^\infty(T)$ does not contain interior points (equivalently, if the inclusion $R^\infty(T) \subseteq \mathbb{R}^n$ is strict) there is nothing to prove, so we may assume that

$$R^\infty(T) = \mathbb{R}^n\,. \tag{1.1.15}$$

Property (1.1.15) is independent of $T > 0$; in fact $R^\infty(t)$ is independent of $t$. To see why, let $t > 0$ and assume there exists $z \in \mathbb{R}^n$ such that $\langle z, y\rangle = 0$ for all $y \in R^\infty(t)$. This is equivalent to

$$\begin{aligned}\int_0^t \langle B^*S(t-\sigma)^*z, u(\sigma)\rangle d\sigma &= \int_0^t \langle z, S(t-\sigma)Bu(\sigma)\rangle d\sigma \\ &= \left\langle z, \int_0^t S(t-\sigma)Bu(\sigma)d\sigma \right\rangle = 0\end{aligned}$$

for all $u(\cdot) \in L^\infty(0,t;\mathbb{R}^m)$, which is in turn equivalent to $B^*S(t-\sigma)^*z = 0$ or $B^*S(\sigma)^*z = 0$ for $0 \le \sigma \le t$. Now, $B^*S(\sigma)^*z$ is analytic, hence this statement is independent of $t$. In particular, (1.1.15) is equivalent to

$$B^*S(\sigma)^*z = 0 \ \ (\sigma \ge 0) \quad \Longrightarrow \ z = 0\,. \tag{1.1.16}$$

Assuming that (1.1.15) (or, equivalently, (1.1.16)) holds, we check that, for any $t > 0$ the expression

$$\|y\|_{R^\infty(t)} = \inf\left\{\|u(\cdot)\|_{L^\infty(0,t;\mathbb{R}^m)}\,;\ \int_0^t S(t-\sigma)Bu(\sigma)d\sigma = y\right\} \tag{1.1.17}$$

defines a norm in $R^\infty(t) = \mathbb{R}^n$, which (as any other norm) must be equivalent to the norm (1.1.3) of $\mathbb{R}^n$. In particular, this means there exists a constant $C(t)$ such that, for every $y \in \mathbb{R}^n$ we can find $u(\cdot) \in L^\infty(0,t;\mathbb{R}^m)$ with

$$y = \int_0^t S(t-\sigma)Bu(\sigma)d\sigma, \quad \|u(\cdot)\|_{L^\infty(0,t;\mathbb{R}^m)} \le C(t)\|y\|\,. \tag{1.1.18}$$

Let $t' > t$. Then we have

$$\int_0^t S(t-\sigma)Bu(\sigma)d\sigma = \int_{t'-t}^{t'} S(t'-\sigma)Bu(\sigma-(t'-t))d\sigma\,,$$

thus, if

$$y = \int_0^{t'} S(t'-\sigma)Bv(\sigma)d\sigma, \quad \|v(\cdot)\|_{L^\infty(0,t';\mathbb{R}^m)} \le C(t')\|y\|$$

is the version of (1.1.18) for $t'$, we have

$$C(t') \le C(t) \quad (t' \ge t)\,. \tag{1.1.19}$$

Assume $\bar{y} - S(T)\zeta$ is not a boundary point of $B_1^\infty(T)$. Then, arguing as in the proof of $(b)$ we deduce that there exists $u(\cdot) \in L^\infty(0, T; I\!R^m)$, $\|u(\cdot)\|_{L^\infty(0,T;I\!R^m)} \le r < 1$ such that

$$\bar{y} - S(T)\zeta = \int_0^T S(T-\sigma)Bu(\sigma)d\sigma\,. \tag{1.1.20}$$

Take $t < T$ and rewrite (1.1.20) as

$$\begin{aligned}\bar{y} - S(t)\zeta &= \int_0^t S(t-\sigma)Bu(\sigma)d\sigma - S(t)\zeta + S(T)\zeta \\ &\quad + \int_0^t (S(T-\sigma) - S(t-\sigma))Bu(\sigma)d\sigma + \int_t^T S(T-\sigma)Bu(\sigma)d\sigma \\ &= \int_0^t S(t-\sigma)Bu(\sigma)d\sigma + y(t,T)\,. \end{aligned}\tag{1.1.21}$$

We use the fact that $y(t,T) \to 0$ as $t \to T$, take $t$ sufficiently near $T$ and use (1.1.18) to construct a control $v(\cdot)$ such that $\|v(\cdot)\|_{L^\infty(0,t;I\!R^m)} \le 1 - r$ and[3]

$$\int_0^t S(t-\sigma)Bv(\sigma)d\sigma = y(t,T)\,. \tag{1.1.22}$$

Putting together (1.1.21) and (1.1.22) we obtain

$$\bar{y} = S(t)\zeta + \int_0^t S(T-\sigma)B(u(\sigma) + v(\sigma))d\sigma\,,$$

so we can drive from $\zeta$ to $\bar{y}$ by means of the admissible control $u(\cdot) + v(\cdot)$ in time $t < T$. This contradicts the fact that $T$ is the optimal time, and ends the proof of $(a)$, always assuming that (1.1.15) holds.

The case where $R^\infty(T) \ne I\!R^n$ is trivial; here (1.1.16) must fail, thus there exists $z \ne 0$ with $B^*S(\sigma)^*z$ identically zero. With this $z$ every control, optimal or not, satisfies the necessary conditions (1.1.6) or (1.1.7). This completes the proof of Theorem 1.1.1.

Condition (1.1.16) and the fact that $z \ne 0$ implies that (1.1.6) is nontrivial, that is, it gives information on the optimal control $\bar{u}(t)$ for almost all $t$ (although this information may not determine $\bar{u}(t)$ uniquely; see **1.2**).

Conditions (1.1.15) $\Leftrightarrow$ (1.1.16) also imply that the assumptions in Theorem 1.1.1 are *sufficient* (with additional conditions in the time optimal case). See Theorem 1.1.3 below.

It follows from (1.1.6) that if condition (1.1.16) is satisfied then time optimal controls satisfy not only the bound (1.1.2) but

$$\|u(t)\| = 1 \quad \text{a. e. in } 0 \le t \le T\,. \tag{1.1.23}$$

---

[3] Here it is essential that the constant $C(t)$ in (1.1.18) does not increase as we move $t$ to the right; this is (1.1.19).

Entirely similar considerations apply to norm optimal controls, where (1.1.23) becomes

$$\|u(t)\| = \rho \quad \text{a. e. in } 0 \le t \le T \,. \tag{1.1.24}$$

Under stronger conditions, we can say more.

**Theorem 1.1.2.** *Assume that, for arbitrary* $z \in I\!R^n$, $z \neq 0$ *the vector* $B^*S(t)^*z$ *is not identically perpendicular to any of the sides of the unit cube* $|y_j| \le 1$, $j = 1, 2, \ldots, m$, *that is, is not identically parallel to any of the vectors*[4]

$$e_j = (0, \ldots, 1, \ldots, 0) \in I\!R^m \quad (1 \textit{ in the } j^{\text{th}} \textit{ place})\,.$$

*Then, if the admissible control* $\bar{u}(t)$ *is time optimal there exists a finite sequence* $\{t_k\}$, $0 = t_0 \le t_1 < \ldots < t_{p-1} < t_p \le T$ *such that*

$$\bar{u}(t) = \textit{one of the vertices of the cube} \tag{1.1.25}$$

*in each interval* $t_k \le t \le t_{k+1}$. *If* $\bar{u}(t)$ *is norm optimal the same conclusion holds for the cube* $|y_j| \le \rho =$ *optimal norm,* $j = 1, 2, \ldots, m$.

*Proof.* Let $P_j : I\!R^m \to I\!R$ be the operator of projection on the $j^{\text{th}}$ coordinate, $z$ the multiplier in (1.1.6). Since $z \neq 0$, $B^*S^*(\sigma)z$ is not identically parallel to $e_j$, so that each analytic function

$$B^*S(\sigma)^*z - P_jB^*S(\sigma)^*z\,, \quad j = 1, \ldots, m$$

is not identically zero. Hence there exists $\{t_k\}$, $0 \le t_1 < \ldots < t_p \le T$ such that

$$t \neq t_k \implies P_jB^*S(\sigma)^*z \neq 0 \quad \text{for all } j\,.$$

Accordingly, in any interval $t_k \le t \le t_{k+1}$ the maximum principle (1.1.6) selects a vertex of the unit cube. This ends the proof for time optimal controls: the proof for the norm optimal case is the same.

Controls satisfying (1.1.25) are called *bang-bang.* We shall extend this name to controls satisfying only (1.1.23) or (1.1.24).

**Theorem 1.1.3.** *Assume* (1.1.16) *holds. Then* (*a*) *Let* $\bar{u}(t)$ *satisfy* (1.1.7) *for a multiplier* $z \neq 0$ *with* $\rho = \|\bar{u}(\cdot)\|_{L^\infty(0,T;I\!R^m)}$. *Then* $\bar{u}(t)$ *is norm optimal.* (*b*) *Let* $\bar{u}(t)$ *satisfy* (1.1.6) *for a multiplier* $z \neq 0$, *and let*

$$\zeta = 0 \quad \textit{or} \quad \bar{y} = 0\,. \tag{1.1.26}$$

*Then* $\bar{u}(t)$ *is time optimal.*

---

[4] For algebraic conditions on $A, B$ that guarantee this, or just condition (1.1.16), see Pontryagin *et al.* [1961].

*Proof:* For $(a)$, note that the maximum principle (1.1.7) is equivalent to (1.1.12), which implies that

$$\bar{y} - S(T)\zeta = \int_0^T S(T-\sigma)B\bar{u}(\sigma)d\sigma$$

is a boundary point of $B_\rho^\infty(T)$. If $\bar{u}(\sigma)$ is not norm optimal there exists another control $u(\cdot) \in L^\infty(0,T;\mathbb{R}^m)$, $\|u(\cdot)\|_{L^\infty(0,T;\mathbb{R}^m)} = r < \rho$ with

$$\bar{y} - S(T)\zeta = \int_0^T S(T-\sigma)Bu(\sigma)d\sigma\,.$$

Multiplying both sides by $\rho/r$ we obtain

$$\frac{\rho}{r}(\bar{y} - S(T)\zeta) \in B_\rho^\infty(T)\,.$$

Since $\rho/r > 1$, this means that $\bar{y} - S(T)\zeta$ is an interior point of the segment joining the origin and $(\rho/r)(\bar{y} - S(T)\zeta)$. On the other hand, equivalence of the norms (1.1.3) and (1.1.17) in $\mathbb{R}^n$ implies that the origin is is an interior point of $B_\rho^\infty(T)$, thus $\bar{y} - S(T)\zeta$ is an interior point of $B_\rho(T)$, a contradiction.

To prove $(b)$, assume $\bar{u}(\cdot)$ is not time optimal. Then there exists $t < T$ and an admissible control $\tilde{u}(\cdot) \in L^\infty(0,t;\mathbb{R}^m)$ that drives $\zeta$ to $\bar{y}$ in time $t < T$. If $\bar{y} = 0$, the control

$$u(\sigma) = \begin{cases} \tilde{u}(\sigma) & (0 \le \sigma \le t) \\ 0 & (t < \sigma \le T) \end{cases}$$

drives $\zeta$ to $\bar{y}$ in time $T$. Again, the maximum principle (1.1.6) is equivalent to (1.1.12), thus

$$-S(T)\zeta = \int_0^T S(T-\sigma)Bu(\sigma)d\sigma = \int_0^T S(T-\sigma)B\bar{u}(\sigma)d\sigma \tag{1.1.27}$$

with $-S(T)\zeta = \bar{y} - S(T)\zeta$ a boundary point of $B_\rho^\infty(T)$. It follows that $u(\sigma)$ itself must satisfy (1.1.6), a contradiction with (1.1.23) since $u(\sigma)$ is zero in an interval.

On the other hand, if $\zeta = 0$ we drive from $\zeta$ to $\bar{y}$ by means of [5]

$$v(\sigma) = \begin{cases} 0 & (0 \le \sigma < T-t) \\ \tilde{u}(\sigma - (T-t)) & (T-t \le \sigma \le T) \end{cases}$$

and a similar argument applies; here, (1.1.27) becomes

$$\bar{y} = \int_0^T S(T-\sigma)Bv(\sigma)d\sigma = \int_0^T S(T-\sigma)B\bar{u}(\sigma)d\sigma\,.$$

[5] With $u(\sigma)$ we get to $\bar{y}$ at time $t$ and then do nothing until time $T$; with $v(\sigma)$ we do nothing until time $T-t$ and then drive to $\bar{y}$.

**Corollary 1.1.4.** *Under conditions* (1.1.16) *and* (1.1.26) *norm optimality* (*with* $\rho = 1$) *is equivalent to time optimality.*

**Example 1.1.5.** Assumption (1.1.26) for sufficiency of the maximum principle for time optimality can be relaxed to $\|A\zeta\| < 1$ or $\|A\bar{y}\| < 1$ (for the proof see Theorem 2.5.7) but no further. Consider the 1-dimensional system

$$y'(t) = -ay(t) + u(t), \quad y(0) = \zeta \tag{1.1.28}$$

with $a > 0$. We have $S(t) = e^{-at} = S(t)^*$, thus controls satisfying the maximum principle (1.1.6) with $z \neq 0$ are

$$\bar{u}(t) = \begin{cases} 1 & \text{if } z > 0\,, \\ -1 & \text{if } z < 0\,. \end{cases} \tag{1.1.29}$$

For initial condition and target $\zeta = 1/a = \bar{y}$ we have

$$\int_0^T S(T-\sigma)\cdot 1\, d\sigma = \frac{1 - e^{-aT}}{a} = \bar{y} - S(T)\zeta\,,$$

so that the first control in (1.1.29) drives $\zeta$ to $\bar{y}$ in any time $T \geq 0$; in other words, $y(T, \zeta, \bar{u}) = \bar{y}$ for all $T \geq 0$. None of these drives is time optimal except for $T = 0$. Note, however, that Theorem 1.1.3 ($a$) says that all the drives are norm optimal. For another situation where the need for restrictions on the size of $\|A\zeta\|$ or $\|A\bar{y}\|$ is put in evidence, see Example 2.5.10.

**Theorem 1.1.6.** (*The optimality principle for the time optimal problem*). *Let* $\bar{u}(\sigma)$ *be a time optimal control in the interval* $[0, T]$. *Then* $\bar{u}(\sigma)$ *is time optimal in any interval* $[0, t]$, $t < T$.

*Proof.* Assume $\bar{u}(\sigma)$ does not drive $\zeta$ to $y(t, \zeta, \bar{u})$ time optimally. Then there exists an admissible control $v(\sigma)$ that drives $\zeta$ to $y(t, \zeta, \bar{u})$ in time $t - \delta < t$, that is

$$y(t - \delta, \zeta, v) = y(t, \zeta, \bar{u})\,.$$

It follows that the control

$$u(\sigma) = \begin{cases} v(\sigma) & 0 \leq \sigma \leq t - \delta \\ \bar{u}(\sigma + \delta) & t - \delta \leq \sigma \leq T - \delta \end{cases}$$

drives $\zeta$ to $y(t, \zeta, \bar{u})$ in time $t - \delta$ and then proceeds to the target $\bar{y} = y(T, \zeta, \bar{u})$, where it arrives at time $T - \delta$. In other words,

$$y(T - \delta, \zeta, u) = y(T, \zeta, \bar{u})\,,$$

which shows that $\bar{u}(\sigma)$ is not time optimal. Obviously, the interval $[0, t]$ in the statement of the theorem can be replaced by any interval $[s, t]$, $0 \leq s < t \leq T$. Note also that the proof uses only raw time optimality, not any additional hypotheses like (1.1.16).

Under (1.1.16), the optimality principle for the norm optimal problem is as well true. To see this, let the control $\bar{u}(\sigma)$ be norm optimal in $0 \leq \sigma \leq T$, $\rho$ its $L^\infty(0,T;I\!R^m)$ norm. Assume that the drive from $\zeta$ to $y(t,\zeta,\bar{u})$, $t < T$ is not norm optimal, so that there exists $v(\cdot) \in L^\infty(0,t;I\!R^m)$ with

$$y(t,\zeta,v) = y(t,\zeta,\bar{u})\,, \quad \|v(\cdot)\|_{L^\infty(0,t;I\!R^m)} < \rho\,.$$

Then the control

$$u(\sigma) = \begin{cases} v(\sigma) & (0 \leq \sigma \leq t) \\ \bar{u}(\sigma) & (t < \sigma \leq T) \end{cases}$$

is norm optimal but doesn't satisfy (1.1.24), absurd.

**Miscellaneous notes.** The equivalence between (1.1.15) and (1.1.16) was given in Kalman, Ho and Narendra [1963], who call a system *completely controllable in time* $t$ if, given $\zeta, \bar{y} \in I\!R^n$ there exists $u(\cdot) \in L^\infty(0,t;I\!R^m)$ such that

$$S(t)\zeta + \int_0^t S(t-\sigma)Bu(\sigma)d\sigma = \bar{y}\,. \tag{1.1.30}$$

Complete controllability in time $t$ implies $R^\infty(t) = I\!R^n$ (setting $\zeta = 0$ in (1.1.30)). Conversely, if $R^\infty(t) = I\!R^n$ we can find $u(\cdot) \in L^\infty(0,t;I\!R^m)$ such that

$$\bar{y} - S(t)\zeta = \int_0^t S(t-\sigma)Bu(\sigma)d\sigma \tag{1.1.31}$$

which is (1.1.30). The results in this section imply that complete controllability is equivalent to (1.1.16). The definition of complete controllability is not sharply dependent on the controls; for instance we may use as controls continuous functions[6] or even polynomials with vector coefficients in $I\!R^m$.

For a discussion of (not necessarily linear) finite dimensional systems that includes controllability, optimality and other questions see Pontryagin *et al.* [1961], Krasovski [1968], Conti [1974], Barnett - Cameron [1985] and Hocking [1991].

**1.2. Finite dimensional systems: existence and uniqueness.** Existence results for optimal controls for

$$y'(t) = Ay(t) + Bu(t), \quad y(0) = \zeta \tag{1.2.1}$$

[6] Complete controllability puts no bounds on the controls. For controllability with bounded controls see Antosiewicz [1963] for finite dimensional systems, the author [1969] for infinite dimensional systems.

are based on duality and weak convergence. The space $L^1(0, T; \mathbb{R}^m)$ consists of all vector functions $f(t) = (f_1(t), \ldots, f_m(t))$ with components in $L^1(0, T)$, equipped with the norm

$$\|f(\cdot)\|_{L^1(0,T;\mathbb{R}^m)} = \int_0^T \|f(\sigma)\|_1 d\sigma \,, \tag{1.2.2}$$

where, for a $m$-vector $y = (y_1, \ldots, y_m)$ the norm $\|y\|_1$ is given by

$$\|y\|_1 = |y_1| + \ldots + |y_m| \,.$$

We have the duality relation

$$L^\infty(0, T; \mathbb{R}^m) = L^\infty(0, T; \mathbb{R}^m)^* \,, \tag{1.2.3}$$

elements $u(\cdot) \in L^\infty(0, T; \mathbb{R}^m)$ acting on elements $f(\cdot) \in L^1(0, T; \mathbb{R}^m)$ as follows:

$$\langle u(\cdot), f(\cdot)\rangle = \int_0^T \langle f(\sigma), u(\sigma)\rangle d\sigma \,. \tag{1.2.4}$$

The identification (1.2.3)-(1.2.4) is algebraic and metric.

*Weak convergence* (or, more precisely, $L^1(0, T; \mathbb{R}^m)$-weak convergence) of a sequence $\{u_n(\cdot)\} \subset L^\infty(0, T; \mathbb{R}^m)$ to a $u(\cdot) \in L^\infty(0, T; \mathbb{R}^m)$ means

$$\langle u_n(\cdot), f(\cdot)\rangle \to \langle u(\cdot), f(\cdot)\rangle$$

for all $f(\cdot) \in L^1(0, T; \mathbb{R}^m)$. If we restrict the sequences to a ball

$$\mathcal{B}^\infty_\rho(0, T; \mathbb{R}^m) = \{u(\cdot) \in L^\infty(0, T; \mathbb{R}^m) \ ; \|u(\cdot)\|_{L^\infty(0,T;\mathbb{R}^m)} \le \rho\}$$

weak convergence is defined by a metric and the metric space $\mathcal{B}^\infty_\rho(0, T; \mathbb{R}^m)$ is compact in the weak topology.[7] This implies that, in dealing with compactness, we can avoid generalized sequences; ordinary sequences are enough.

**Theorem 1.2.1.** *(a) Assume there exists a control $u(\cdot) \in L^\infty(0, T; \mathbb{R}^m)$ such that*

$$y(T, \zeta, u) = \bar{y} \,.$$

*Then there exists a norm optimal control $\bar{u}(\cdot)$ in $0 \le t \le T$. (b) Assume there exists a control $u(\cdot) \in L^\infty(0, t; \mathbb{R}^m)$ satisfying*

$$\|u(\cdot)\|_{L^\infty(0,t;\mathbb{R}^m)} \le 1 \,, \quad y(t, \zeta, u) = \bar{y}$$

*for some $t > 0$. Then there exists a time optimal control $\bar{u}(\cdot)$.*

---

[7] This is due to the fact that $L^1(0, T; \mathbb{R}^m)$ is separable. See Dunford - Schwartz [1958, p. 426, Theorem 1].

*Proof.* (*a*) Under the assumptions we have a sequence $\{u_n(\cdot)\} \subset L^\infty(0, T; \mathbb{R}^m)$ of controls all driving $\zeta$ to $\bar{y}$,

$$\bar{y} - S(T)\zeta = \int_0^T S(T - \sigma)Bu_n(\sigma)d\sigma\,, \tag{1.2.5}$$

and such that

$$\begin{aligned} &\|u_1(\cdot)\|_{L^\infty(0,T;\mathbb{R}^m)} \geq \|u_2(\cdot)\|_{L^\infty(0,T;\mathbb{R}^m)} \geq \cdots\,, \\ &\|u_n(\cdot)\|_{L^\infty(0,T;\mathbb{R}^m)} \to \rho = \inf\left\{\|u(\cdot)\|_{L^\infty(0,T;\mathbb{R}^m)}\right\}, \end{aligned} \tag{1.2.6}$$

the infimum taken over all controls $u(\cdot)$ driving $\zeta$ to $\bar{y}$ in time $T$. Since the sequence $\{u_n(\cdot)\}$ is bounded in $L^\infty(0, T; \mathbb{R}^m)$ we can select a subsequence (named in the same way) such that $\{u_n(\cdot)\}$ is $L^1(0, T; \mathbb{R}^m)$-weakly convergent to an element $\bar{u}(\cdot) \in L^\infty(0, T; \mathbb{R}^m)$. We obtain from (1.2.5) that

$$\begin{aligned} \langle z, \bar{y} - S(T)\zeta\rangle &= \left\langle z, \int_0^T S(T - \sigma)Bu_n(\sigma)d\sigma\right\rangle \\ &= \int_0^T \langle B^*S(T - \sigma)^*z, u_n(\sigma)\rangle d\sigma \end{aligned}$$

for any $z \in \mathbb{R}^n$. As $B^*S(T - \,\cdot\,)^*z \in L^1(0, T; \mathbb{R}^m)$ we can take limits using weak convergence and obtain

$$\begin{aligned} \langle z, \bar{y} - S(T)\zeta\rangle &= \int_0^T \langle B^*S(T - \sigma)^*z, \bar{u}(\sigma)\rangle d\sigma \\ &= \left\langle z, \int_0^T S(T - \sigma)B\bar{u}(\sigma)d\sigma\right\rangle \end{aligned}$$

which, due to the fact that $z$ is arbitrary implies

$$\bar{y} - S(T)\zeta = \int_0^T S(T - \sigma)B\bar{u}(\sigma)d\sigma\,,$$

hence the control $\bar{u}(\sigma)$ drives $\zeta$ to $\bar{y}$. Now, each ball $\mathcal{B}^\infty_{\rho_k}(0, T; \mathbb{R}^m)$ of radius $\rho_k = \|u_k(\cdot)\|_{L^\infty(0,T;\mathbb{R}^m)}$ is closed and convex, thus weakly closed, hence it follows from (1.2.6) that $\|u_n(\cdot)\|_{L^\infty(0,T;\mathbb{R}^m)} \leq \rho_k$ $(n \geq k)$. Accordingly, $\|\bar{u}(\cdot)\|_{L^\infty(0,T;\mathbb{R}^m)} \leq \rho_k$ for all $k$, thus

$$\|\bar{u}(\cdot)\|_{L^\infty(0,T;\mathbb{R}^m)} = \rho$$

and $\bar{u}(\cdot)$ is norm optimal.

For (*b*), the hypotheses imply that there exists a decreasing sequence $\{t_n\}$ with $t_n \to T =$ optimal time and a sequence $\{u_n(\cdot)\}$, $u_n(\cdot) \in L^\infty(0, t_n; \mathbb{R}^m)$ of controls all driving $\zeta$ to $\bar{y}$ in time $t_n$,

$$\bar{y} - S(t_n)\zeta = \int_0^{t_n} S(t_n - \sigma)Bu_n(\sigma)d\sigma\,, \tag{1.2.7}$$

and such that

$$\|u_n(\cdot)\|_{L^\infty(0,t_n;\mathbb{R}^m)} \le 1\,. \tag{1.2.8}$$

Take $s > t_1$ and extend each $u_n(\cdot)$ to $t_n \le \sigma \le s$ setting $u_n(\sigma) = 0$ there. Then select a subsequence of the extended $\{u_n(\cdot)\}$ (denoted with the same name) such that $\{u_n(\cdot)\}$ is $L^1(0,s;\mathbb{R}^m)$-weakly convergent to an element $\bar{u}(\cdot) \in L^\infty(0,s;\mathbb{R}^m)$; clearly, $\bar{u}(\sigma) = 0$ for $t > T$. Next, pick $z \in \mathbb{R}^n$ and define

$$\chi_n(\sigma) = \begin{cases} B^*S(t_n-\sigma)^*z & (0 \le \sigma \le t_n) \\ 0 & (t_n < \sigma \le s) \end{cases}$$

and

$$\chi(\sigma) = \begin{cases} B^*S(T-\sigma)^*z & (0 \le \sigma \le T) \\ 0 & (T < \sigma \le s)\,. \end{cases}$$

Then,

$$\chi_n(\cdot) \to \chi(\cdot) \quad \text{in } L^1(0,s;\mathbb{R}^m)\,.$$

Now, (1.2.7) implies

$$\begin{aligned}\langle z, \bar{y} - S(t_n)\zeta\rangle &= \left\langle z, \int_0^{t_n} S(t_n-\sigma)Bu_n(\sigma)d\sigma\right\rangle \\ &= \int_0^s \langle \chi_n(\sigma), u_n(\sigma)\rangle d\sigma\,.\end{aligned}$$

Taking limits,

$$\begin{aligned}\langle z, \bar{y} - S(T)\zeta\rangle &= \int_0^s \langle \chi(\sigma), u(\sigma)\rangle d\sigma \\ &= \left\langle z, \int_0^T S(T-\sigma)B\bar{u}(\sigma)d\sigma\right\rangle,\end{aligned}$$

which again implies

$$\bar{y} - S(T)\zeta = \int_0^T S(T-\sigma)Bu(\sigma)d\sigma\,,$$

so that $\bar{u}(\sigma)$ drives $\zeta$ to $\bar{y}$ in optimal time $T$.

**Remark 1.2.2.** All results in **1.1** and in this section hold if one replaces the norm (1.1.3) in $\mathbb{R}^m$ by any other, for instance the $p$-norm

$$\|u\|_p = +\sqrt[p]{|u_1|^p + \ldots + |u_m|^p} \quad (1 \le p < \infty)\,. \tag{1.2.9}$$

Of course, the nicest things happen for $p = 2$, where the geometry is that of a Hilbert space. For the norm $\|\cdot\|_2$, (1.1.7) is equivalent to

$$\bar{u}(t) = \rho\,\frac{B^*S(T-t)^*z}{\|B^*S(T-t)^*z\|_2} \tag{1.2.10}$$

outside of the set $\subseteq [0, T]$ where $B^*S(T-t)^*z = 0$. Under the complete controllability condition (1.1.16) this set is finite, so that (1.2.10) shows that optimal controls $\bar{u}(t)$ are piecewise analytic in $0 \le t \le T$. Condition (1.2.10) implies the bang-bang properties (1.1.23) and (1.1.24). Piecewise analyticity and the bang-bang properties can be extended to the $p$-norm (1.2.9) ($p > 1$); moreover, (1.1.23) and (1.1.24) respectively imply uniqueness of time and norm optimal controls. In fact, we have

**Theorem 1.2.3.** *Let $\|\cdot\|$ be a norm in $\mathbb{R}^m$ such that*

$$\|u\| = \|v\| = 1, \quad u \neq v \quad \Longrightarrow \quad \|u+v\| < 2\,. \tag{1.2.11}$$

*Then time and norm optimal controls are unique.*

*Proof.* Let $\bar{u}(t)$, $\bar{v}(t)$ be two optimal controls, that is, two admissible controls driving $\zeta$ to $\bar{y}$ in optimal time $T$. Define

$$w(t) = \frac{1}{2}\big(\bar{u}(t) + \bar{v}(t)\big)\,. \tag{1.2.12}$$

Then $w(t)$ is admissible and drives $\zeta$ to $\bar{y}$ in time $T$, thus it is also optimal and must satisfy the bang-bang property (1.1.23). If $\bar{u}(t) \neq \bar{v}(t)$ in a set of positive measure $e$ then by (1.2.11) $\|w(t)\| < 1$ in $e$, which contradicts the bang-bang property (1.1.23). For norm optimal controls we use (1.1.24).

All the norms $\|\cdot\|_p$ with $1 < p < \infty$ satisfy (1.2.11); in fact, these norms are *uniformly convex* (Adams [1975, p. 7]), which is a stronger property than (1.2.11). On the other hand, uniqueness results for the maximum norm (1.1.3) and for the $\|\cdot\|_1$ norm require special assumptions on the control system (for the norm (1.1.3) the same that guarantee the true bang-bang property in Theorem 1.1.2).

**Theorem 1.2.4.** *Assume that, for arbitrary $z \in \mathbb{R}^n$, $z \neq 0$ the vector $B^*S(t)^*z$ is not identically perpendicular to any of the sides of the unit cube $|y_j| \le 1$, $j = 1, 2, \ldots, m$, that is, is not identically parallel to any of the vectors*

$$e_j = (0, \ldots, 1, \ldots, 0) \in \mathbb{R}^m \quad (1 \text{ in the } j^{\text{th}} \text{ place})\,.$$

*Then time and norm optimal controls are unique.*

*Proof.* If $\bar{u}(t)$, $\bar{v}(t)$ are two time optimal controls in $0 \le t \le T$ then both must obey the conclusions of Theorem 1.1.2. If the two controls don't coincide, there is a nonempty interval $a < t < b$ where they take values in different vertices of the unit cube, which in turn means that the optimal control $w(t)$ in (1.2.12) does not take values in a vertex in $a < t < b$; this is a contradiction. The same argument works for the norm optimal problem.

Theorems 1.1.2 and 1.2.4 have analogues for the $\|\cdot\|_1$ norm. In Theorem 1.1.2 we only have to replace the unit cube $|y_j| \le 1$ by the cube $|y_1| + \ldots + |y_m| \le 1$;

same for Theorem 1.2.4. When the assumptions on either unit cube don't hold, uniqueness collapses, as seen below.

**Example 1.2.5.** The system is

$$y'(t) = Ay(t) + u(t)\,, \quad y(0) = \zeta \tag{1.2.13}$$

in $\mathbb{R}^2$, with control constraint

$$\|u(t)\| = \max(|u_1(t)|, |u_2(t)|) \le 1 \tag{1.2.14}$$

($\|\cdot\|$ the maximum norm (1.1.3)). We use the diagonal matrix

$$A = \begin{bmatrix} \lambda_1 & 0 \\ 0 & \lambda_2 \end{bmatrix}$$

with $\lambda_1, \lambda_2$ real. We have

$$S(t) = S(t)^* = \begin{bmatrix} e^{\lambda_1 t} & 0 \\ 0 & e^{\lambda_2 t} \end{bmatrix},$$

thus, if $z = (z_1, z_2) \in \mathbb{R}^2$,

$$S(t)z = S(t)^* z = (e^{\lambda_1 t} z_1, e^{\lambda_2 t} z_2)\,. \tag{1.2.15}$$

In this system $B = I$, so that $B^* = I$ and the complete controllability condition (1.1.16) is trivially satisfied. The eigenvectors of $A$ are $z^1 = (1, 0)$ and $z^2 = (0, 1)$, with eigenvalues $\lambda_1, \lambda_2$. The assumptions in Theorem 1.2.4 are not satisfied; if $z = z^1$ we have

$$B^* S(t)^* z = S(t)^* z = (e^{\lambda_1 t}, 0)\,,$$

so that $B^* S(t)^* z$ is identically parallel to $e_1 = (1, 0)$. Now, for the same $z$

$$S(T-t)^* z = (e^{\lambda_1 (T-t)}, 0)\,,$$

and (1.1.6) for a control $\bar{u}(t) = (\bar{u}_1(t), \bar{u}_2(t))$ is

$$\begin{aligned} &\langle S(T-t)^* z, \bar{u}(t)\rangle = e^{\lambda_1 (T-t)} \bar{u}_1(t) \\ &= \max_{\|u\| \le 1} \langle S(T-t)^* z, u\rangle = \max_{|u_1|,\, |u_2| \le 1} e^{\lambda_1 (T-t)} u_1\,. \end{aligned} \tag{1.2.16}$$

This gives

$$\bar{u}_1(t) = 1\,, \quad |\bar{u}_2(t)| \le 1 \tag{1.2.17}$$

(the condition on $\bar{u}_2(t)$ is simply (1.2.14), as the maximum principle (1.2.16) doesn't concern $\bar{u}_2(t)$). In view of Theorem 1.1.3 any control $\bar{u}(t)$ satisfying (1.2.17) drives $(0,0)$ time and norm optimally to $y(T,0,\bar{u})$. Select a measurable function $\bar{u}_2(t)$ satisfying the second condition (1.2.17) and such that

$$u_2(\cdot) \neq 0\,, \qquad \int_0^T e^{\lambda_2(T-\sigma)}u_2(\sigma)d\sigma = 0\,,$$

and consider the two (distinct) controls

$$\bar{u}(t) = (1,0)\,, \quad \bar{v}(t) = (1, u_2(t))\,. \tag{1.2.18}$$

Then, using (1.2.15),

$$\begin{aligned} y(0,T,\bar{v}) &= \left( \int_0^T e^{\lambda_1(T-\sigma)}d\sigma\,,\ \int_0^T e^{\lambda_2(T-\sigma)}u_2(\sigma)d\sigma \right) \\ &= \left( \int_0^T e^{\lambda_1(T-\sigma)}d\sigma\,,\ 0 \right) = y(T,0,\bar{u})\,. \end{aligned}$$

It follows that both controls drive time and norm optimally the origin (0, 0) to the same target.

**Example 1.2.6.** To obtain a nonuniqueness example for the control constraint

$$\|u(t)\|_1 = |u_1(t)| + |u_2(t)| \leq 1 \tag{1.2.19}$$

we use the system (1.2.13) with matrix

$$A = \begin{bmatrix} 1 & a \\ a & 1 \end{bmatrix}$$

$(a \neq 0)$. The eigenvectors are $z^1 = (1,1)$, $z^2 = (1,-1)$ corresponding to eigenvalues $\lambda_1 = 1 + a$, $\lambda_2 = 1 - a$. We have

$$S(t) = S(t)^* = \frac{1}{2}\begin{bmatrix} e^{\lambda_1 t} + e^{\lambda_2 t} & e^{\lambda_1 t} - e^{\lambda_2 t} \\ e^{\lambda_1 t} - e^{\lambda_2 t} & e^{\lambda_1 t} + e^{\lambda_2 t} \end{bmatrix},$$

hence, if $z = (z_1, z_2) \in I\!\!R^2$,

$$\begin{aligned} &S(t)z = S(t)^* z \\ &= \left( e^{\lambda_1 t}\frac{z_1+z_2}{2} + e^{\lambda_2 t}\frac{z_1-z_2}{2}\,,\ e^{\lambda_1 t}\frac{z_1+z_2}{2} + e^{\lambda_2 t}\frac{z_2-z_1}{2} \right). \end{aligned} \tag{1.2.20}$$

Again, the complete controllability condition (1.1.16) is trivially satisfied but the assumptions in the version of Theorem 1.2.4 for the norm $\|\cdot\|_1$ are not (for these assumptions, see the comments after Theorem 1.2.4). In fact, for the vector $z = z^1 = (1,1)$ we have

$$B^*S(t)^*z = S(t)^*z = \left(e^{\lambda_1(T-t)}\,,\, e^{\lambda_1(T-t)}\right),$$

which is identically parallel to the vector $z = (1,1)$, thus identically perpendicular to the side of the cube $|y_1| + |y_2| \leq 1$ lying in the positive quadrant. For this $z$

$$S(T-t)^*z = \left(e^{\lambda_1(T-t)}\,,\, e^{\lambda_1(T-t)}\right),$$

and (1.1.6) for a control $\bar{u}(t) = (\bar{u}_1(t), \bar{u}_2(t))$ is

$$\begin{aligned} &\langle S(T-t)^*z, \bar{u}(t)\rangle = e^{\lambda_1(T-t)}(\bar{u}_1(t) + \bar{u}_2(t)) \\ &= \max_{\|u\|_1 \leq 1} \langle S(T-t)^*z, u\rangle = \max_{|u_1|+|u_2|\leq 1} e^{\lambda_1(T-t)}(u_1 + u_2)\,. \end{aligned} \tag{1.2.21}$$

This gives

$$\bar{u}_1(t) \geq 0\,,\ \bar{u}_2(t) \geq 0 \quad \bar{u}_1(t) + \bar{u}_2(t) = 1\,. \tag{1.2.22}$$

Select a measurable function $\phi(\cdot) \neq 0$ such that

$$|\phi(t)| \leq \frac{1}{2}\,, \qquad \int_0^T e^{\lambda_2(T-\sigma)}\phi(\sigma)d\sigma = 0\,,$$

and consider the two (distinct) admissible controls

$$\bar{u}(t) = \left(\frac{1}{2}\,,\, \frac{1}{2}\right), \qquad \bar{v}(t) = \left(\frac{1}{2} + \phi(t)\,,\, \frac{1}{2} - \phi(t)\right). \tag{1.2.23}$$

Both satisfy (1.2.22), thus both are norm and time optimal by Theorem 1.1.3. Using (1.2.20) we obtain

$$\begin{aligned} y(0,T,\bar{v}) &= \left(\int_0^T e^{\lambda_1(T-\sigma)}\frac{1}{2}d\sigma + \int_0^T e^{\lambda_2(T-\sigma)}\phi(\sigma)d\sigma\,,\right. \\ &\qquad \left.\int_0^T e^{\lambda_1(T-\sigma)}\frac{1}{2}d\sigma - \int_0^T e^{\lambda_2(T-\sigma)}\phi(\sigma)d\sigma\right) \\ &= \left(\int_0^T e^{\lambda_1(T-\sigma)}\frac{1}{2}d\sigma\,,\, \int_0^T e^{\lambda_1(T-\sigma)}\frac{1}{2}d\sigma\right) = y(0,T;\bar{u})\,, \end{aligned}$$

hence both controls drive time and norm optimally $(0,0)$ to the same target.

**Miscellaneous notes.** The existence and uniqueness arguments in Theorems 1.2.1 and 1.2.3 follow Bellman *et al.* [1956], who do only the time optimal case.

### 1.3. Infinite dimensional systems.

Let $A$ be the infinitesimal generator of a strongly continuous semigroup in a Banach space $E$. Beginning with the work of Hille, Feller, Yosida and others, the equation

$$y'(t) = Ay(t) + Bu(t)\,, \quad y(0) = \zeta \tag{1.3.1}$$

has been used for more than sixty years to model evolution partial differential equations

$$\frac{\partial y(t,x)}{\partial t} = Ay(t,x) + Bu(t,x)\,, \quad \beta y(t,x) = 0\,, \quad u(0,x) = \zeta(x) \tag{1.3.2}$$

in a domain $\Omega$ of $m$-dimensional Euclidean space $I\!R^m$, where $A$ (resp. $\beta$) is a differential operator acting on the space variables $x = (x_1, \ldots, x_m)$ in the domain $\Omega$ (resp. on the boundary $\Gamma$). In the the early 1960s this setup was found to be suitable for the study of partial differential equations like (1.3.2) as control systems. In the abstract model (1.3.1) controls take values in a *control set* $U \subseteq F$, where the *control space* $F$ is a second Banach space and $B : F \to E$ is bounded.[8] The *full control* particular case of (1.3.1) is

$$y'(t) = Ay(t) + u(t)\,, \quad y(0) = \zeta \tag{1.3.3}$$

where $F = E$ and $B = I$; here, controls take values in the same space as the solutions $y(t)$. The system (1.3.3) is a model for the equation

$$\frac{\partial y(t,x)}{\partial t} = Ay(t,x) + u(t,x)\,, \quad \beta y(t,x) = 0\,, \quad u(0,x) = \zeta(x)\,. \tag{1.3.4}$$

The way in which the partial differential equations (1.3.2) or (1.3.4) are reduced to the "abstract differential equations" (1.3.1) or (1.3.3) depends on the choice of control space and also on the choice of the *state space* $E$ where solutions live. To fix ideas, we look at the one-dimensional heat equation with Dirichlet boundary conditions.

**Example 1.3.1.** The system is

$$\frac{\partial y(t,x)}{\partial t} = \frac{\partial^2 y(t,x)}{\partial x^2} + u(t,x)\,, \quad y(t,a) = y(t,b) = 0\,, \quad u(0,x) = \zeta(x) \tag{1.3.5}$$

where $\Omega = (a,b)$. If the bound on the controls is

$$\int_a^b |u(t,x)|^2 dx \le C \tag{1.3.6}$$

then the "natural" control space is $F = L^2(a,b)$ and, for simplicity, we may take $E = L^2(a,b)$ as well. The operator $A$ in (1.3.3) is

$$A_2 u(x) = u''(x) \tag{1.3.7}$$

---

[8] The case $B$ unbounded is also of interest for the modelling of boundary control systems. We only deal with $B$ bounded in these notes.

with domain $D(A_2)$ consisting of all $u(\cdot) \in L^2(a,b)$ such that $u''(\cdot)$ exists in the sense of distributions, belongs to $L^2(a,b)$ and[9]

$$u(a) = u(b) = 0\,. \tag{1.3.8}$$

Note that we are incorporating the boundary conditions into the definition of the domain of $A_2$. A similar approach works when the $L^2$ constraint (1.3.6) is replaced by a $L^p$ constraint, $1 \le p < \infty$. On the other hand, if the bound on the controls is

$$|u(t,x)| \le C \tag{1.3.9}$$

the "natural" control space is $F = L^\infty(a,b)$. As for the state space $E$ we may again take $E = F$. However, $L^\infty(a,b)$ is not separable and has a complicated dual, and it's easier to model using $E = C[a,b]$, the space of continuous functions in $[a,b]$ equipped with the supremum norm. This time, the boundary conditions (1.3.8) must be incorporated in the definition of the space.[10] We define $E = C_0[a,b]$ as the subspace of all $u(\cdot) \in C[a,b]$ satisfying (1.3.8). The operator $A_c$ is defined by

$$A_c u(x) = u''(x)\,,$$

and the domain of $A_c$ is the set of all twice continuously differentiable $u(\cdot)$ such that $u(\cdot), u''(\cdot) \in C_0[a,b]$. With this choice of space, the instantaneous values $u(t,\cdot)$ of the control don't belong to $C_0[a,b]$; however this can be obviated using the smoothing properties of the semigroup generated by $A_c$.

The treatment of (1.3.4) in arbitrary space dimension as the abstract control system (1.3.3) in $E = L^p(\Omega)$, $1 < p < \infty$ is done in Chapters 2 and 3 of this monograph. Chapter 4 deals with the state space $C(\overline{\Omega})$. The $L^1(\Omega)$ setting is in Chapter 5; due to existence needs, controls must take values in measure spaces rather than in $L^1(\Omega)$.

**Example 1.3.2.** The system is

$$\frac{\partial y(t,x)}{\partial t} = \frac{\partial^2 y(t,x)}{\partial x^2} + b(x)u(t)\,, \quad y(t,a) = y(t,b) = 0, \quad u(0,x) = \zeta(x) \tag{1.3.10}$$

where the control is a scalar function $u(t)$. We use the model (1.3.1) with $F = \mathbb{R}$, $Bu = b(x)u$. Just as for the system (1.3.4), we can model (1.3.10) with $E = L^p(a,b)$ or $E = C[a,b]$; in this example, the control and state spaces are different. The definition of the infinitesimal generators is the same.

[9] This is equivalent to: $u(\cdot)$ is continuously differentiable with absolutely continuous derivative and $u''(\cdot) \in L^2(a,b)$.

[10] If we take $E = C[a,b]$ and incorporate the boundary conditions (1.3.8) in $D(A)$, then $A$ is not densely defined in $E$. This precludes application of classical semigroup theory.

We go back to the general model (1.3.1). To make it work we require that the controls $u(\cdot)$ in (1.3.1) be *strongly measurable* and *a.e. bounded,* that is, that they belong to the space $L^\infty(0,T;F)$. Under these assumptions, the variation-of-constants integral

$$y(t,\zeta,u) = S(t)\zeta + \int_0^t S(t-\sigma)Bu(\sigma)d\sigma \tag{1.3.11}$$

exists in the sense of Lebesgue - Bochner (Hille - Phillips [1957]) since the integrand is strongly measurable and bounded in any interval $0 \le \sigma \le t$. It is a particular case of Theorem 1.3.3 below that $y(t,\zeta,u)$ is continuous. We *define*

$$y(t) = y(t,\zeta,u)$$

as the *solution* of the initial value problem (1.3.1). It should be pointed out that $y(t)$ is just a *weak* solution; it may not be a solution of (1.3.3) in the strong sense, that is, $y(t)$ may not belong to $D(A)$ for any $t$, and/or the derivative $y'(t)$ may not exist as the limit of the quotient of increments in $E$. This turns out to be totally irrelevant in the control theory of (1.3.1); only continuity of $y(t)$ is needed. It guarantees that the evaluation $y(t)$ at a given $t$ is well defined, which allows us to hit point targets. It can be shown (see for instance the author [1983, Theorem 2.4.6, p. 89]) that the function (1.3.11) is a solution of the initial value problem (1.3.1) in the sense of (vector valued) distributions. This plays no role or has any application in the results.

We recall (Hille - Phillips [1957]) that there exist constants $C$, $\omega$ such that

$$\|S(t)\| \le Ce^{\omega t} \quad (t \ge 0)\,.$$

**Theorem 1.3.3.** *Assume that $u(\cdot) \in L^1(0,T;F)$. Then the function $t \to y(t,\zeta,u)$ is continuous in $0 \le t \le T$ in the norm of $E$.*

*Proof.* If $t < t'$ we have

$$\begin{aligned} y(t',\zeta,u) - y(t,\zeta,u) = &\int_t^{t'} S(t'-\sigma)Bu(\sigma)d\sigma \\ &+ \int_0^t (S(t'-\sigma) - S(t-\sigma))Bu(\sigma)d\sigma\,, \end{aligned}$$

so that

$$\begin{aligned} \|y(t',\zeta,u) - y(t,\zeta,u)\| \le &\int_t^{t'} \|S(t'-\sigma)Bu(\sigma)\|d\sigma \\ &+ \int_0^t \|(S(t'-\sigma) - S(t-\sigma))Bu(\sigma)\|d\sigma\,. \end{aligned}$$

The first integral tends to zero since the integrand is $\leq C\|u(\sigma)\|$; the integrand of the second integral shares this bound and tends to zero due to strong continuity of the semigroup, which makes possible application of the dominated convergence theorem.[11]

Solutions $y(t, \zeta, u)$ are also called *trajectories.* We say that a control $u(t)$ *drives* the initial condition $\zeta \in E$ to $\bar{y} \in E$ in time $T$ if the *target condition*

$$y(T, \zeta, u) = S(T)\zeta + \int_0^T S(T - \sigma)Bu(\sigma)d\sigma = \bar{y} \tag{1.3.12}$$

is satisfied; $\bar{y}$ is the *target.* The time optimal problem and the norm optimal problem are set up in the same way as in **1.1.**

This monograph deals with generalizations of the finite dimensional results in **1.1** and **1.2** to the infinite dimensional control systems (1.3.1) and (1.3.3) (mostly the latter). Unlike in finite dimension, where the approaches to (1.3.1) and (1.3.3) are essentially the same, the results are very different in infinite dimension; a lot more is known about the fully controlled equation (1.3.3) than about (1.3.1). Also, some sharp differences appear between the norm optimal and the time optimal problem, which problems are equivalent in finite dimension if (1.1.16) is satisfied. Here is a preview of some of the results.

EXISTENCE. This is the one area where finite dimensional results generalize with few, if any, unexpected complications. If the control space $F$ is the dual of another Banach space $X$, all existence arguments in **1.2** generalize automatically if

$$L^\infty(0, T; F) = L^1(0, T; X)\,. \tag{1.3.13}$$

This happens whenever $X$ is reflexive or $F$ separable. This setup works well for modeling (1.3.5) in $L^2(a, b)$ where $F = X = L^2(a, b)$, or more generally in $L^p(a, b)$ with $1 < p < \infty$. For infinite dimensional existence theory based on (1.3.13) see **3.1**, especially Theorem 3.1.2.

If the control constraint, however, is (1.3.9), modifications are needed. We outline these modifications for the system (1.3.5). The state space is $C[a, b]$, with control space $F = L^\infty(a, b)$ and $X = L^1(a, b)$. Equality (1.3.13) is no longer true. However, it can be repaired if the space $L^\infty(0, T; F)$ on the left is replaced by $L^\infty_w(0, T; F)$ defined in the same way but using $X$-weakly measurable (instead of strongly measurable) functions. This approach is used in Chapter 4 (to model heat

[11] When (1.3.5) is modeled as an equation in the space $E = C_0[a, b]$ with controls in $L^\infty((0, T) \times (a, b))$ then $u(\sigma)$ in the integral (1.3.12) is not strongly measurable (in fact, it doesn't even take values in $E$). However, (1.3.12) can be extended and produces a strongly continuous $y(t, \zeta, u)$. The same happens for the multidimensional equation (1.3.4). This is the setting in Chapters 4 and 5.

propagation processes in arbitrary space dimension). The corresponding existence result for the $m$-dimensional system (1.3.4) is Theorem 4.8.1.

If the bound on the controls is

$$\int_a^b |u(t,x)|dx \leq C$$

the natural state space for (1.3.5) is $L^1(a,b)$. It doesn't make it as a control space, since it is not the dual of any other space. This motivates the choice $F = \Sigma[a,b]$, the space of all finite Borel measures in $a \leq x \leq b$ equipped with the total variation norm. Equality (1.3.13) is still in force if we replace $L^\infty(0,T;F)$ by $L^\infty_w(0,T;F)$. This is the setup in the study of controlled diffusions, and the pertinent existence result (in arbitrary dimension) is Theorem 5.5.1.

COMPLETE CONTROLLABILITY. In finite dimension, (1.1.16) is equivalent to complete controllability (see the end of **1.2**). The equivalence vanishes in infinite dimension; (1.1.16) only implies *approximate controllability* (we can drive from the origin to a dense subspace). The time needed to drive to $\bar{y}$ may depend on $\bar{y}$.

THE OPTIMALITY PRINCIPLE. Theorem 1.1.6 holds (without changes in the proof) for the time optimal problem. In finite dimension, the optimality principle is as well true for the norm optimal problem under (1.1.16) (as pointed out after Theorem 1.1.6) but it does not hold in infinite dimension: the optimality principle depends on the bang-bang property (1.1.24), which may not hold for certain norm optimal controls.

FULL CONTROL. This refers to the general system (1.3.1). If the semigroup $S(t)$ is a *group* or, more generally, if

$$S(t)E = E \quad (t \geq 0) \tag{1.3.14}$$

then the maximum principle (Theorem 1.1.1) generalizes without changes: every time optimal control $\bar{u}(t)$ satisfies

$$\langle S(T-t)^*z, \bar{u}(t)\rangle = \max_{\|u\|\leq 1} \langle S(T-t)^*z, u\rangle \,, \tag{1.3.15}$$

and every norm optimal control satisfies

$$\langle S(T-t)^*z, \bar{u}(t)\rangle = \max_{\|u\|\leq \rho} \langle S(T-t)^*z, u\rangle \tag{1.3.16}$$

with $\rho = \|\bar{u}(\cdot)\|_{L^\infty(0,T;E)}$; in both cases $z \in E^* =$ dual space of $E$, $z \neq 0$. The proofs use the same separation scheme in Theorem 1.1.1. However, if condition (1.3.14) is not satisfied, the set

$$B^\infty_\rho(T) = \left\{ \int_0^T S(T-\sigma)u(\sigma)d\sigma;\ \|u(\cdot)\|_{L^\infty(0,T;E)} \leq \rho \right\}$$

does not contain interior points in $E$ (Theorem 2.2.3) so that it may not be possible to separate boundary points of $B_\rho^\infty(T)$ from $B_\rho^\infty(T)$ by means of functionals in $E^*$. This difficulty is (partly) bypassed doing the separation in the space

$$R^\infty(T) = \left\{ y = \int_0^T S(T-\sigma)u(\sigma)d\sigma;\ u(\cdot) \in L^\infty(0,T;E) \right\}$$

equipped with the norm

$$\|y\|_{R^\infty(T)} = \left\{ \inf \|u(\cdot)\|_{L^\infty(0,T;F)};\ \int_0^T S(T-\sigma)u(\sigma)d\sigma = y \right\}.$$

In this space, $B_\rho^\infty(T)$ is (contained in) the sphere of center $0$ and radius $\rho$. To do the separation in $R^\infty(T)$ of course requires the identification[12] of the dual $R^\infty(T)^\star$ and produces the following result: if

$$\bar{y} \in \overline{D(A)} \tag{1.3.17}$$

(the bar indicates closure in $R^\infty(T)$) then the maximum principle (1.3.15) (resp. (1.3.16)) holds for time (resp. norm) optimal controls with $z$ in a space $Z$ larger than the dual $E^*$ (Theorem 2.5.3). The dual semigroup $S(t)^*$ is defined in $Z$ for $t > 0$ and satisfies $S(t)^*Z \subseteq E^*$ and[13]

$$\int_0^T \|S(\sigma)^* z\| d\sigma < \infty \tag{1.3.18}$$

((1.3.18) is essentially the definition of $Z$). On the other hand, under (1.3.18), (1.3.15) (resp. (1.3.16) is a sufficient condition for time (resp. norm) optimality; we don't need (1.3.17) here. See Theorems 2.5.5. and 2.5.7.

Necessary conditions collapse without (1.3.17) and the collapse is different for the norm and the time optimal problem. Under a condition which is roughly the opposite of (1.3.14) we can produce norm optimal controls that don't satisfy (1.3.16) even if $z$ is taken in a space larger than $Z$ (Theorem 2.8.6). On the other hand, (at least for analytic semigroups in Hilbert spaces) we can show that time optimal controls will always satisfy (1.3.15) with $z$ in a space much larger than $Z$, where (1.3.18) may not hold (Theorem 3.2.3). For certain nonanalytic semigroups

---

[12] "Identification" is overly optimistic. In fact, for many spaces and semigroups, most of the functionals in $R^\infty(T)^\star$ are *singular functionals* (definition in (2.3.36)). Not much is known about these functionals except that separation with them doesn't produce the maximum principle. In fact, much of the work in this book consists in finding ways to avoid singular functionals.

[13] A linear space $\mathcal{Z} \supseteq E^*$ where $S(t)^*$ can be defined in such a way that $S(t)^*\mathcal{Z} \subseteq E^*$ for $t > 0$ is called a *multiplier space.*

we can produce examples of time optimal controls that do not satisfy (1.3.15) for any multiplier $z$ (Theorem 2.8.4).

There are known facts and open questions about the role of (1.3.18) in the maximum principle. One outstanding open question is: what growth conditions on $\|S(t)^*z\|$ as $t \to 0$ are necessary (or sufficient) for a control $\bar{u}(t)$ satisfying (1.3.15) to be time optimal? norm optimal? We know that (1.3.18) is sufficient, but it is not necessary: there exists an example of a (time, norm) optimal control $\bar{u}(t)$ that satisfies (1.3.15) with

$$\|S(t)^*z\| \approx \frac{c}{t}, \tag{1.3.19}$$

so that (1.3.18) does not hold (Theorem 3.2.4). On the other hand, it is not known if (1.3.19) is necessary; all that has been proved (in Hilbert space, self adjoint generator) is that if $\|S(t)^*z\|$ increases "fast enough" as $t \to 0$ (for instance, exponentially) a control satisfying (1.3.15) will not be time or norm optimal. This is proved in Lemma 3.5.9 (Example 3.5.11 shows that the rate of growth required in Lemma 3.5.9 actually appears in "real life").

As all of the above implies, under condition (1.3.17) we have a pretty tidy theory of the time and norm optimal problems. On the other hand, if (1.3.17) is discarded not much remains standing, especially necessary conditions.

How demanding is (1.3.17)? The inclusion

$$\overline{D(A)} \subseteq R^{\infty}(T) \tag{1.3.20}$$

is strict (except in very special situations), so that (1.3.17) leaves some targets out, that is, excludes some optimal trajectories.

Since the bang-bang property

$$\|\bar{u}(t)\| = 1 \tag{1.3.21}$$

for time optimal controls follows from the maximum principle (1.3.15) and there are so many caveats in the results leading to (1.3.15), one may expect that the same caveats would apply to (1.3.21). Surprisingly, this is not the case; (1.3.21) *always* holds for time optimal controls with no restriction on the space, the semigroup or the target (Theorem 2.1.3). On the other hand, if (1.3.17) is given up, (1.3.21) may not hold for norm optimal controls; examples in Theorems 2.6.1 and 2.8.6.

UNIQUENESS AND CONTINUOUS DEPENDENCE. Uniqueness of time optimal controls follows from (1.3.21) in strictly convex spaces under no conditions on the target (Theorem 2.1.7). In the $C(\overline{\Omega})$ or $L^1(\Omega)$ setting in Chapters 4 and 5 (1.3.21) does not imply uniqueness. Uniqueness in $C(\overline{\Omega})$ is still attainable under a condition corresponding to (1.3.17) (Theorem 4.8.4), but in $L^1(\Omega)$ uniqueness collapses even under that condition (Theorem 5.5.6). It is interesting to note that the geometry of the nonuniqueness example roughly corresponds to that of Example 1.2.6. For the norm optimal problem, there are no uniqueness results without a condition of the type of (1.3.17); counterexamples make an appearance in Theorem 2.6.1, Theorem 3.1.7 and other places.

Roughly speaking, each time uniqueness can be proved a continuous dependence statement falls into place. These statements can be found in **3.6** and **4.8**. In the $L^1(\Omega)$ setting lack of uniqueness prevents continuous dependence, but there are results on "continuous dependence of supports" (Theorem 5.5.4).

THE GENERAL CASE. Not unexpectedly, the treatment of (1.3.1) is much more difficult than that of (1.3.3) and results are far from comprehensive. The example below gives an idea of the complications.

**Example 1.3.4.** Let $E$ be a Hilbert space, $\{\phi_n\}$ a complete orthonormal sequence and $A$ the self adjoint operator

$$A\Big(\sum_{n=1}^{\infty} c_n\phi_n\Big) = -\sum_{n=1}^{\infty} \lambda_n c_n \phi_n\,, \quad 0 < \lambda_1 < \lambda_2 < \ldots, \quad \lambda_n \to \infty\,. \tag{1.3.22}$$

If

$$\sum_{n=1}^{\infty} \frac{1}{\lambda_n} = \infty \tag{1.3.23}$$

then there exists $b \in E$ such that the system

$$y'(t) = Ay(t) + bu(t) \tag{1.3.24}$$

with one-dimensional control space $F = I\!R$ is *rigid* in the sense that, for every $T > 0$ the map

$$L^\infty(0,T) \ni u(\cdot) \to y(T,0,u) = \int_0^T S(T-\sigma)bu(\sigma)d\sigma \in E \tag{1.3.25}$$

is one-to-one. This means, if a control $u(\cdot)$ drives $\zeta$ to $\bar{y}$ in time $T$,

$$\bar{y} - S(T)\zeta = \int_0^T S(T-\sigma)bu(\sigma)d\sigma\,,$$

no other control can do the same job in the same time, so that *every* $u(\cdot) \in L^\infty(0,T)$ *is norm optimal.* It can be shown that $B^*S(\sigma)^*z = \langle b, S(\sigma)^*z\rangle \neq 0$ except in a sequence $\{t_n\}$, so that (1.3.16) is

$$\bar{u}(t) = \rho\,\mathrm{sign}\langle b, S(\sigma)^*z\rangle \tag{1.3.26}$$

which is no longer a necessary condition for optimality. It happens to be a sufficient condition, but an empty one; every control satisfying (1.3.26) is norm optimal, but then, so is any other control. Thus, we have a complete breakdown of the necessity and sufficiency of the maximum principle for norm optimal controls.

A similar breakdown occurs for the time optimal problem. Assume that $u(t)$, $\|u(\cdot)\|_{L^\infty(0,T)} \leq 1$ drives 0 to $\bar{y}$ in time $T$,

$$\bar{y} = \int_0^T S(T-\sigma)bu(\sigma)d\sigma \,. \tag{1.3.27}$$

If there exists another control $v(t)$, $\|v(\cdot)\|_{L^\infty(0,T)} \leq 1$ driving 0 to $\bar{y}$ in a smaller time $T - \delta < T$ we have

$$\bar{y} = \int_0^{T-\delta} S(T-\delta-\sigma)bv(\sigma)d\sigma = \int_\delta^T S(T-\sigma)bv(\sigma-\delta)d\sigma \,. \tag{1.3.28}$$

Comparing (1.3.27) with (1.3.28) and using the fact that the map (1.3.25) is 1-1 we deduce that the integrands in (1.3.27) and (1.3.28) are the same, thus

$$u(\sigma) = 0 \quad \text{a. e. in} \quad 0 \leq t \leq \delta \,.$$

Hence, *a control* $u(\cdot)$ *that does not vanish identically in any interval* $0 \leq \sigma \leq \delta$ *drives* 0 *to* $y(T, 0, u)$ *time optimally.* Again, (1.3.15) collapses as a necessary condition for time optimality. It is valid (but useless) as a sufficient condition.

With the operator (1.3.22), the situation is totally different if

$$\sum_{n=1}^{\infty} \frac{1}{\lambda_n} < \infty \tag{1.3.29}$$

which is the case for the system (1.3.10) in any interval $[a, b]$; for $[a, b] = [0, \pi]$, $\lambda_n = n^2$. It can be shown that if the target $\bar{y}$ belongs to a space $K$ (defined by very fast decay of Fourier coefficients) then (1.3.15) is a necessary condition for optimality. However the condition that $\bar{y} \in K$ leaves out many (perhaps most) targets. For references and additional information on the system (1.3.24) under (1.3.29) and other assumptions see the author [1999:2].

**Miscellaneous notes.** There exist numerous treatments of semigroup theory that include results on the variation-of-constants formula (1.3.11); see for instance the author [1983], Pazy [1983], Goldstein [1985]. For a treatment specifically oriented to control applications, see the author [1999:2]. In the abstract parabolic case (analytic semigroup $S(t)$) there is much recent work on regularity properties of $y(t, \zeta, u)$; see Amann [1995].

Example 1.3.4 is a particular case of the results in the author [1966:1], where $A$ need not be self adjoint in Hilbert space; the growth condition (1.3.23) on the eigenvalues is the same. The rigidity property (1-1 character of the map (1.3.25)) also appears when $A$ does not have pure point spectrum. Rigidity of the system (1.3.24) is a generic property; it holds for any $b$ outside of a set of the first category in $E$. For generalizations of these ideas from 1-dimensional to finite dimensional controls see the author [1966/7].

The theory of infinite dimensional systems had its beginnings in three different fields at about the same time. The first field was that of functional differential equations and equations with memory, where the state of the system is the restriction of the solution (in a suitable function space) to a finite or infinite time interval (Jaratishvili [1961], Friedman [1964], Halanay [1968], Banks - Jacobs [1970]). For additional references see Bensoussan - Da Prato - Delfour - Mitter [1992], [1993]. The second field was that of differential equations in Banach spaces, where Yu. B. Egorov [1962], [1964] proved an infinite dimensional generalization of Pontryagin's maximum principle. For linear equations, semigroup theory was applied in the author [1964], [1966:1], [1966:2], Balakrishnan [1965], Friedman [1967], [1968] and Conti [1968]. The third field was that of partial differential equations; some of the early works are Yu. B. Egorov [1963:1], [1963:2], Butkovski [1965], A. I. Egorov [1967], Lions [1966], [1968], Russell [1966], [1967:1], [1967:2] and Galchuk [1968].

Of course, these three areas are not disjoint; differential equations in Banach spaces are used for modeling both functional and partial differential equations, an approach that has proved very fruitful in matters related to Pontryagin's maximum principle for evolution partial differential equations, either parabolic or hyperbolic. In other subjects, such as optimal problems for elliptic equations or the relation of the maximum principle with domains of influence and dependence for hyperbolic equations, methods that deal directly with the partial differential equation in question have been useful.

Further history of infinite dimensional control theory is outside the scope of this book, and can be found in treatises on general (linear as well as nonlinear) infinite dimensional systems such as Li-Yong [1995] or the author [1999:2]. We include, however in **6.3** some additional references on various subjects not covered in this book.

CHAPTER 2

# SYSTEMS WITH STRONGLY MEASURABLE CONTROLS, I

**2.1. The reachable space and the bang-bang property.** The system is

$$y'(t) = Ay(t) + u(t)\,, \quad y(0) = \zeta \tag{2.1.1}$$

under the conditions set up in **1.3**; this means $A$ is the infinitesimal generator of a strongly continuous semigroup $S(t)$ in a Banach space $E$ and $u(\cdot) \in L^\infty(0, T; E)$. The solution of (2.1.1) is, by definition,

$$y(t, \zeta, u) = S(t)\zeta + \int_0^t S(t-\sigma)u(\sigma)d\sigma\,.$$

Given $t > 0$, the *reachable space* $R^\infty(t)$ (at time $t$) consists of all

$$y = y(t, 0, u) = \int_0^t S(t-\sigma)u(\sigma)d\sigma\,, \quad u(\cdot) \in L^\infty(0, t; E)\,, \tag{2.1.2}$$

and is equipped with the norm

$$\|y\|_{R^\infty(t)} = \inf\left\{\|u\|_{L^\infty(0,t;E)};\ \int_0^t S(t-\sigma)u(\sigma)d\sigma = y\right\}. \tag{2.1.3}$$

Every strongly continuous semigroup $S(t)$ satisfies

$$\|S(t)\| \le Ce^{\omega t} \quad (t \ge 0)$$

for suitable constants $C, \omega$, hence

$$M(t) = \max_{0\le\sigma\le t} \|S(\sigma)\|_{(E,E)} \tag{2.1.4}$$

is finite ($M(t) \le Ce^{\omega t}$ if $\omega \ge 0$, $M(t) \le C$ if $\omega \le 0$). We have

$$\|y\|_E \le tM(t)\|u(\cdot)\|_{L^\infty(0,T;E)}$$

for every $y \in E$, $u(\cdot) \in L^\infty(0,T;E)$ that satisfy (2.1.2); taking infimum on the right side,

$$\|y\|_E \le tM(t)\|y\|_{R^\infty(t)}, \tag{2.1.5}$$

hence we have the imbedding $R^\infty(t) \hookrightarrow E$.

**Lemma 2.1.1.** *The space $R^\infty(t)$ is complete.*

*Proof.* The space $R^\infty(t)$ coincides algebraically and metrically with the quotient space $L^\infty(0,t;E)/\mathcal{N}(t)$, $\mathcal{N}(t)$ the closed subspace of $L^\infty(0,t;E)$ defined by

$$\int_0^t S(t-\sigma)u(\sigma)d\sigma = 0 .$$

Accordingly, $R^\infty(t)$ is complete. (Dunford - Schwartz [1958 p. 72]). For a direct proof, let $\{y_n\}$ be a Cauchy sequence in $R^\infty(t)$. Obviously, we can select inductively a subsequence, denoted again $\{y_n\}$, with

$$\|y_{n+1} - y_n\|_{R^\infty(t)} \le \frac{1}{2^n},$$

and then, using the definition (2.1.3) of the norm of $R^\infty(t)$, select inductively a subsequence $\{u_n(\cdot)\} \in L^\infty(0,t;E)$, denoted again $\{u_n(\cdot)\}$, such that

$$y_n = \int_0^t S(t-\sigma)u_n(\sigma)d\sigma, \quad \|u_{n+1}(\cdot) - u_n(\cdot)\|_{L^\infty(0,t;E)} \le \frac{1}{2^{n-1}}. \tag{2.1.6}$$

It follows that

$$y_{n+k} - y_n = \int_0^t S(t-\sigma)(u_{n+k}(\sigma) - u_n(\sigma))d\sigma,$$
$$\|u_{n+k}(\cdot) - u_n(\cdot)\|_{L^\infty(0,t;E)} \le \sum_{k=n}^{\infty} \frac{1}{2^{n-1}} = \frac{1}{2^{n-2}},$$

the second line implying that $\{u_n(\cdot)\} \subset L^\infty(0,t;E)$ is Cauchy, hence convergent to $u(\cdot) \in L^\infty(0,t;E)$. Taking limits in the first line as $k \to \infty$ we see that $y_{n+k}$ is convergent to some $y \in E$. Using the definition of $y_n$ in (2.1.6),

$$y = \int_0^t S(t-\sigma)u(\sigma)d\sigma, \quad y - y_n = \int_0^t S(t-\sigma)(u(\sigma) - u_n(\sigma))d\sigma,$$
$$\|u(\cdot) - u_n(\cdot)\|_{L^\infty(0,t;E)} \le \frac{1}{2^{n-2}},$$

so that $\|y - y_n\|_{R^\infty(T)} \to 0$. A Cauchy sequence having a convergent subsequence is itself convergent (to the same limit), thus we are through.

The equality

$$\int_0^s S(s-\sigma)u(\sigma)d\sigma = \int_{t-s}^t S(t-\sigma)u(\sigma-(t-s))d\sigma$$

shows that

$$R^\infty(s) \subseteq R^\infty(t)\,, \quad \|y\|_{R^\infty(s)} \le \|y\|_{R^\infty(t)} \quad (0 < s \le t)\,. \tag{2.1.7}$$

On the other hand,

$$\begin{aligned}\int_0^t S(t-\tau)u(\tau)d\tau &= S(s)\int_0^{t-s} S(t-s-\tau)u(\tau)d\tau \\ &\quad + \int_{t-s}^t S(t-\tau)u(\tau)d\tau \\ &= \int_0^s S(s-\sigma)\Big(\frac{S(\sigma)}{s}\int_0^{t-s} S(t-s-\tau)u(\tau)d\tau\Big)d\sigma \\ &\quad + \int_0^s S(s-\sigma)u(\sigma+t-s)d\sigma\,,\end{aligned} \tag{2.1.8}$$

hence $R^\infty(t) \subseteq R^\infty(s)$ and

$$\|y\|_{R^\infty(s)} \le \Big(1+\frac{t-s}{s}M(s)\,M(t-s)\Big)\|y\|_{R^\infty(t)} \quad (0 < s \le t)\,. \tag{2.1.9}$$

Combined with (2.1.7), (2.1.8) shows that $R^\infty(s) = R^\infty(t)$ for all $s, t > 0$; all that can be reached in time $s > 0$ can be reached in any other time $t > 0$. Moreover, (2.1.7) and (2.1.9) show that all norms $\|\cdot\|_{R^\infty(t)}$ are equivalent to each other, thus "at time $t$" can be dropped from the definition (2.1.2) of the reachable space. Finally,

$$S(t)y = \int_0^t S(t-\sigma)\frac{S(\sigma)y}{t}d\sigma\,, \tag{2.1.10}$$

hence

$$S(t)E \subseteq R^\infty(t)\,, \quad \|S(t)y\|_{R^\infty(t)} \le \frac{M(t)}{t}\|y\|_E\,. \tag{2.1.11}$$

Fix $T > 0$. In view of (2.1.10), (2.1.11) and the equivalence of the norms $\|\cdot\|_{R^\infty(t)}$ and $\|\cdot\|_{R^\infty(T)}$, $S(t) \in \mathcal{L}(E, R^\infty(T))$ for $t > 0$.[1] In view of (2.1.5), $S(t) \in \mathcal{L}(R^\infty(T), R^\infty(T))$ for $t > 0$. It follows that

$$\mathbf{S}(t)y = S(t)y$$

[1] $\mathcal{L}(X,Y)$ is the space of all linear bounded operators from the Banach space $X$ into the Banach space $Y$ equipped with the operator norm.

defines a semigroup in $R^\infty(T)$. We have

$$
\begin{aligned}
&(\mathbf{S}(t+h) - \mathbf{S}(t)) \int_0^T S(T-\sigma)u(\sigma)d\sigma \\
&= (S(t+h) - S(t)) \int_0^T S(T-\sigma)u(\sigma)d\sigma \\
&= \int_0^T S(T-\sigma)(S(t+h) - S(t))u(\sigma)d\sigma\,,
\end{aligned}
$$

so that, if $S(t)$ is continuous in the norm of $\mathcal{L}(E, E)$ for $t > 0$, $\mathbf{S}(t)$ is continuous in the norm of $\mathcal{L}(R^\infty(T), R^\infty(T))$ for $t > 0$. Continuity at $t = 0$ does not follow, since $S(h)u(\cdot)$ does not in general converge to $u(\cdot)$ in $L^\infty(0, T; E)$ as $h \to 0$ except in the trivial case where $A$ is bounded.

The domain $D(A)$ of $A$ is a Banach space equipped with the graph norm $\|y\|_{D(A)} = \|y\| + \|Ay\|$. If $y \in D(A)$, integration by parts gives

$$
y = \int_0^T S(T-\sigma)\frac{1}{T}(y - \sigma Ay)d\sigma\,, \tag{2.1.12}
$$

thus

$$
D(A) \subseteq R^\infty(T)\,, \quad \|y\|_{R^\infty(T)} \le \frac{\|y\| + T\|Ay\|}{T} \le \frac{\max(1,T)}{T}\|y\|_{D(A)}\,. \tag{2.1.13}
$$

Combining with (2.1.5) we obtain the two imbeddings

$$
D(A) \hookrightarrow R^\infty(T) \hookrightarrow E\,. \tag{2.1.14}
$$

If $y \in D(A)$ then $t \to S(t)y$ is continuous in the norm of $D(A)$ in $t \ge 0$, thus, by the first imbedding (2.1.14), $t \to S(t)y$ is continuous in $t \ge 0$ in the norm of $R^\infty(T)$ for any $T > 0$.

**Theorem 2.1.2.** *Let $v(\cdot) \in L^\infty(0, t; E)$, $\|v(\cdot)\|_{L^\infty(0,t;E)} = 1$ drive $\zeta$ to $\bar{y}$ in time $t$. Assume $v(\cdot)$ is not norm optimal. Then $v(\cdot)$ is not time optimal.*

*Proof.* "Not norm optimal" means there is another control $u(\cdot) \in L^\infty(0, t; E)$ driving $\zeta$ to $\bar{y}$ in the same interval,

$$
\bar{y} = S(t)\zeta + \int_0^t S(t-\sigma)u(\sigma)d\sigma\,, \tag{2.1.15}
$$

and such that

$$
\|u(\cdot)\|_{L^\infty(0,t;E)} = \rho < 1\,. \tag{2.1.16}
$$

We take $s < t$, rewrite (2.1.15) in the form

$$\begin{aligned}\bar{y} &= S(s)\zeta + (S(t) - S(s))\zeta + \int_0^t S(t-\sigma)u(\sigma)d\sigma \\ &= S(s)\zeta + S(s)(S(t-s) - I)\zeta + \int_0^t S(t-\sigma)u(\sigma)d\sigma\,,\end{aligned}$$

and use (2.1.10) (resp. (2.1.8)) in the second (resp. third) term on the right side, obtaining

$$\begin{aligned}\bar{y} &= S(s)\zeta + \int_0^s S(s-\sigma)\frac{S(\sigma)}{s}(S(t-s) - I)\zeta\, d\sigma \\ &\quad + \int_0^s S(s-\sigma)\Big(\frac{S(\sigma)}{s}\int_0^{t-s} S(t-s-\tau)u(\tau)d\tau\Big)d\sigma \\ &\quad + \int_0^s S(s-\sigma)u(\sigma+t-s)d\sigma \\ &= S(s)\zeta + \int_0^s S(s-\sigma)w(\sigma)d\sigma \end{aligned} \tag{2.1.17}$$

where

$$\begin{aligned}w(\sigma) &= \frac{S(\sigma)}{s}(S(t-s) - I)\zeta \\ &\quad + \frac{S(\sigma)}{s}\int_0^{t-s} S(t-s-\tau)u(\tau)d\tau + u(\sigma+t-s)\,,\end{aligned}$$

thus

$$\begin{aligned}\|w(\sigma)\| &\le \frac{M(s)}{s}\|(S(t-s)\zeta - \zeta\| \\ &\quad + \Big(1 + \frac{t-s}{s}M(s)\,M(t-s)\Big)\rho \quad (0 \le \sigma \le s)\,. \end{aligned} \tag{2.1.18}$$

According to (2.1.17) the control $w(\cdot)$ drives $\zeta$ to $\bar{y}$ in time $s$ and it follows from (2.1.16) and (2.1.18) that, if $s$ is sufficiently close to $t$ then $\|w(\cdot)\|_{L^\infty(0,s;E)} \le 1$, which shows that $t$ is not the optimal driving time. This ends the proof.

**Theorem 2.1.3.** *Let $\bar{u}(t)$ be a time optimal control in $0 \le t \le T$. Then*

$$\|\bar{u}(t)\| = 1 \quad \textit{a. e. in } \; 0 \le t \le T\,. \tag{2.1.19}$$

The proof requires some auxiliary results. Denote by $|e|$ the Lebesgue measure of a measurable set $e \subseteq I\!R$. A point $t \in e$ is a *density point* of $e$ if

$$\lim_{h\to 0}\frac{|(t-h, t+h)\cap e|}{2h} = 1\,.$$

**Lemma 2.1.4.** *Almost every point of a measurable set $e$ is a density point.*

For a proof see Natanson [1955, p. 261]. A point $t \in e$ is a *left density point* of $e$ if

$$\lim_{h \to 0} \frac{|(t-h,t) \cap e|}{h} = 1 .$$

Since

$$\frac{|(t-h,t+h) \cap e|}{h} = \frac{|(t-h,t) \cap e|}{h} + \frac{|(t,t+h) \cap e|}{h}$$

where both terms on the sum on the right are $\leq 1$, every density point is as well a left density point.

**Lemma 2.1.5.** *Let $e$ be a set of positive measure in $I\!R$. Then, given $\rho$ with $0 < \rho < 1$ there exists $c > 0$ such that: for almost every $t \in e$ there exists a sequence $\{t_n\} \subset e$, $t_1 < t_2 < \ldots < t$, $t_n \to t$ such that*

$$|(t_n, t_{n+1}) \cap e| \geq \rho(t_{n+1} - t_n) , \qquad \frac{t_n - t_{n-1}}{t_{n+1} - t_n} \leq c \qquad (n = 1, 2, \ldots) . \tag{2.1.20}$$

*Proof.* Let $e_m \subset e$ be the set

$$e_m = \Big\{ t \in e;\ |(t-h,t) \cap e| \geq \rho h = \rho |(t-h,t)| ,\ h \leq \frac{1}{m} \Big\} . \tag{2.1.21}$$

If $t$ is a left density point of $e$ then $t \in e_m$ for $m$ large enough, so that

$$e = \Big( \bigcup_{m=1}^{\infty} e_m \Big) \cup b , \quad |b| = 0 .$$

Moreover, if $d_m \subseteq e_m$ is the set of left density points of $e_m$, each $d_m$ has full measure in $e_m$, thus we have

$$e = \Big( \bigcup_{m=1}^{\infty} d_m \Big) \cup b' , \quad |b'| = 0 .$$

Hence, it is enough to prove existence of the sequence $\{t_n\}$ for arbitrary $m$ and $t \in d_m$. Since $t$ is a left density point of $e_m$ we can choose an arbitrary sequence such that

$$t_n \in e_m , \quad t - \frac{1}{m} \leq t_1 < t_2 < \ldots < t , \quad t_n \to t . \tag{2.1.22}$$

Taking into account that $t_{n+1} \in e_m$ and that

$$t_{n+1} - t_n \leq t - t_1 \leq \frac{1}{m} ,$$

we conclude from the definition of $e_m$ that

$$|(t_n, t_{n+1}) \cap e| \geq \rho(t_{n+1} - t_n) .$$

This takes care of the first condition (2.1.20) under the only requirement that the choice of the sequence be made under (2.1.22). To realize the second condition (2.1.20) , we have to pick the $t_n$ with more care. Select $r$ with

$$\frac{2}{3} < r < 1 . \tag{2.1.23}$$

The fact that $t$ is a left density point of $e_m$ implies that if necessary, moving $t_1$ closer to $t$ we have

$$|(s, t) \cap e_m| \geq r(t - s) \quad (t_1 \leq s \leq t) \tag{2.1.24}$$

(recall that $t$ is a left density point of $e_m$). We divide the interval $I_1 = (t_1, t)$ in three equally long disjoint subintervals; $I_1 = I_{11} \cup I_{12} \cup I_{13}$.

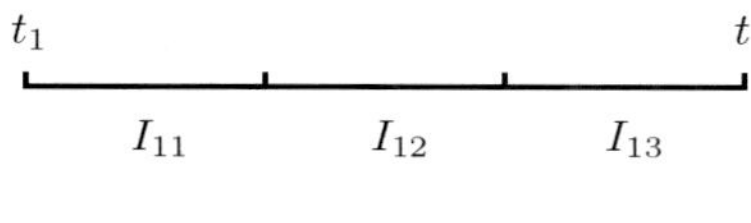


Figure 2.1.1

If the intersection $I_{12} \cap e_m$ is of measure zero we have

$$|(t_1, t) \cap e_m| \leq \frac{2}{3}(t - t_1)$$

which contradicts (2.1.23)-(2.1.24). Accordingly, we can select $t_2 \in I_{12} \cap e_m$ and we have

$$\frac{1}{3}(t - t_1) \leq t_2 - t_1 \leq \frac{2}{3}(t - t_1) , \qquad \frac{t - t_1}{t - t_2} \leq 3 .$$

Next, we divide $I_2 = (t_2, t)$ in three equally long subintervals; $I_2 = I_{21} \cup I_{22} \cup I_{23}$, pick $t_3 \in I_{22} \cap e_m$ such that

$$\frac{1}{3}(t - t_2) \leq t_3 - t_2 \leq \frac{2}{3}(t - t_2) , \qquad \frac{t - t_2}{t - t_3} \leq 3 \ldots$$

and so on. The sequence $\{t_n\}$ so chosen satisfies

$$\frac{1}{3}(t - t_n) \leq t_{n+1} - t_n \leq \frac{2}{3}(t - t_n) , \qquad \frac{t - t_n}{t - t_{n+1}} \leq 3 ,$$

so that

$$\frac{t_n - t_{n-1}}{t_{n+1} - t_n} \leq \frac{(2/3)(t - t_{n-1})}{(1/3)(t - t_n)} = 2 \, \frac{t - t_{n-1}}{t - t_n} \leq 6 .$$

This ends the proof.

Given a measurable set $e \subseteq I\!R$ we denote by $R^\infty(t; e)$ the subspace of $R^\infty(t)$ corresponding to controls $u(\cdot) \in L^\infty((0, t) \cap e; E)$ (or, more precisely, controls in $L^\infty(0, t; E)$ supported by $e$). We give $R^\infty(t; e)$ the norm

$$\|y\|_{R^\infty(t;e)} = \inf \|u(\cdot)\|_{L^\infty((0,t)\cap e;E)}\,, \tag{2.1.25}$$

the infimum taken over all $u(\cdot) \in L^\infty((0, t) \cap e; E)$ such that

$$y = \int_0^t S(t - \sigma)u(\sigma)d\sigma\,. \tag{2.1.26}$$

Obviously,

$$R^\infty(t; e) \subseteq R^\infty(t)\,, \quad \|y\|_{R^\infty(t)} \le \|y\|_{R^\infty(t;e)} \tag{2.1.27}$$

(in other words, $R^\infty(t; e) \hookrightarrow R^\infty(t)$). In general, $R^\infty(t; e)$ may be smaller than $R^\infty(t)$; for instance, if $e = (0, s)$, $s < t$, the equality

$$\int_0^s S(t - \sigma)u(\sigma)d\sigma = S(t - s)\int_0^s S(s - \sigma)u(\sigma)d\sigma$$

shows that $R^\infty(t; e) = S(t - s)R^\infty(t)$, so that $R^\infty(t; e)$ is *much* smaller than $R^\infty(t)$ if, say, $S(t)$ is analytic with unbounded infinitesimal generator. However, we have

**Lemma 2.1.6.** *Let $|e| > 0$. Then for almost all $t \in e$*

$$R^\infty(t; e) = R^\infty(t) \tag{2.1.28}$$

*with equivalence of norms.*

*Proof.* Let $t \in e$ be such that the sequence $\{t_n\}$ in Lemma 2.1.5 exists, and let $y \in R^\infty(t)$. We have

$$\begin{aligned} y &= \int_0^t S(t - \sigma)u(\sigma)d\sigma = S(t - t_1)\int_0^{t_1} S(t_1 - \tau)u(\tau)d\tau \\ &\quad + \int_{t_1}^t S(t - \sigma)u(\sigma)d\sigma \\ &= \int_{t_1}^t S(t - \sigma)\left(\frac{S(\sigma - t_1)}{t - t_1}\int_0^{t_1} S(t_1 - \tau)u(\tau)d\tau\right)d\sigma \\ &\quad + \int_{t_1}^t S(t - \sigma)u(\sigma)d\sigma\,, \end{aligned}$$

so we may assume from the beginning that the integral in (2.1.26) is taken in $t_1 \le \sigma \le t$ with $t_1 < t$ as close to $t$ as we want. In other words, we have

$$\begin{aligned} R^\infty(t) &= R^\infty(t; (t_1, t))\,, \\ \|y\|_{R^\infty(t;(t_1,t))} &\le \left(1 + \frac{t_1 M(t - t_1)M(t_1)}{t - t_1}\right)\|y\|_{R^\infty(t)}\,. \end{aligned} \tag{2.1.29}$$

Now,

$$\begin{aligned} y &= \int_{t_1}^{t} S(t-\sigma)u(\sigma)d\sigma \\ &= \sum_{n=1}^{\infty} S(t-t_{n+1}) \int_{t_n}^{t_{n+1}} S(t_{n+1}-\tau)u(\tau)d\tau \\ &= \sum_{n=1}^{\infty} \int_{(t_{n+1},t_{n+2})\cap e} S(t-\sigma) \\ &\quad \times \left( \frac{S(\sigma - t_{n+1})}{|(t_{n+1},t_{n+2})\cap e)|} \int_{t_n}^{t_{n+1}} S(t_{n+1}-\tau)u(\tau)d\tau \right) d\sigma \\ &= \int_{t_1}^{t} S(t-\sigma)v(\sigma)d\sigma\,, \end{aligned} \tag{2.1.30}$$

where

$$\begin{aligned} v(\sigma) &= \frac{S(\sigma - t_{n+1})}{|(t_{n+1},t_{n+2})\cap e)|} \int_{t_n}^{t_{n+1}} S(t_{n+1}-\tau)u(\tau)d\tau \\ &\qquad (\sigma \in (t_{n+1},t_{n+2})\cap e,\ n = 1,2,\ldots) \\ v(\sigma) &= 0 \qquad \text{elsewhere,} \end{aligned} \tag{2.1.31}$$

so that $v(\cdot)$ has support in $(t_2, t)\cap e$. Moreover, we have

$$\begin{aligned} \|v(\sigma)\| &\le \frac{M(t_{n+2}-t_{n+1})M(t_{n+1}-t_n)(t_{n+1}-t_n)}{|(t_{n+1},t_{n+2})\cap e)|}\|u(\cdot)\|_{L^\infty(0,t;E)} \\ &\le \frac{M(t_{n+2}-t_{n+1})M(t_{n+1}-t_n)(t_{n+1}-t_n)}{\rho(t_{n+2}-t_{n+1})}\|u(\cdot)\|_{L^\infty(0,t;E)} \\ &\le \frac{cM(t_{n+2}-t_{n+1})M(t_{n+1}-t_n)}{\rho}\|u(\cdot)\|_{L^\infty(0,t;E)} \\ &\le C\|u(\cdot)\|_{L^\infty(0,t;E)} \quad (t_n \le \sigma \le t_{n+1})\,, \end{aligned} \tag{2.1.32}$$

where $c$, $\rho$ and the sequence $\{t_n\}$ come from Lemma 2.1.5 and $M(t)$ from (2.1.4). Combining this estimation with (2.1.29) deduce that for almost all $t \in e$ there exists a constant $C(t)$ such that

$$\|y\|_{R^\infty(t;e)} \le C(t)\|y\|_{R^\infty(t)}\,. \tag{2.1.33}$$

This estimate, combined with (2.1.27) ends the proof of Lemma 2.1.6. We can obtain an "explicit" expression for $C(t)$ from (2.1.29), (2.1.30) and (2.1.32), but this wouldn't serve any purpose now.

*Proof of Theorem* 2.1.3. Assume that the set $\subseteq (0,t)$ where $\|u(t)\| < 1$ has positive measure. Then there exists $\varepsilon > 0$ and a set $e$ of positive measure where

$$\|u(t)\| \le 1-\varepsilon \quad (t \in e)\,.$$

Using Lemma 2.1.6 we know that there exists $s \in e$ such that $R^\infty(s) = R^\infty(s;e)$ so we can construct a control $v(\cdot) \in L^\infty((0,s)\cap e; E)$ such that

$$\begin{aligned} y(s,\zeta,v) &= S(s)\zeta + \int_0^s S(s-\sigma)v(\sigma)d\sigma \\ &= S(s)\zeta + \int_0^s S(s-\sigma)u(\sigma)d\sigma = y(s,\zeta,u)\,. \end{aligned} \tag{2.1.34}$$

Let $0<\delta<1$, $v_\delta(\sigma) = (1-\delta)u(\sigma) + \delta v(\sigma)$. In view of (2.1.34),

$$y(s,\zeta,v_\delta) = (1-\delta)y(s,\zeta,u) + \delta y(s,\zeta,v) = y(s,\zeta,u)$$

for any $\delta$. If $C = \|v(\cdot)\|_{L^\infty(0,T;E)}$ ,

$$\|v_\delta(\sigma)\| \le (1-\delta)(1-\varepsilon) + C\delta \le 1-\varepsilon + C\delta \le 1 - \frac{\varepsilon}{2} \quad (\sigma \in e)\,,$$

the last inequality for $\delta > 0$ sufficiently small; on the other hand, outside of $e$ we have $\|v_\delta(\sigma)\| = (1-\delta)\|u(\sigma)\| \le 1-\delta$, so that $\|v_\delta(\cdot)\|_{L^\infty(0,s;E)} < 1$ and $v_\delta(\cdot)$ is not norm optimal in $(0,s)$, thus not time optimal in the same interval (Theorem 2.1.2). This means that $\zeta$ can be driven to $y(s,\zeta,v_\delta) = y(s,\zeta,u)$ in time $< s$, which implies that $u(\sigma)$ is not time optimal in the subinterval $(0,s)$. It then follows from the optimality principle (see Theorem 1.1.6 and **1.3**) that $u(\sigma)$ is not time optimal in the entire interval $(0,t)$. This completes the proof of Theorem 2.1.3.

There is *no* analogue of Theorem 2.1.3 for the norm optimal problem except under additional conditions on the semigroup $S(t)$ or on the target $\bar{y}$.

A Banach space $E$ is *strictly convex* if

$$\|u\| = \|v\| = 1, \quad u \ne v \quad \Longrightarrow \quad \|u+v\| < 2\,. \tag{2.1.35}$$

**Theorem 2.1.7.** *Let $E$ be strictly convex. Then time optimal controls are unique.*

The proof is based on the bang-bang property (2.1.19) and is the same as that of the time optimal part of Theorem 1.2.3. Assume two controls $\bar{u}(t)$, $\bar{v}(t)$ drive $\zeta$ to $\bar{y}$ time optimally. Then $w(t) = (\bar{u}(t)+\bar{v}(t))/2$ drives $\zeta$ to $\bar{y}$ in the same optimal time. If $\bar{u}(t) \ne \bar{v}(t)$ in a set $e$ of positive measure, (2.1.35) implies that $\|w(t)\| < 1$ in $e$, which contradicts Theorem 2.1.3.

**Miscellaneous notes.** The results are taken from the author [1964].

**2.2. Reversible systems.** Assuming that $S(t)$ satisfies

$$S(t)E = E \quad (t \ge 0)\,, \tag{2.2.1}$$

the reachable space $R^\infty(t)$ (at any time $t>0$) of all

$$y = y(t,0,u) = \int_0^t S(t-\sigma)u(\sigma)d\sigma\,, \quad u(\cdot) \in L^\infty(0,t;E)$$

coincides with $E$;

$$R^\infty(t) = E \quad (t > 0)\,. \tag{2.2.2}$$

In fact, every $y \in E$ can be written $y = S(t)z$, hence

$$y = S(t)z = \int_0^t S(t-\sigma)\frac{S(\sigma)z}{t}d\sigma\,.$$

It follows from (2.1.5), (2.2.2) and the closed graph theorem that the norm

$$\|y\|_{R^\infty(t)} = \inf\left\{\|u\|_{L^\infty(0,t;E)};\int_0^t S(t-\sigma)u(\sigma)d\sigma = y\right\}$$

is equivalent to the original norm $\|\cdot\|$ of the space, so that the dual $R^\infty(t)^*$ of $R^\infty(t)$ and the dual $E^*$ of $E$ are the same; the norms of $R^\infty(t)^*$ and $E^*$ are equivalent.

Condition (2.2.1) needs to be satisfied only for a single $t_0 > 0$. In fact, since $E = S(t_0)E = S(t)S(t_0 - t)E$ we have $S(t)E \supseteq S(t)S(t_0 - t)E = E$ for $0 \le t \le t_0$. For large $t$ we write $S(t) = S(t/n)^n$ for $n$ large enough.

Also, assuming (2.2.1) does *not* imply that $S(t)$ is a group. Consider the left translation semigroup

$$S(t)y(x) = y(x+t) \quad (t \ge 0) \tag{2.2.3}$$

in $E = L^2(0,\infty)$ ($y(x+t)$ is restricted to $x \ge 0$ after translation). Here, (2.2.1) is satisfied but $S(t)$ is not one-to-one (hence not invertible) for any $t > 0$.

We call *reversible* any semigroup satisfying (2.2.1). This is the situation where the finite dimensional theory extends the easiest, but there are some complications. One is that the adjoint semigroup may not be strongly continuous or even strongly measurable. To see this consider again the semigroup (2.2.3), now in the space $E = L^1(-\infty,\infty)$, where it is a group. The adjoint is the right translation group

$$S(t)^*y(x) = y(x-t) \tag{2.2.4}$$

in $L^\infty(\infty,\infty)$, which is not strongly continuous. In fact, if $\chi(x)$ is the characteristic function of $[0,\infty)$ we have $\|S(t')^*\chi - S(t)^*\chi\|_{L^\infty(-\infty,\infty)} = 1$ for $t \ne t'$. For a group, strong measurability implies strong continuity (this follows from Hille - Phillips [1957, Theorem 10.2.3, p. 305]), hence $S(t)^*$ is not strongly measurable.

We may avoid complications such as these restricting $E$; if $E$ is reflexive or if $E^*$ is separable then $S(t)^*$ is strongly continuous. However, we try for more generality and we only assume (for the moment) that $E$ is *separable*.[2]

---

[2] All the results in this chapter are true in the nonseparable case. See Miscellaneous Notes, in particular Lemma 2.2.10.

Let $E$ be an arbitrary Banach space. We say that a $E^*$-valued function $f(\cdot)$ defined in $[0,T]$ is *$E$-weakly measurable* if $t \to \langle f(t), y\rangle$ is measurable for all $y \in E$. Under the separability assumption we have

$$\|f(t)\|_{E^*} = \sup |\langle f(t), u_n\rangle| \,, \tag{2.2.5}$$

where $\{u_n\}$ is a countable dense subset of the unit ball of $E$, so that $\|f(t)\|_{E^*}$ is measurable.

Assume the function $\|f(t)\|_{E^*}$ is integrable in $0 \le t \le T$. If $u(\cdot)$ is a function in $L^\infty(0,T;E)$ then there exists a sequence $u_n(t) = \sum \chi_{nk}(t)u_{nk}$ of countably valued functions (the $\chi_k(\cdot)$ characteristic functions of a finite or countable partition of $[0,T]$ in measurable sets) such that $\|u_n(t) \to u(t)\|_E \to 0$ and $\|u_n(t)\| \le C$ a. e. The function

$$\langle f(t), u_n(t)\rangle = \sum \chi_{nk}(t)\langle f(t), u_{nk}\rangle$$

is measurable and $\langle f(t), u_n(t)\rangle \to \langle f(t), u(t)\rangle$, $|\langle f(t), u_n(t)\rangle| \le C\|f(t)\|_{E^*}$ a. e. This implies that the function $\langle f(t), u(t)\rangle$ is integrable in $0 \le t \le T$.

**Lemma 2.2.1.** *Assume the $E^*$-valued function $f(\cdot)$ is $E$-weakly measurable. Then*

$$\sup_{\|u(\cdot)\|_{L^\infty(0,T;E)}\le 1} \int_0^T \langle f(t), u(t)\rangle dt = \int_0^T \|f(t)\|_{E^*} dt$$

*if the right side is finite.*

It is enough to prove

$$\sup_{\|u(\cdot)\|_{L^\infty(0,T;E)}\le 1} \int_0^T |\langle f(t), u(t)\rangle| dt = \int_0^T \|f(t)\|_{E^*} dt \,. \tag{2.2.6}$$

In fact, if $e$ is the set where $\langle f(t), u(t)\rangle \ge 0$, $\chi_e(t)$ is the characteristic function of $e$ and $v(t) = \chi_e(t)u(t) - (1-\chi_e(t))u(t)$ we have $\|v(\cdot)\|_{L^\infty(0,T;E)} = \|u(\cdot)\|_{L^\infty(0,T;E)}$ and $|\langle f(t), u(t)\rangle| = \langle f(t), v(t)\rangle$.

The proof of Lemma 2.2.1 uses Lemma 2.2.2 below. We say that a family $\mathcal{A}$ of nonnegative measurable functions $\alpha(\cdot), \beta(\cdot), \dots$ defined in $0 \le t \le T$ is *directed* if, given any two functions $\alpha(\cdot), \beta(\cdot) \in \mathcal{A}$ there exists $\gamma(\cdot) \in \mathcal{A}$ with

$$\gamma(t) \ge \max(\alpha(t), \beta(t)) \ \text{ a. e.}$$

**Lemma 2.2.2.** *Let $\mathcal{A}$ be a directed family of functions in $L^1(0,T)$ such that*

$$\sup_{\alpha\in\mathcal{A}} \int_0^T \alpha(t) dt = M < \infty \,. \tag{2.2.7}$$

*Then there exists $\eta(\cdot) \in L^1(0,T)$ such that*

$$\alpha(t) \le \eta(t) \ \text{a. e. in } 0 \le t \le T \text{ for all } \ \alpha(\cdot) \in \mathcal{A} \,, \quad \int_0^T \eta(t) dt = M \,. \tag{2.2.8}$$

*Proof.* Call a sequence $\{\alpha_n(\cdot)\} \subseteq \mathcal{A}$ *maximizing* if

$$\alpha_n(t) \le \alpha_{n+1}(t) \text{ a. e.,} \quad \int_0^T \alpha_n(t)dt \to M \quad \text{as } n \to \infty\,. \tag{2.2.9}$$

That sequences $\{\beta_n(\cdot)\}$ that satisfy the second condition (2.2.9) always exist is a consequence of the definition (2.2.7) of $M$, but these sequences may not satisfy the first condition (2.2.9). Since the family $\mathcal{A}$ is directed, we may obtain a maximizing sequence $\{\alpha_n(\cdot)\}$ from $\{\beta_n(\cdot)\}$ choosing elements $\alpha_n(\cdot) \in \mathcal{A}$ such that

$$\alpha_1(\cdot) = \beta_1(\cdot)\,, \quad \alpha_2(t) \ge \max(\beta_2(t), \alpha_1(t))\,, \quad \alpha_3(t) \ge \max(\beta_3(t), \alpha_2(t))\,, \ldots$$

and so on, all inequalities a. e. If $\{\alpha_n(\cdot)\}$ is a maximizing sequence, then it follows from Beppo Levi's theorem (Natanson [1955, p. 141]) that the *maximizing function*

$$\alpha_{\max}(t) = \max_{n\ge1} \alpha_n(t) = \lim_{n\to\infty} \alpha_n(t) \tag{2.2.10}$$

(defined a. e.) is finite a. e. and belongs to $L^1(0,T)$, with

$$\int_0^T \alpha_{\max}(t)dt = M\,. \tag{2.2.11}$$

If $\{\beta_n(\cdot)\}$ is another maximizing sequence, we deduce in the same way that the maximizing function $\beta_{\max}(\cdot)$ (defined from the $\beta_n(\cdot)$ as in (2.2.10)) satisfies

$$\int_0^T \beta_{\max}(t)dt = M\,. \tag{2.2.12}$$

Construct a third maximizing sequence $\{\gamma_n(\cdot)\}$ taking $\gamma_n(t) \ge \max(\alpha_n(t), \beta_n(t))$ a. e. For the corresponding maximizing function $\gamma_{\max}(t)$ we have

$$\gamma_{\max}(t) \ge \alpha_{\max}(t),\ \gamma_{\max}(t) \ge \beta_{\max}(t) \text{ a. e.,} \qquad \int_0^T \gamma_{\max}(t)dt = M\,.$$

Comparing with (2.2.11) and (2.2.12) we deduce that

$$\alpha_{\max}(t) = \beta_{\max}(t) = \gamma_{\max}(t) \quad \text{a. e.}$$

so that the maximizing function doesn't depend on the particular choice of the maximizing sequence. We then define

$$\eta(t) = \alpha_{\max}(t)$$

and the first statement in (2.2.8) follows from the fact that any $\alpha(t)$ can be made the first element of a maximizing sequence.

*Proof of Lemma* 2.2.1. We apply Lemma 2.2.2 to the family

$$\mathcal{A} = \left\{ |\langle f(\cdot), u(\cdot)\rangle| \, ; u(\cdot) \in L^\infty(0,T;E), \ \|u(\cdot)\|_{L^\infty(0,T;E)} \le 1 \right\}.$$

This family is directed. In fact, if $|\langle f(\cdot), u(\cdot)\rangle|$, $|\langle f(\cdot), v(\cdot)\rangle|$ are two members, $e = \{t \in [0,T]; |\langle f(t), u(t)\rangle| \le |\langle f(t), v(t)\rangle|\}$, $\chi(\cdot)$ is the characteristic function of $e$ and $w(t) = \chi(t)v(t) + (1-\chi(t))u(t)$ then

$$|\langle f(t), w(t)\rangle| \ge \max\left(|\langle f(t), u(t)\rangle|, |\langle f(\cdot), v(\cdot)\rangle|\right).$$

Now, for each $u(\cdot) \in L^\infty(0,T;E)$, $\|u(\cdot)\|_{L^\infty(0,T;E)} \le 1$ we have

$$|\langle f(t), u(t)\rangle| \le \|f(t)\|_{E^*} \quad \text{a. e.}$$

Since $\|f(\cdot)\| \in L^1(0,T)$, each member of the family $\mathcal{A}$ belongs to $L^1(0,T)$ and

$$M = \sup_{\|u(\cdot)\|_{L^\infty(0,T;E)} \le 1} \int_0^T |\langle f(t), u(t)\rangle| dt \le \int_0^T \|f(t)\|_{E^*} dt \, .$$

For the opposite inequality, let $\eta(t)$ be the function provided by Lemma 2.2.2 for the family $\mathcal{A}$. Then

$$\int_0^T \eta(t) dt = M \, ,$$

and we only have to show that

$$\|f(t)\|_{E^*} \le \eta(t) \quad \text{a. e.} \tag{2.2.13}$$

This follows from (2.2.5), as each of the functions $|\langle f(t), u_n\rangle|$ belongs to $\mathcal{A}$. This concludes the proof.

Separability of the space $E$, which gives $\|f(t)\|_{E^*}$ as the supremum (2.2.5) of the *countable* set of functions $|\langle f(t), u_n\rangle|$ is essential in passing to (2.2.13); the set where (2.2.13) does not hold is the union of the countable family of sets where $|\langle f(t), u_n\rangle| \le \eta(t)$ does not hold.

**Theorem 2.2.3.** *Assume* $R^\infty(T) = E$ *for some* $T > 0$. *Then* (2.2.1) *holds.*

*Proof.* Consider the operator $\Lambda : L^\infty(0,T;E) \to E$ given by

$$\Lambda u(\cdot) = \int_0^T S(T-\sigma)u(\sigma)d\sigma \, .$$

Its adjoint $\Lambda^* : E^* \to L^\infty(0,T;E)^*$ is[3]

$$\Lambda^* y^* = S(T - \, \cdot)^* y^*.$$

[3] Fortunately, we don't need a characterization of $L^\infty(0,T;E)^*$; we are just using part of this dual space.

Since $S(T - \cdot)^* y^*$ is $E$-weakly measurable, Lemma 2.2.1 says that the the norm of $S(T - \cdot)^* y^*$ as an element of $L^\infty(0, T; E)^*$ is

$$\|S(T - \cdot)^* y^*\|_{L^\infty(0,T;E)^*} = \sup_{\|u(\cdot)\|_{L^\infty(0,T;E)} \le 1} \int_0^T \langle S(T-t)^* y^*, u(t) \rangle dt$$
$$= \int_0^T \|S(T-t)^* y^*\| dt = \|S(T - \cdot)^* y^*\|_{L^1(0,T)} . \tag{2.2.14}$$

**Theorem 2.2.4.** *Let $X$, $Y$ be Banach spaces, $\Lambda : X \to Y$ a bounded operator. Then $\Lambda X = Y$ if and only if there exists $m > 0$ such that $\|\Lambda^* y^*\|_{X^*} \ge m \|y^*\|_{Y^*}$ $(y^* \in Y^*)$.*

Theorem 2.2.4 is Banach [1932, Théorème 1, p.146] (see also Kantorovich - Akilov [1964, Theorem 2*, (2.XII), p. 480]).

*End of proof of Theorem* 2.2.3. It follows from the assumptions, from (2.2.14) and from Theorem 2.2.4 that there exists $m > 0$ such that

$$\|S(T - \cdot)^* y^*\|_{L^1(0,T)} \ge m \|y^*\|_{E^*} \quad (y^* \in E^*) . \tag{2.2.15}$$

Let $C$ be a bound for $\|S(\sigma)^*\|$ in $0 \le \sigma \le T$, let $0 < t < T$, $r > 0$, and assume there exists $y^* \in E^*$, $y^* \ne 0$ with $\|S(t)^* y^*\| \le r \|y^*\|$ . Then we have

$$\|S(T-\sigma)^* y^*\| \le \|S(T-t-\sigma)^*\| \, \|S(t) y^*\| \le Cr \|y^*\| \quad (0 \le \sigma \le T-t) .$$

We bound $\|S(T-\sigma)^* y^*\|$ in the rest of the interval $0 \le \sigma \le T$ by

$$\|S(T-\sigma)^* y^*\| \le C \|y^*\| \quad (0 \le \sigma \le T) .$$

Using the first bound in $0 \le \sigma \le T-t$ and the second in $T-t < \sigma \le T$ we obtain

$$\|S(T - \cdot) y^*\|_{L^1(0,T)} = \int_0^{T-t} \|S(T-\sigma) y^*\| d\sigma + \int_{T-t}^T \|S(T-\sigma) y^*\| d\sigma$$
$$\le C(r(T-t) + t) \|y^*\| < C(rT + t) \|y^*\| ,$$

which contradicts (2.2.15) if $t \le t_0 = m/2C$, $r \le m/2CT$, since in this case

$$C(rT + t) \le C\Big(\frac{m}{2C} + \frac{m}{2C}\Big) = m .$$

We have then shown that

$$\|S(t)^* y^*\| \ge r \|y^*\| \quad (y^* \in E^*, \ 0 \le t \le t_0) .$$

Application of Theorem 2.2.4 to $S(t)$ produces (2.2.1) for $0 \le t \le t_0$. For large $t$ we write $S(t) = S(t/n)^n$ .

**Theorem 2.2.5.** *Let (2.2.1) hold. Then, $\bar{u}(t)$ is norm optimal*[4] *in $0 \le t \le T$ if and only if there exists $z \in E^*$, $z \neq 0$ such that*

$$\langle S(T-t)^* z, \bar{u}(t)\rangle = \max_{\|u\| \le \rho} \langle S(T-t)^* z, u\rangle \quad a.\ e.\ in \ \ 0 \le t \le T \tag{2.2.16}$$

*with $\rho = \|\bar{u}(\cdot)\|_{L^\infty(0,T;E)}$.*

*Proof.* It $\bar{u}(\cdot)$ is norm optimal,

$$\bar{y} - S(T)\zeta = \int_0^T S(T-\sigma)\bar{u}(\sigma)d\sigma$$

is a boundary point of the ball $B_\rho^\infty(T)$ of center 0 and radius $\rho$ in $R^\infty(T)$. We separate $\bar{y} - S(T)\zeta$ from $B_\rho^\infty(T)$ with a nonzero functional $z \in E^*$. The separation inequality $\langle z, y\rangle \le \langle z, \bar{y} - S(T)\zeta\rangle$ $(y \in B_\rho^\infty(T))$ is equivalent to

$$\left\langle z\,, \int_0^T S(T-\sigma)u(\sigma)d\sigma \right\rangle \le \left\langle z\,, \int_0^T S(T-\sigma)\bar{u}(\sigma)d\sigma \right\rangle$$
$$u(\cdot) \in L^\infty(0,T;E)\,, \quad \|u(\cdot)\|_{L^\infty(0,T;E)} \le \rho\,. \tag{2.2.17}$$

In fact, that (2.2.17) is a consequence of the separation inequality is obvious. For the opposite implication, let $y \in R^\infty(T)$, $\|y\|_{R^\infty(T)} \le \rho$. By definition of the norm of $R^\infty(T)$, for every $\varepsilon > 0$ we can write[5]

$$y = \int_0^T S(T-\sigma)u_\varepsilon(\sigma)d\sigma\,, \quad \|u_\varepsilon(\cdot)\|_{L^\infty(0,T;E)} \le \rho + \varepsilon\,.$$

If (2.2.17) holds we have

$$\begin{aligned} \frac{\rho}{\rho+\varepsilon}\langle z, y\rangle &= \left\langle z\,, \int_0^T S(T-\sigma)\frac{\rho}{\rho+\varepsilon}u_\varepsilon(\sigma)d\sigma \right\rangle \\ &\le \left\langle z\,, \int_0^T S(T-\sigma)\bar{u}(\sigma)d\sigma \right\rangle \le \langle z, \bar{y} - S(T)\zeta\rangle\,, \end{aligned}$$

which inequality (as $\varepsilon$ is arbitrary and $y$ does not depend on $\varepsilon$) implies the separation inequality. We can rearrange (2.2.17) as

$$\int_0^T \langle S(T-\sigma)^* z, u(\sigma)\rangle d\sigma \le \int_0^T \langle S(T-\sigma)^* z, \bar{u}(\sigma)\rangle d\sigma\,. \tag{2.2.18}$$

[4] In the necessity part of Theorem 2.2.5 "$\bar{u}(t)$ is norm optimal" means $\bar{u}(t)$ drives *some* initial condition $\zeta$ norm optimally to the target $\bar{y} = y(T, \zeta, \bar{u})$. In the sufficiency part, "$\bar{u}(t)$ is norm optimal" means $\bar{u}(t)$ drives an *arbitrary* initial condition $\zeta$ norm optimally to the target $\bar{y} = y(T, \zeta, \bar{u})$.

[5] It is not necessarily true that we may take $\|u(\cdot)\|_{L^\infty(0,T;E)} = \|y\|_{R^\infty(T)}$; this needs an existence theorem for norm optimal controls, which in turn needs assumptions on $E$. See **3.1**, especially Corollary 3.1.3.

We take the supremum of the left side over $u(\cdot) \in B^\infty_\rho(T)$ using Lemma 2.2.1:

$$\rho \int_0^T \|S(T-\sigma)^* z\| d\sigma = \int_0^T \langle S(T-\sigma)^* z, \bar{u}(\sigma)\rangle d\sigma \,, \tag{2.2.19}$$

which, since $\langle S(T-\sigma)^* z, \bar{u}(\sigma)\rangle \le \rho \|S(T-\sigma)^* z\|$, implies

$$\langle S(T-\sigma)^* z, \bar{u}(\sigma)\rangle = \rho \|S(T-\sigma)^* z\| \text{ a. e.}$$

This is equivalent to (2.2.16). Conversely, assume that (2.2.16) holds with $z \in E^*$, $z \neq 0$. Arguing as before we deduce

$$\langle z, \bar{y} - S(T)\zeta\rangle = \int_0^T \langle S(T-\sigma)^* z, \bar{u}(\sigma)\rangle d\sigma = \rho \int_0^T \|S(T-\sigma)^* z\| d\sigma \,. \tag{2.2.20}$$

If $\bar{u}(t)$ is not norm optimal there exists $u(\cdot) \in L^\infty(0,T;E)$ with $\|u(\cdot)\|_{L^\infty(0,T;E)} = \rho' < \rho$ and $y(T,\zeta,u) = y(T,\zeta,\bar{u})$. This means

$$\langle z, \bar{y} - S(T)\zeta\rangle = \int_0^T \langle S(T-\sigma)^* z, u(\sigma)\rangle d\sigma \le \rho' \int_0^T \|S^*(T-\sigma) z\| d\sigma$$

which contradicts (2.2.20). This completes the proof.

We note for the record that

$$z \neq 0 \implies S(t)^* z \neq 0 \quad \text{in} \quad 0 \le t < \infty \tag{2.2.21}$$

since $\langle S(t)^* z, y\rangle = \langle z, S(t)y\rangle$ and (2.2.1) is in force.

**Theorem 2.2.6.** *Let* (2.2.1) *hold. Then, if* $\bar{u}(t)$ *is time optimal in* $0 \le t \le T$ *there exists* $z \in E^*$, $z \neq 0$ *such that*

$$\langle S(T-t)^* z, \bar{u}(t)\rangle = \max_{\|u\| \le 1} \langle S(T-t)^* z, u\rangle \quad \textit{a. e. in} \quad 0 \le t \le T \,. \tag{2.2.22}$$

*Conversely, assume that* (2.2.22) *holds and that, either*

$$\begin{aligned} &(a) \quad \bar{y} = y(T,\zeta,\bar{u}) \in D(A), \quad \|A\bar{y}\| < 1 \,, \ \textit{or} \\ &(b) \qquad\qquad\qquad\quad \zeta \in D(A), \quad \|A\zeta\| < 1 \,. \end{aligned} \tag{2.2.23}$$

*Then* $\bar{u}(t)$ *is time optimal.*[6]

[6] In the necessity part of Theorem 2.2.6 "$\bar{u}(t)$ is time optimal" means $\bar{u}(t)$ drives *some* initial condition $\zeta$ time optimally to the target $\bar{y} = y(T,\zeta,\bar{u})$. In the sufficiency part, "$\bar{u}(t)$ is time optimal" means $\bar{u}(t)$ drives an initial condition $\zeta$ time optimally to the target $\bar{y} = y(T,\zeta,\bar{u})$ as long as one of the two conditions (2.2.23) is satisfied.

*Proof.* The necessity part is obvious, since time optimality $\Rightarrow$ norm optimality (Theorem 2.1.2). The sufficiency proof is a particular case of Theorem 2.5.7 thus is omitted. See Remark 2.5.8 for details and for comments on condition (2.2.23).

**Remark 2.2.7.** The only nontrivial case known to us where "everything can be computed explicitly" is that of a semigroup $S(t)$ in a Hilbert space $E$ such that

$$S(t)S(t)^* = I\,. \tag{2.2.24}$$

Condition (2.2.24) implies (2.2.1). It also implies that $S(t)^*$ is *isometric;*

$$\|S(t)^*y\|^2 = \langle S(t)^*y, S(t)^*y\rangle = \langle y, S(t)S(t)^*y\rangle = \langle y, y\rangle = \|y\|^2\,.$$

It does *not* imply that the semigroup is unitary; if $S(t)$ is the left translation semigroup (2.2.3) in $E = L^2(0,\infty)$ its adjoint $S(t)^*$ is the right translation semigroup (2.2.4) in $E$ ($y(x)$ is extended to $y(x) = 0$ for $x < 0$) and condition (2.2.24) holds; on the other hand, $S(t)^*S(t)y(x) = \chi_t(x)y(x)$, $\chi_t(x)$ the characteristic function of the interval $[t,\infty)$.

The maximum principle (2.2.16) gives

$$\bar{u}(t) = \frac{S(T-t)^*z}{\|S(T-t)^*z\|} = \frac{\rho}{\|z\|}S(T-t)^*z \tag{2.2.25}$$

due to isometry of $S(t)^*$. We then have (assuming, as we may, that $\|z\| = 1$)

$$\int_0^T S(T-\sigma)\bar{u}(\sigma)d\sigma = \rho\int_0^T S(T-\sigma)S(T-\sigma)^*zd\sigma = \rho\int_0^T zd\sigma = \rho Tz$$

so that the control (2.2.25) drives $\zeta$ to $\bar{y}$ in time $T$ if and only if $T$ solves the equation

$$\rho Tz = \bar{y} - S(T)\zeta\,. \tag{2.2.26}$$

Taking norms, this equation implies the scalar equation

$$\rho T = \|S(T)\zeta - \bar{y}\|\,. \tag{2.2.27}$$

Conversely, if $T > 0$ is a solution of (2.2.27) it is clear that (2.2.26) holds with

$$z = \frac{\bar{y} - S(T)\zeta}{\|\bar{y} - S(T)\zeta\|}\,. \tag{2.2.28}$$

The control (2.2.25) drives $\zeta$ to $\bar{y}$ norm optimally (Theorem 2.2.5). For $\rho = 1$ it drives $\zeta$ to $\bar{y}$ time optimally if one of the two conditions (2.2.23) holds (Theorem 2.2.6). All possible norm or time optimal controls are of the form (2.2.25).

**Remark 2.2.8.** Theorem 2.2.6 gives the bang-bang property $\|\bar{u}(t)\| = 1$ a. e. for time optimal controls, although the result is just a very particular case of the bang-bang property shown in Theorem 2.1.3. Theorem 2.2.5 takes care of norm optimal controls:

$$\|\bar{u}(t)\| = \rho = \|\bar{u}(\cdot)\|_{L^\infty(0,T;E)} \quad \text{a. e. in } 0 \le t \le T\,. \tag{2.2.29}$$

As seen later, for the norm optimal problem the implication maximum principle $\Rightarrow$ bang-bang property may not hold without special assumptions.

Uniqueness of time optimal controls in strictly convex spaces has been proved in full generality in Theorem 2.1.7. For the norm optimal problem we have

**Theorem 2.2.9.** *Let $E$ be strictly convex and let* (2.2.1) *be satisfied. Then norm optimal controls are unique.*

*Proof.* Let $\bar{u}(t)$ be two different norm optimal controls in the interval $0 \le t \le T$ corresponding to the initial condition $\zeta$ and the target $\bar{y}$. Then the control $w(t) = (\bar{u}(t) + \bar{v}(t))/2$ drives $\zeta$ to $\bar{y}$ in $0 \le t \le T$ and

$$\|w(\cdot)\|_{L^\infty(0,T;E)} \le \frac{1}{2}\Big(\|\bar{u}(\cdot)\|_{L^\infty(0,T;E)} + \|\bar{v}(\cdot)\|_{L^\infty(0,T;E)}\Big) = \rho$$

(" $\le$ " because of the triangle inequality, " $=$ " because $\rho$ is the optimal value of the norm). It follows that $w(\cdot)$ is norm optimal as well. If $\bar{u}(t) \ne \bar{v}(t)$ in a set $e$ of positive measure then $\|w(t)\| < \rho$ $(t \in e)$ which contradicts (2.2.29).

**Miscellaneous notes.** One of the ideas in this in this section – when $S(t)$ satisfies (2.2.1) $B_\rho^\infty(T)$ contains interior points, which implies the maximum principle as a necessary condition – is in Friedman [1967], [1968]. That (2.2.1) is a necessary condition for existence of interior points is in the author [1974/75] in the Hilbert space case; for the general case see the author [2001:3]. The sufficient conditions are in the author [2001:3] (more on this in the coming sections).

The proofs of Lemma 2.2.1 and Lemma 2.2.2 are from Ionescu Tulcea [1969]. To remove the separability assumption on $E$ in Lemma 2.2.1 requires a quantum leap in sophistication; in fact, we need to wield the Dunford - Pettis theorem in full force (for details see Ionescu Tulcea [1969] or the author [1999:2, Section 12.2]) and deal with problems such as $\|f(t)\|_{E^*}$ not being necessarily measurable. However, all the generality we need is to deduce (2.2.19) from (2.2.18) in the case where $f(t) = S(T-t)^*z$; here, $\langle f(t), u\rangle = \langle z, S(T-t)u\rangle$ $(u \in E)$ so that $f(t)$ is $E$-weakly continuous. The only place where separability is used in Lemma 2.2.1 is in showing (2.2.13); in fact, if $E$ is not separable (2.2.5) must be replaced by

$$\|f(t)\|_{E^*} = \sup_{\|u\|_E \le 1} \langle f(t), u\rangle \tag{2.2.30}$$

and the fact that

$$|\langle f(t), u\rangle| \le \eta(t) \quad \text{a. e.} \tag{2.2.31}$$

("a. e" depending on $u$) no longer guarantees (2.2.13) (see the comments after the proof of Lemma 2.2.1).

Let $\ell^\infty$ be the space of all real bounded sequences $s = \{s_1, s_2, \ldots\}$ equipped with the supremum norm $\|s\| = \sup |s_j|$. A *Banach limit* is a linear functional (denoted $\mathrm{LIM}_{n\to\infty}$) in $\ell^\infty$ having the following properties:

$$\operatorname*{LIM}_{n\to\infty} s_n = \operatorname*{LIM}_{n\to\infty} s_{n+1}\,, \quad \operatorname*{LIM}_{n\to\infty} s_n \ge 0 \ \text{ if } s_n \ge 0\,, \quad \operatorname*{LIM}_{n\to\infty} 1 = 1\,.$$

Linearity and all conditions above imply

$$\liminf_{n\to\infty} s_n \le \operatorname*{LIM}_{n\to\infty} s_n \le \limsup_{n\to\infty} s_n$$

for any $\{s_j\} \subset \ell^\infty$; thus, in particular,

$$\operatorname*{LIM}_{n\to\infty} s_n = \lim_{n\to\infty} s_n$$

whenever the latter exists; a Banach limit extends the ordinary limit. That Banach limits exist is a direct consequence of the Hahn - Banach theorem; see Banach [1932, p. 34], Dunford - Schwartz [1958, p. 73] or the author [1999:2, p. 173].

Let $B(0,T)$ be the space of all bounded functions in $0 \le t \le T$ equipped with the maximum norm. We define an operator $\mathcal{L} : L^\infty(0,T) \to B(0,T)$ by

$$\mathcal{L}\phi(t) = \operatorname*{LIM}_{n\to\infty} \phi_n(t) = \operatorname*{LIM}_{n\to\infty} \frac{1}{h_n(t)} \int_{I_n(t)} \phi(\sigma)d\sigma$$

where $I_n(t) = (t - 1/n, t + 1/n) \cap [0,T]$ and $h_n(t)$ is the length of $I_n(t)$. Since $\lim_{n\to\infty} \phi_n(t)$ exists for almost all $t$ and equals $\phi(t)$, the function $\mathcal{L}\phi(\cdot)$ belongs to the equivalence class of $\phi(\cdot)$ in $L^\infty(0,T)$. Moreover, $\phi(t) \le \psi(t)$ a. e. implies $\mathcal{L}\phi(t) \le \mathcal{L}\psi(t)$ everywhere, and $\mathcal{L}\phi(t) = \phi(t)$ everywhere in $0 < t < T$ if $\phi(t)$ is continuous in $0 < t < T$.

To prove Lemma 2.2.1 for $f(t)$ $E$-weakly continuous note first that (2.2.30) gives $\|f(t)\|_{E^*}$ as the supremum of the family of continuous functions $|\langle f(t), u\rangle|$, thus $\|f(t)\|_{E^*}$ is lower semicontinuous, hence measurable (for a proof see Lemma 2.3.4). We assume first that $f(t)$ is bounded. Applying $\mathcal{L}$ to both sides of (2.2.31) we obtain

$$|\langle f(t), u\rangle| = \mathcal{L}|\langle f(t), u\rangle| \le \mathcal{L}\eta(t) \quad \text{everywhere in } 0 < t < T,$$

thus, taking supremum in $\|u\| \le 1$ on the left side,

$$\|f(t)\|_{E^*} \le \mathcal{L}\eta(t) = \eta(t) \quad \text{a. e. in } 0 \le t \le T,$$

which is (2.2.13), so that (2.2.6) holds.

To remove the assumption that $f(t)$ is bounded we apply (2.2.6) to $\chi_n(t)f(t)$, where $\chi_n(t)$ is the characteristic function of the set $\{t \in [0,T]; \|f(t)\| \le n\}$ and obtain

$$\begin{aligned} &\sup_{\|u(\cdot)\|_{L^\infty(0,T;E)}\le 1} \int_0^T |\langle f(t), u(t)\rangle|\, dt \\ \ge &\sup_{\|u(\cdot)\|_{L^\infty(0,T;E)}\le 1} \int_0^T |\langle \chi_n(t) f(t), u(t)\rangle|\, dt = \int_0^T \|\chi_n(t)f(t)\|_{E^*}\, dt\,. \end{aligned}$$

We then get (2.2.6) letting $n \to \infty$ and applying the dominated convergence theorem on the right side. We have proved

**Lemma 2.2.10.** *Lemma* 2.2.1 *remains true even if $E$ is not separable as long as $f(t)$ is $E$-weakly continuous in $0 < t < T$. Thus, all results in this section remain true without assuming $E$ separable.*

**2.3. The reachable space and its dual, I.** When $R^\infty(T)$ doesn't have interior points in $E$ (that is, when (2.1) does not hold) the separation argument in **1.2** collapses, and the only way out is to separate with a functional in the dual space[7] $R^\infty(T)^\star$. We identify (to a certain extent) this dual in this section and the next, with no restriction on $E$ or on the semigroup $S(t)$.

Let $C$, $\omega$ be such that

$$\|S(t)\| \le Ce^{\omega t} \quad (t \ge 0)\,. \tag{2.3.1}$$

It follows from the Hille - Yosida - Phillips theorem that $R(\mu; A) = (\mu I - A)^{-1}$ exists and is bounded for $\mu > \omega$, with

$$\|R(\mu; A)\| \le \frac{C}{\mu - \omega} \quad (\mu > \omega)\,. \tag{2.3.2}$$

We don't require $E^*$ to be reflexive, thus $D(A^*)$ may not be dense in $E^*$; note, however, that the assumptions on $A$ and adjoint theory imply that

$$R(\mu; A^*) = (\mu I^* - A^*)^{-1} = ((\mu I - A)^{-1})^* = R(\mu; A)^*\,,$$

($I^*$ the identity operator in $E^*$) thus (2.3.2) yields

$$\|\mu R(\mu; A^*)\| \le \frac{C\mu}{\mu - \omega} \quad (\mu > \omega)\,. \tag{2.3.3}$$

We have

$$\mu R(\mu; A^*)y^* - y^* = R(\mu; A^*)A^*y^* \quad (y^* \in D(A^*))\,,$$

---

[7] The dual of $R^\infty(T)$ is marked by a different star $^\star$ to emphasize the difference with the duality of $E$ and $E^*$.

so that, in view of (2.3.3),[8]

$$\lim_{\mu\to\infty} \mu R(\mu; A^*)y^* = y^* \quad (y^* \in D(A^*))\,. \tag{2.3.4}$$

Let $\mu > \omega$ and $y \in D(A)$. Applying $R(\mu; A)$ to $(\mu I - A)y$ and using (2.3.2) we obtain $\|y\| \le C(\mu - \omega)^{-1}\|(\mu I - A)y\|$, thus the norm

$$\|y\|_{D(A),\mu} = \|(\mu I - A)y\| \tag{2.3.5}$$

in $D(A)$ is equivalent to the graph norm $\|y\|_{D(A)} = \|y\| + \|Ay\|$; all norms (2.3.5) are equivalent for $\mu > \omega$. We have shown (see (2.1.14)) that

$$D(A) \hookrightarrow R^\infty(T) \hookrightarrow E \tag{2.3.6}$$

for the graph norm, thus the imbedding holds for any of the norms (2.3.6).

Let $E^*_{-1}(\mu)$ be the completion of $E^*$ in the norm

$$\|y^*\|_{E^*_{-1},\mu} = \|R(\mu; A)^* y^*\|_{E^*}\,. \tag{2.3.7}$$

We have

$$R(\lambda; A^*) - R(\mu; A^*) = (\mu - \lambda)R(\lambda; A^*)R(\mu; A^*)\,,$$

so all the norms $\|\cdot\|_{E^*_{-1},\mu}$ are equivalent for $\mu > \omega$ and define the same space $E^*_{-1} = E^*_{-1}(\mu)$. It follows from the definition that

$$E^* \hookrightarrow E^*_{-1}\,. \tag{2.3.8}$$

We extend $R(\mu; A^*) = R(\mu; A)^*$ to an operator $R(\mu; A)^\bullet$ in $E^*_{-1}$ as follows. If $\psi \in E^*_{-1}(\mu)$ then there exists $\{y^*_n\} \subset E^*$ such that

$$\|\psi - y^*_n\|_{E^*_{-1}(\mu)} \to 0\,. \tag{2.3.9}$$

This means $\{R(\mu; A)^* y^*_n\}$ is Cauchy in $E^*$; we define

$$R(\mu; A)^\bullet \psi = \lim_{n\to\infty} R(\mu; A)^* y^*_n\,. \tag{2.3.10}$$

The definition doesn't depend on the particular $\{y^*_n\}$; this can be seen alternating the members of any two sequences used in (2.3.9)-(2.3.10). If $y^* \in E^*$ we take $\{y^*_n\}$ with $y^*_n = y^*$ and show that $R(\mu; A)^\bullet y^* = R(\mu; A)^* y^*$, justifying the first equality below:

$$R(\mu; A)^\bullet|_{E^*} = R(\mu; A)^*\,, \qquad R(\mu; A)^\bullet E^*_{-1} = \overline{D(A^*)}\,. \tag{2.3.11}$$

[8] The convergence relation (2.3.4) can be extended to $y^* \in \overline{D(A^*)}$ using (2.3.3).

The inclusion $\subseteq$ in the second equality follows from the definition (2.3.9)-(2.3.10) of $R(\mu; A)^\bullet$ and from the fact that each $R(\mu; A)^* y_n^* = R(\mu; A^*) y_n^*$ belongs to $D(A^*)$. To show the opposite inclusion $\supseteq$ we note that if $z^* \in \overline{D(A^*)}$ there exists a sequence $\{z_n^*\} \subset D(A^*)$ such that $z_n^* \to z^*$; using $y_n^* = (\mu I^* - A^*) z_n^*$ in (2.3.9)-(2.3.10) we deduce that $z^* \in R(\mu; A)^\bullet E_{-1}^*$.

It is obvious from the definition that the operator

$$R(\mu; A)^\bullet : (E_{-1}^*, \|\cdot\|_{E_{-1}^*,\mu}) \to (\overline{D(A^*)}, \|\cdot\|_{E^*}) \tag{2.3.12}$$

is an isometry onto,[9] thus it has an isometric inverse $(R(\mu; A)^\bullet)^{-1}$. We extend $A^*$ to $\overline{D(A^*)}$ by

$$A^\bullet = \mu I^* - (R(\mu; A)^\bullet)^{-1} . \tag{2.3.13}$$

If $y^* \in D(A^*)$ then, using the first equality (2.3.11) we have $y^* = R(\mu; A^*) z^* = R(\mu; A)^\bullet z^*$ for $z^* = \mu y^* - A^* y^*$, thus the definition (2.3.13) gives

$$\begin{aligned} A^\bullet y^* &= \mu y^* - (R(\mu; A)^\bullet)^{-1} y^* \\ &= \mu y^* - (R(\mu; A)^\bullet)^{-1} R(\mu; A)^\bullet z^* = \mu y^* - z^* = A^* y^* , \end{aligned}$$

justifying the first equality

$$A^\bullet|_{D(A^*)} = A^*, \qquad R(\mu; A^\bullet) = (\mu I^* - A^\bullet)^{-1} = R(\mu; A)^\bullet . \tag{2.3.14}$$

The second follows writing (2.3.13) in the form $\mu I^* - A^\bullet = (R(\mu; A)^\bullet)^{-1}$.

**Lemma 2.3.1.** *$D(A^*)$ is dense in $E_{-1}^*$.*

*Proof.* It is enough to show that $D(A^*)$ is dense in $E^*$ in (any of) the norm(s) of $E_{-1}^*$. This means: given $y^* \in E^*$ there exists a sequence $\{y_n^*\} \subset D(A^*)$ such that

$$R(\mu; A^*) y^* = \lim_{n \to \infty} R(\mu; A) y_n^* .$$

To show this we use (2.3.4) with $y_n^* = \lambda_n R(\lambda_n; A^*) y^*$, $\lambda_n \to \infty$. In fact, since $R(\mu; A^*) y^* \in D(A^*)$,

$$\begin{aligned} \lim_{n \to \infty} R(\mu; A^*) y_n^* &= \lim_{n \to \infty} R(\mu; A^*) \lambda_n R(\lambda_n; A^*) y^* \\ &= \lim_{n \to \infty} \lambda_n R(\lambda_n; A^*) R(\mu; A^*) y^* = R(\mu; A^*) y^* . \end{aligned}$$

**Lemma 2.3.2.** *$A^\bullet$ does not depend on $\mu > \omega$.*

[9] In (2.3.12) and other statements, $(E, \|\cdot\|)$ means "the space $E$ equipped with the norm $\|\cdot\|$." We only use this notation when there are several norms at play in $E$. If the norm $\|\cdot\|_{E_{-1}^*,\mu}$ is replaced by the equivalent norm $\|\cdot\|_{E_{-1}^*,\lambda}$ with $\lambda \neq \mu$ "isometry onto" in (2.3.12) becomes "bounded operator with bounded inverse".

*Proof.* (2.3.12) and the fact that $\mu I^* - A^\bullet$ is the inverse of $R(\mu; A)^\bullet$ imply that

$$\mu I^* - A^\bullet : (\overline{D(A^*)}, \|\cdot\|_{E^*}) \to (E^*_{-1}, \|\cdot\|_{E^*_{-1},\mu}) \tag{2.3.15}$$

is an isometry onto, in particular an operator in $\mathcal{L}(\overline{D(A^*)}, E^*_{-1})$; then, the first equality (2.3.11) shows that $\mu I^* - A^\bullet$ is the unique extension in $\mathcal{L}(\overline{D(A^*)}, E^*_{-1})$ of $\mu I^* - A^*$, hence $A^\bullet$ is the unique extension in $\mathcal{L}(\overline{D(A^*)}, E^*_{-1})$ of $A^*$. The last statement does not concern $\mu$, so the proof is complete.

Figure 2.3.1 shows all the spaces and maps at play. Horizontal arrows are isometries; vertical arrows are imbeddings.

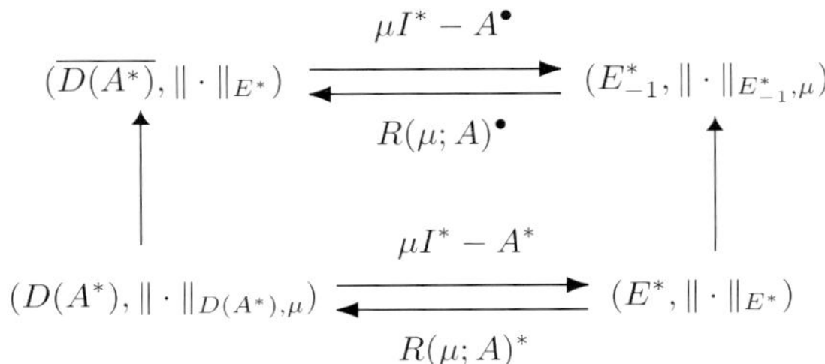


Figure 2.3.1

We extend $S(t)^*$ to $E^*_{-1}$ by

$$S(t)^\bullet \psi = (\mu I^* - A^\bullet) S(t)^* R(\mu; A)^\bullet \psi \,. \tag{2.3.16}$$

In view of the second equality (2.3.11) and of the fact that $A^\bullet$ is only defined in $\overline{D(A^*)}$, (2.3.16) makes sense if

$$S(t)^* \overline{D(A^*)} \subseteq \overline{D(A^*)} \,.$$

To see this, note that $S(t)^*$ and $A^*$ commute, so that $S(t)^* D(A^*) \subseteq D(A^*)$; the inclusion then follows taking limits and using the fact that $S(t)^*$ is bounded.

Boundedness of all three operators on the right side of (2.3.16) implies that $S(t)^\bullet \in \mathcal{L}(E^*_{-1}, E^*_{-1})$ for $t > 0$ with

$$\begin{aligned} &\|S(t)^\bullet\|_{\mathcal{L}(E^*_{-1}, E^*_{-1})} \\ &\le \|\mu I^* - A^\bullet\|_{\mathcal{L}(E^*, E^*_{-1})} \|S(t)^*\|_{\mathcal{L}(E^*, E^*)} \|R(\mu; A)^\bullet\|_{\mathcal{L}(E^*_{-1}, E^*)} \,. \end{aligned} \tag{2.3.17}$$

On the other hand, the first equality (2.3.11) and the first equality (2.3.14) imply

$$\begin{aligned} S(t)^\bullet y^* &= (\mu I^* - A^\bullet) S(t)^* R(\mu; A)^\bullet y^* \\ &= (\mu I^* - A^*) S(t)^* R(\mu; A)^* y^* = S(t)^* y^* \quad (y^* \in E^*) \,, \end{aligned}$$

so that

$$S(t)^\bullet|_{E^*} = S(t)^*\,, \tag{2.3.18}$$

and we see that $S(t)^\bullet$ is the unique extension in $\mathcal{L}(E^*_{-1}, E^*_{-1})$ of $S(t)^*$; in particular,

**Lemma 2.3.3.** *$S(t)^\bullet$ does not depend on $\mu > \omega$.*

The space $\mathcal{Z} \subseteq E^*_{-1}$ consists of all $\psi \in E^*_{-1}$ such that

$$S(t)^\bullet \psi \in E^* \quad (t > 0)\,. \tag{2.3.19}$$

The definition (2.3.16) of $S(t)^\bullet$ gives

$$\begin{aligned} S(s+t)^\bullet\psi &= (\mu I^* - A^\bullet)S(s+t)^* R(\mu; A)^\bullet \psi \\ &= (\mu I^* - A^\bullet)S(s)^* R(\mu; A)^\bullet (\mu I^* - A^\bullet)S(t)^* R(\mu; A)^\bullet \psi \\ &= S(s)^\bullet S(t)^\bullet \psi\,, \end{aligned}$$

so that, if $S(h)^\bullet\psi \in E^*$ and $t \geq h$ we obtain, using (2.3.18),

$$S(t)^\bullet\psi = S(t-h)^* S^\bullet(h)\psi \in E^*\,. \tag{2.3.20}$$

This means, condition (2.3.19) only has to be required in an arbitrarily small interval $0 \leq t \leq h$ (or, more generally, in a set accumulating at 0).

An $E^*$-valued function $f(t)$ is *$E$-weakly continuous* in an open set $e$ if $\langle f(t), y\rangle$ is continuous in $e$ for every $y \in E$.

**Lemma 2.3.4.** *Assume $f(t)$ is $E$-weakly continuous in $e$. Then $\|f(t)\|_{E^*}$ is lower semicontinuous in $e$.*

*Proof.* We have

$$\|f(t)\|_{E^*} = \sup_{\|u\|_E \leq 1} \langle f(t), u\rangle\,.$$

Let $t_0 \in e$, $\epsilon > 0$. Select $u \in E$, $\|u\| = 1$ such that $\langle f(t_0), u\rangle \geq \|f(t_0)\|_{E^*} - \epsilon$. Since $\langle f(t), u\rangle$ is continuous there exists $\delta > 0$ such that $\langle f(t), u\rangle \geq \|f(t_0)\|_{E^*} - 2\epsilon$ in $|t - t_0| \leq \delta$. Accordingly $\|f(t)\|_{E^*} \geq \langle f(t), u\rangle \geq \|f(t_0)\|_{E^*} - 2\epsilon$ in $|t - t_0| \leq \delta$, and the proof is complete.

For simplicity, if $\psi \in \mathcal{Z}$ we write

$$S(t)^*\psi = S(t)^\bullet\psi\,. \tag{2.3.21}$$

The space[10] $Z_w(T) \subseteq \mathcal{Z} \subseteq E^*_{-1}$ consists of all $z \in \mathcal{Z}$ such that

$$\|z\|_{Z_w(T)} = \int_0^T \|S(t)^* z\| dt < \infty\,. \tag{2.3.22}$$

[10] The subscript $w$ is a reminder that $S(t)^*z$ may not be strongly measurable.

It follows from (2.3.20) that if $z \in \mathcal{Z}$ we have

$$\langle S(t)^{\bullet} z, y \rangle = \langle S(h)^{\bullet} z, S(t-h) y \rangle \quad (t \geq h)$$

for $h > 0$ arbitrarily small, thus $S(t)^* z$ is $E$-weakly continuous in $t > 0$. Lemma 2.3.4 then says that the the integrand of (2.3.22) is lower semicontinuous in $t > 0$, hence measurable. It is a consequence of (2.3.20) that $\|S(t)^* z\|_{E^*}$ is bounded in $h \leq t \leq T$ for any $h > 0$, thus the summability condition (2.3.22) only bears on the behavior of $\|S(t)^* z\|_{E^*}$ near zero. This means, all spaces $Z_w(T)$ coincide:

$$Z_w(T) = Z_w \quad (T > 0)\,. \tag{2.3.23}$$

Moreover, all norms $\|\cdot\|_{Z_w(T)}$ are equivalent. In fact, we have $\|z\|_{Z_w(T)} \leq \|z\|_{Z_w(T+h)}$; on the other hand,

$$\begin{aligned}\int_0^{T+h} \|S(t)^* z\| dt &= \int_0^T \|S(t)^* z\| dt + \int_T^{T+h} \|S(t)^* z\| dt \\ &\leq \int_0^T \|S(t)^* z\| dt + \|S(T)^*\| \int_0^h \|S(t)^* z\| dt \\ &\leq (1 + \|S(T)^*\|)\|z\|_{Z_w(T)}\end{aligned}$$

if $h \leq T$; if $h > T$ we take $n$ with $h/n \leq T$ and apply the above inequality repeatedly.

The first imbedding (2.3.6) implies that every bounded linear functional in $R^\infty(T)$ is a bounded linear functional in $D(A)$, so we begin with the identification of the dual space $D(A)^*$.

**Lemma 2.3.5.** *Every bounded linear functional* $\Phi$ *in* $D(A)$ *is given by*

$$\Phi(y) = \langle y^*, (\mu I - A) y \rangle \quad (y \in D(A)) \tag{2.3.24}$$

*for some* $y^* \in E^*$. *Moreover, if* $D(A)$ *is given the norm* (2.3.5) *then*

$$\|\Phi\|_{D(A)^*} = \|y^*\|_{E^*}\,. \tag{2.3.25}$$

*Proof.* Let $y^* \in E^*$. Then (2.3.24) defines a bounded linear functional in $D(A)$. That (2.3.25) holds follows from the fact that $(\mu I - A)D(A) = E$. Conversely, let $\Phi$ be a bounded linear functional in $D(A)$. Then $\Phi(R(\mu; A)y)$ is a bounded linear functional in $E$, so that $\Phi(R(\mu; A)y) = \langle y^*, y \rangle$. This is equivalent to (2.3.24).

The second imbedding (2.3.6) implies that every bounded linear functional in $E$ is a bounded linear functional in $R^\infty(T)$. On the other hand, $D(A)$ is dense in $E$, hence $R^\infty(T)$ is dense in $E$. Accordingly, an element of $E^*$ that vanishes in $R^\infty(T)$ must vanish in $E$, thus

$$E^* \hookrightarrow R^\infty(T)^\star.$$

Elements of $R^\infty(T)^\star$ are named $\xi, \eta, \ldots$ and the duality of $R^\infty(T)$ and $R^\infty(T)^\star$ is indicated by $\langle\langle y, \xi \rangle\rangle$ or $\langle\langle \xi, y \rangle\rangle$. Given $z \in Z_w(T)$ we define a functional $\xi_z$ in $R^\infty(T)$ through the formula

$$\left\langle\!\!\left\langle \xi_z\,, \int_0^T S(T-\sigma)u(\sigma)d\sigma \right\rangle\!\!\right\rangle = \int_0^T \langle S(T-\sigma)^* z, u(\sigma)\rangle d\sigma \tag{2.3.26}$$

but (2.3.26) requires explanation, as $S(T-\sigma)^* z$ is merely $E$-weakly measurable. The fact that the integrand $\langle S(T-\sigma)^* z, u(\sigma)\rangle$ is measurable is proved as in the comments following (2.2.5). We have

$$|\langle S(T-\sigma)^* z, u(\sigma)\rangle| \le \|S(T-\sigma)^* z\|_{E^*} \|u(\sigma)\|_E\,,$$

thus the integral (2.3.26) exists and

$$\begin{aligned} \left| \int_0^T \langle S(T-\sigma)^* z, u(\sigma)\rangle d\sigma \right| &\le \int_0^T \|S(T-\sigma)^* z\|_{E^*} d\sigma \cdot \|u(\cdot)\|_{L^\infty(0,T;E)} \\ = \int_0^T \|S(\sigma)^* z\|_{E^*} d\sigma \cdot \|u(\cdot)\|_{L^\infty(0,T;E)} &= \|z\|_{Z_w(T)} \|u(\cdot)\|_{L^\infty(0,T;E)}\,. \end{aligned} \tag{2.3.27}$$

If (2.3.26) is to produce a functional in $R^\infty(T)$ we must check that the definition heeds the equivalence relation in $R^\infty(T)$, that is,

$$\int_0^T S(T-\sigma)u(\sigma)d\sigma = 0 \implies \left\langle\!\!\left\langle \xi_z\,, \int_0^T S(T-\sigma)u(\sigma)d\sigma \right\rangle\!\!\right\rangle = 0\,. \tag{2.3.28}$$

For this we use the operator $\lambda R(\lambda; A) = \lambda(\lambda I - A)^{-1}$ $(\lambda > \omega)$. Inequality (2.3.2) implies

$$\|\lambda R(\lambda; A)\| \le \frac{C\lambda}{\lambda - \omega}\,, \tag{2.3.29}$$

and arguing as in the proof of (2.3.4) we show that

$$\lim_{\lambda\to\infty} \lambda R(\lambda; A)y = y \tag{2.3.30}$$

first for $y \in D(A)$ and then (using (2.3.29) and the denseness of $D(A)$) for $y \in E$.

Assume $y^* \in \overline{D(A^*)}$ is such that $(\mu I^* - A^\bullet)y^* \in E^*$. Using the second equality (2.3.14) and the first equality (2.3.11) we obtain

$$\begin{aligned} y^* &= R(\mu; A^\bullet)(\mu I^* - A^\bullet)y^* = R(\mu; A)^\bullet(\mu I^* - A^\bullet)y^* \\ &\in R(\mu; A)^\bullet E^* = R(\mu; A)^* E^* = R(\mu; A^*)E^* = D(A^*)\,. \end{aligned}$$

Accordingly, $y^* \in D(A^*)$ and

$$(\mu I^* - A^\bullet)y^* = (\mu I^* - A^*)y^*\,.$$

We use this in the definition (2.3.16) of $S(t)^\bullet$ for $\psi$ restricted to the space $\mathcal{Z}$; since, by definition of $\mathcal{Z}$ we have $(\mu I^* - A^\bullet)S(t)^* R(\mu; A)^\bullet \psi \in E^*$, we deduce that $y^* = S(t)^* R(\mu; A)^\bullet \psi \in D(A^*)$ and

$$S(t)^\bullet \psi = (\mu I^* - A^*)S(t)^* R(\mu; A)^\bullet \psi \quad (\psi \in \mathcal{Z}). \tag{2.3.31}$$

In particular, for $z \in Z_w(T) \subseteq \mathcal{Z}$ we have

$$\begin{aligned} &\langle \lambda R(\lambda; A^*)(\mu I^* - A^*)S(T-\sigma)^* R(\mu; A)^\bullet z, u(\sigma)\rangle \\ &= \langle \lambda R(\lambda; A^*)S(T-\sigma)^\bullet z, u(\sigma)\rangle \\ &= \langle \lambda R(\lambda; A^*)S(T-\sigma)^* z, u(\sigma)\rangle \\ &= \langle S(T-\sigma)^* z, \lambda R(\lambda; A)u(\sigma)\rangle \to \langle S(T-\sigma)^* z, u(\sigma)\rangle \quad (0 \le \sigma < T). \end{aligned} \tag{2.3.32}$$

The equalities in (2.3.32) also imply

$$\begin{aligned} &|\langle \lambda R(\lambda; A^*)(\mu I^* - A^*)S(T-\sigma)^* R(\mu; A)^\bullet z, u(\sigma)\rangle| \\ &\le C\|S(T-\sigma)^* z\|_{E^*} \|u(\cdot)\|_{L^\infty(0,T;E)}, \end{aligned}$$

hence, by the dominated convergence theorem, the right side of (2.3.26) is the limit of

$$\begin{aligned} &\int_0^T \langle \lambda R(\lambda; A^*)(\mu I^* - A^*)S(T-\sigma)^* R(\mu; A)^\bullet z, u(\sigma)\rangle \, d\sigma \\ &= \int_0^T \langle R(\mu; A)^\bullet z\,,\, S(T-\sigma)(\mu I - A)\lambda R(\lambda; A)u(\sigma)\rangle \, d\sigma \\ &= \int_0^T \langle R(\mu; A)^\bullet z\,,\, (\mu I - A)\lambda R(\lambda; A)S(T-\sigma)u(\sigma)\rangle \, d\sigma \\ &= \left\langle R(\mu; A)^\bullet z\,,\ (\mu I - A)\lambda R(\lambda; A)\int_0^T S(T-\sigma)u(\sigma)d\sigma \right\rangle \end{aligned} \tag{2.3.33}$$

as $\lambda \to \infty$, which is zero if

$$\int_0^T S(T-\sigma)u(\sigma)d\sigma = 0\,,$$

proving the implication (2.3.28) and thus showing that the functional (2.3.26) is well defined in $R^\infty(T)$.

It is clear from (2.3.27) that $\|\xi_z\|_{R^\infty(T)^*} \le \|z\|_{Z_w(T)}$. Using Lemma 2.2.1 (or Lemma 2.2.10 if $E$ is not separable) we have

$$\begin{aligned} &\sup_{\|u(\cdot)\|_{L^\infty(0,T;E)} \le 1} \int_0^T \langle S(T-\sigma)^* z, u(\sigma)\rangle d\sigma \\ &= \int_0^T \|S(T-\sigma)^* z\|_{E^*} d\sigma = \|z\|_{Z_w(T)}\,, \end{aligned}$$

so that there exists $u(\cdot) \in L^\infty(0, T; E)$, $\|u(\cdot)\|_{L^\infty(0,T;E)} \le 1$ with

$$\left\langle\!\!\left\langle \xi_z\,, \int_0^T S(T-\sigma)u(\sigma)d\sigma \right\rangle\!\!\right\rangle = \int_0^T \langle S(T-\sigma)^* z, u(\sigma)\rangle d\sigma \ge \|z\|_{Z_w(T)} - \epsilon\,,$$

and we have

$$\left\| \int_0^T S(T-\sigma)u(\sigma)d\sigma \right\|_{R^\infty(T)} \le \|u(\cdot)\|_{L^\infty(0,T;E)} \le 1\,,$$

thus we conclude that

$$\|\xi_z\|_{R^\infty(T)^\star} = \|z\|_{Z_w(T)}\,. \tag{2.3.34}$$

Equality (2.3.34) implies the second imbedding in[11]

$$E^* \hookrightarrow Z_w(T) \overset{i}{\hookrightarrow} R^\infty(T)^\star\,, \tag{2.3.35}$$

the first imbedding obvious since for every $y^* \in E^*$ we have

$$\|y^*\|_{Z_w(T)} = \int_0^T \|S(\sigma)^* y^*\|_{E^*} d\sigma \le C\|y^*\|_{E^*}\,.$$

Functionals $\xi_z$ satisfying (2.3.26) are called *regular*, and we name $\mathcal{R}(T)$ the space of all regular functionals.

If $D(A)$ is not dense in $R^\infty(T)$, by the Hahn - Banach theorem there are nonzero functionals $\xi_s \in R^\infty(T)^\star$ (called *singular*) such that

$$\langle\!\langle \xi_s, y\rangle\!\rangle = 0 \quad (y \in D(A))\,. \tag{2.3.36}$$

If nonzero, a functional satisfying (2.3.36) cannot be regular. In fact, if $\xi_s = \xi_z$ for some $z \in Z_w(T)$ we have

$$\begin{aligned} \left\langle\!\!\left\langle \xi_s\,, \int_0^T S(T-\sigma)u(\sigma)d\sigma \right\rangle\!\!\right\rangle &= \int_0^T \langle S(T-\sigma)^* z, u(\sigma)\rangle d\sigma \\ &= \lim_{\lambda\to\infty} \int_0^T \langle S(T-\sigma)^* z,\, \lambda R(\lambda; A)u(\sigma)\rangle d\sigma \end{aligned} \tag{2.3.37}$$

by the dominated convergence theorem. On the other hand

$$\begin{aligned} &\int_0^T \langle S(T-\sigma)^* z,\, \lambda R(\lambda; A)u(\sigma)\rangle d\sigma = \left\langle\!\!\left\langle \xi_s\,, \int_0^T S(T-\sigma)\lambda R(\lambda; A)u(\sigma)d\sigma \right\rangle\!\!\right\rangle \\ &= \left\langle\!\!\left\langle \xi_s\,,\, \lambda R(\lambda; A)\int_0^T S(T-\sigma)u(\sigma)d\sigma \right\rangle\!\!\right\rangle = 0 \end{aligned}$$

[11] $\overset{i}{\hookrightarrow}$ means isometric imbedding.

so that (2.3.37) says

$$\int_0^T \langle S(T-\sigma)^* z, u(\sigma)\rangle d\sigma = 0 \qquad (u(\cdot) \in L^\infty(0,T;E))$$

and an application of Lemma 2.2.1 (Lemma 2.2.10 when $E$ is not separable) yields $S(T-\sigma)^* z = 0$ $(0 \le \sigma < T)$ which is equivalent to $z = 0$. We can summarize the argument after (2.3.36) writing

$$\mathcal{R}(T) \cap \mathcal{S}(T) = \{0\}\,, \tag{2.3.38}$$

where $\mathcal{S}(T)$ is the space of all singular functionals.

**Miscellaneous notes.** The results in this section are proved in the author [2001:3] under the requirement that $A^{-1}$ exist; this restriction is completely removed here. Functionals in $R^\infty(T)$ that vanish in $D(A)$ make a ghostly appearance in the author [1999:1] (their existence was not proved at that time).

**2.4. The reachable space and its dual, II.** We need the *Phillips adjoint theory* in Hille - Phillips [1957, Chapter 14]. The *Phillips dual* $E^\odot$ of $E$ is

$$E^\odot = \overline{D(A^*)}$$

(closure in $E^*$) and the *Phillips adjoint* $S^\odot(t)$ is

$$S^\odot(t) = S(t)^*|_{E^\odot}\,.$$

The semigroup $S^\odot(t)$ is strongly continuous in $E^\odot$. Its infinitesimal generator is

$$A^\odot y^* = A^* y^*\,, \quad D(A^\odot) = \{y^* \in D(A^*);\ A^* y^* \in E^\odot\}\,.$$

If $\overline{D(A^*)} = E^*$ (in particular, if $E$ is reflexive) then $E^\odot = E^*$ and $S^\odot(t) = S^*(t)$. The space $E^\odot$ is the "strong continuity space" of $S(t)^*$, that is, it consists of all $y^* \in E^*$ for which $S(t)^* y^*$ is continuous in $E^*$ for $t \ge 0$.

**Theorem 2.4.1.** *We have*

$$R^\infty(T)^\star = \mathcal{R}(T) \oplus \mathcal{S}(T) \quad (\textit{Banach direct sum})\,. \tag{2.4.1}$$

*The spaces $\mathcal{R}(T)$, $\mathcal{S}(T)$ are closed.*[12]

[12] *Banach direct sum* means algebraic direct sum of $\mathcal{R}(T)$ and $\mathcal{S}(T)$ plus boundedness of the projections onto both subspaces (it is enough to verify boundedness of one of the projections).

*Proof.* Below, $\mu > \omega$ is fixed ($\omega$ the constant in (2.3.1)), and we take $\lambda$ so large that $\lambda/(\lambda-\omega) \le 2$; $C$ is the constant in (2.3.1). In view of (2.3.2), the restriction on $\lambda$ implies

$$\|\lambda R(\lambda; A)\| = \|\lambda R(\lambda; A^*)\| \le 2C. \tag{2.4.2}$$

Let $\xi \in R(T)^\star$. By the first imbedding (2.3.6) $\xi$ is also a bounded linear functional in $D(A)$, thus Lemma 2.3.5 says that there exists $y^* \in E^*$ such that

$$\left\langle\!\!\left\langle \xi, \int_0^T S(T-\sigma)u(\sigma)d\sigma \right\rangle\!\!\right\rangle = \left\langle y^*, (\mu I - A)\int_0^T S(T-\sigma)u(\sigma)d\sigma \right\rangle \tag{2.4.3}$$

whenever

$$\int_0^T S(T-\sigma)u(\sigma)d\sigma \in D(A). \tag{2.4.4}$$

We can insure (2.4.4) by using the control $\lambda R(\lambda; A)u(\sigma)$ with $u(\cdot) \in L^\infty(0,T;E)$ arbitrary. Since $\xi \in R^\infty(T)^\star$,

$$\begin{aligned}
&\int_0^T \langle (\mu I^* - A^*)\lambda R(\lambda; A^*)S(T-\sigma)^* y^*, u(\sigma)\rangle d\sigma \\
&= \left\langle y^*, (\mu I - A)\int_0^T S(T-\sigma)\lambda R(\lambda; A)u(\sigma)d\sigma \right\rangle \\
&= \left\langle\!\!\left\langle \xi, \int_0^T S(T-\sigma)\lambda R(\lambda; A)u(\sigma)d\sigma \right\rangle\!\!\right\rangle \\
&\le \|\xi\|_{R(T)^\star} \|\lambda R(\lambda; A)u(\cdot)\|_{L^\infty(0,T;E)} \\
&\le 2C\|\xi\|_{R(T)^\star} \|u(\cdot)\|_{L^\infty(0,T;E)},
\end{aligned}$$

thus, by Lemma 2.2.1 (or Lemma 2.2.10 if $E$ is not separable)[13]

$$\int_0^T \|(\mu I^* - A^*)\lambda R(\lambda; A^*)S(T-\sigma)^* y^*\|_{E^*} \le 2C\|\xi\|_{R(T)^\star}. \tag{2.4.5}$$

Now, in view of (2.4.2) we have

$$\int_0^T \|\mu\lambda R(\lambda; A^*)S(T-\sigma)^* y^*\|_{E^*} \le 2C\mu \int_0^T \|S(T-\sigma)^* y^*\|_{E^*} d\sigma,$$

[13] Application of Lemma 2.2.10 is justified since

$$\langle (\mu I^* - A^*)\lambda R(\lambda; A^*)S(T-\sigma)^* y^*, y\rangle = \langle y^*, S(T-\sigma)(\lambda I - A)R(\lambda; A)y\rangle$$

We prove this equality first for $y \in D(A)$ and then extend it to $y \in E$ by continuity of both sides. It implies that $(\mu I^* - A^*)\lambda R(\lambda; A^*)S(T - \cdot)^* y^*$ is $E$-weakly continuous.

hence (2.4.5) implies

$$\int_0^T \|\lambda A^* R(\lambda; A^*) S(T-\sigma)^* y^*\|_{E^*} \le C' . \tag{2.4.6}$$

Select an increasing sequence $\{\lambda_n\}$ such that $\lambda_n \to \infty$. By (2.4.6) and Fatou's theorem,

$$\int_0^T \big( \liminf_{n\to\infty} \|\lambda_n A^* R(\lambda_n; A^*) S(T-\sigma)^* y^*\|_{E^*} \big) d\sigma \le C' , \tag{2.4.7}$$

so that the integrand is finite in a set $e$ of full measure in $[0, T]$. Accordingly, for each $\sigma \in e$ we can find a subsequence $\{\lambda_n(\sigma)\}$ of $\{\lambda_n\}$ depending on $\sigma$ such that $\|\lambda_n(\sigma) A^* R(\lambda_n(\sigma); A^*) S(T-\sigma)^* y^*\|$ is bounded in $E^*$. We have

$$\begin{aligned} &\lambda_n(\sigma) A^* R(\lambda_n(\sigma); A^*) S(T-\sigma)^* y^* \\ &= \lambda_n(\sigma) \big( \lambda_n(\sigma) R(\lambda_n(\sigma); A^*) S(T-\sigma)^* y^* - S(T-\sigma)^* y^* \big) , \end{aligned}$$

which, since $\lambda_n(\sigma) \to \infty$, implies $\lambda_n(\sigma) R(\lambda_n(\sigma); A^*) S(T-\sigma)^* y^* \to S(T-\sigma)^* y^*$ in $E^*$. On the other hand, we have $\lambda_n(\sigma) R(\lambda_n(\sigma); A^*) S(T-\sigma)^* y^* \in D(A^*)$, thus $S(T-\sigma)^* y^* \in \overline{D(A^*)} = E^\odot$. Now,

$$\begin{aligned} &\|S^\odot(t) S(T-\sigma)^* y^* - S(T-\sigma)^* y^*\|_{E^*} \\ &= \lim_{n\to\infty} \|(S^\odot(t) - I) \lambda_n(\sigma) R(\lambda_n(\sigma); A^*) S(T-\sigma)^* y^*\|_{E^*} \\ &= \lim_{n\to\infty} \left\| A^\odot \int_0^t S^\odot(\tau) \lambda_n(\sigma) R(\lambda_n(\sigma); A^*) S(T-\sigma)^* y^* \, d\tau \right\|_{E^*} \\ &= \lim_{n\to\infty} \left\| \int_0^t S^\odot(\tau) \lambda_n(\sigma) A^* R(\lambda_n(\sigma); A^*) S(T-\sigma)^* y^* \, d\tau \right\|_{E^*} \\ &\quad \le t \max_{0\le\tau\le t} \|S^\odot(\tau)\| \sup_n \|\lambda_n(\sigma) A^* R(\lambda_n(\sigma); A^*) S(T-\sigma)^* y^*\|_{E^*} \\ &= O(t) \quad (t \to 0+) , \end{aligned} \tag{2.4.8}$$

and we invoke Butzer - Berens [1967, Corollary 2.1.5, p. 92].

**Theorem 2.4.2.** *Let $y^* \in E^*$. Then $y^* \in D(A^*)$ if and only if*

$$\liminf_{t\to 0+} \frac{\|S(t)^* y^* - y^*\|}{t} < \infty . \tag{2.4.9}$$

Since Theorem 2.4.2 is so basic for our arguments, we give a proof. Assume that $y^* \in D(A^*)$. Then, if $y \in D(A)$,

$$\begin{aligned} \langle S(t)^* y^* - y^*, y \rangle &= \langle y^*, S(t) y - y \rangle \\ &= \left\langle y^* , A \int_0^t S(\sigma) y d\sigma \right\rangle = \int_0^t \langle A^* y^*, S(\sigma) y \rangle d\sigma , \end{aligned}$$

hence

$$|\langle S(t)^* y^* - y^*, y\rangle| \le Ct\,\|A^* y^*\|\|y\| \quad (y \in D(A))\,,$$

which inequality, by denseness of $D(A)$, implies (2.4.9).

Conversely, assume (2.4.9) holds and let $\{t_n\}$ be a positive sequence such that $t_n \to 0$ and $\{(S(t_n)y^* - y^*)/t_n\}$ is bounded in $E^*$. Any closed bounded subset of $E^*$ is $E$-weakly compact, thus there is a subsequence of $\{t_n\}$ (denoted with the same name) such that[14]

$$\frac{S(t_n)^* y^* - y^*}{t_n} \to z^* \in E^* \quad E\text{-weakly}.$$

If $y \in D(A)$ we have

$$\begin{aligned}\langle y^*, Ay\rangle &= \lim_{n\to\infty} \left\langle y^*, \frac{S(t_n)y - y}{t_n}\right\rangle \\ &= \lim_{n\to\infty} \left\langle \frac{S(t_n)^* y^* - y^*}{t_n}, y\right\rangle = \langle z^*, y\rangle\,.\end{aligned}$$

This implies that $y^* \in D(A^*)$ and that $A^* y^* = z^*$, concluding the proof.

Theorem 2.4.2 is applied to $S(T-\sigma)^* y^*$ in (2.4.8); we have already proved that $S(T-\sigma)^* y^* \in E^{\odot}$, thus $S(t)^* S(T-\sigma)^* y^* = S^{\odot}(t) S(T-\sigma)^* y^*$. We conclude that

$$S(T-\sigma)^* y^* \in D(A^*) \tag{2.4.10}$$

for $\sigma \in e$, thus for $0 \le \sigma < T$ since picking $\tau \in e \cap (\sigma, T)$ we may take advantage of the semigroup equation in the form $S(T-\sigma)^* y^* = S(\tau-\sigma)^* S(T-\tau)^* y^*$. That $A^* S(T - \,\cdot\,)^* y^*$ is $E$-weakly continuous in $0 \le \sigma < T$ follows from the semigroup equation;

$$\begin{aligned}&\langle A^* S(T-\sigma)^* y^*, y\rangle = \langle S(T-\sigma-h)^* A^* S(h)^* y^*, y\rangle \\ &= \langle A^* S(h)^* y^*, S(T-\sigma-h)y\rangle \quad (0 \le \sigma < T-h,\ h > 0)\,,\end{aligned}$$

thus (Lemma 2.3.4) $\|A^* S(T - \,\cdot\,)^* y^*\|$ is lower semicontinuous.

Let $\{\lambda_n\}$ be an arbitrary sequence with $\lambda_n \to \infty$, $z^* \in E^*$. Then

$$\langle \lambda_n R(\lambda_n; A^*) z^*, y\rangle = \langle z^*, \lambda_n R(\lambda_n; A) y\rangle \to \langle z^*, y\rangle\,,$$

so that $\lambda_n R(\lambda_n; A^*) z^* \to z^*$ $E$-weakly. Accordingly,

$$\lambda_n R(\lambda_n; A^*)(\mu I^* - A^*) S(T-\sigma)^* y^* \to (\mu I^* - A^*) S(T-\sigma)^* y^* \quad E\text{-weakly}\,,$$

[14] If $E$ is separable then the $E$-weak topology of closed bounded sets in $E^*$ is defined by a metric (Dunford - Schwartz [1958, Theorem 3, p. 434]) and we only need to use sequences and subsequences. If $E$ is not separable, generalized subsequences are needed.

and we deduce that

$$\begin{aligned}&\|(\mu I^* - A^*)S(T-\sigma)^* y^*\|_{E^*}\\ &\quad\le \liminf_{n\to\infty} \|\lambda_n R(\lambda_n; A^*)(\mu I^* - A^*)S(T-\sigma)^* y^*\|_{E^*}\\ &\quad= \liminf_{n\to\infty} \|(\mu I^* - A^*)\lambda_n R(\lambda_n; A^*)S(T-\sigma)^* y^*\|_{E^*}\,,\end{aligned}$$

which inequality, combined with (2.4.5) and Fatou's theorem yields

$$\begin{aligned}&\int_0^T \|(\mu I^* - A^*)S(\sigma)^* y^*\|_{E^*}\\ &\quad= \int_0^T \|(\mu I^* - A^*)S(T-\sigma)^* y^*\|_{E^*} d\sigma \le 2C\|\xi\|_{R(T)^*}\,. \qquad (2.4.11)\end{aligned}$$

We show that (2.4.11) implies

$$(\mu I^* - A^*)S(\sigma)^* y^* = S(\sigma)^* z = S(\sigma)^\bullet z \qquad (2.4.12)$$

for some $z \in Z_w(T)$ (see (2.3.21)). To this end, we define

$$z = (\mu I^* - A^\bullet)y^* \in E^*_{-1}. \qquad (2.4.13)$$

By (2.3.16) and (2.3.31) we have

$$\begin{aligned}S(\sigma)^* z &= S(\sigma)^\bullet z = (\mu I^* - A^\bullet)S(\sigma)^* R(\mu; A)^\bullet z\\ &= (\mu I^* - A^*)S(\sigma)^* R(\mu; A)^\bullet z = (\mu I^* - A^*)S(\sigma)^* y^*\,. \qquad (2.4.14)\end{aligned}$$

That $z \in Z_w(T)$ follows from $S(\sigma)^\bullet z \in E^*$ ($\sigma > 0$) and from (2.4.14) and the integral estimate (2.4.11).

We define a regular functional $\xi_z$ by

$$\left\langle\!\!\left\langle \xi_z\,, \int_0^T S(T-\sigma)u(\sigma)d\sigma \right\rangle\!\!\right\rangle = \int_0^T \langle S(T-\sigma)^* z, u(\sigma)\rangle d\sigma\,. \qquad (2.4.15)$$

It was shown in **2.3** (see (2.3.33)) that the right side is the limit of

$$\left\langle R(\mu; A)^\bullet z\,,\ (\mu I - A)\lambda R(\lambda; A) \int_0^T S(T-\sigma)u(\sigma)d\sigma \right\rangle \qquad (2.4.16)$$

as $\lambda \to \infty$. If (2.4.4) holds we switch the operators $(\mu I - A)$ and $\lambda R(\lambda; A)$ on the right side of (2.4.16) and use (2.4.13), whereupon (2.4.16) becomes

$$\left\langle y^*\,,\ \lambda R(\lambda; A)(\mu I - A) \int_0^T S(T-\sigma)u(\sigma)d\sigma \right\rangle. \qquad (2.4.17)$$

a restriction that affects generality but simplifies the treatment; in particular, the multiple norms $\|\cdot\|_{D(A),\mu}$, and $\|\cdot\|_{E^*_{-1},\mu}$ need not be used. Some of the results were proved earlier in less general versions (for instance, Theorem 2.4.3 (*a*) is in the author [1999:1], Theorem 2.4.3 (*b*) in the author [2000]). The constructs leading to (2.4.10) are in the author [1974/75]; for more on this see the Miscellaneous Notes in next section **2.5.**

The results in this and the previous section fall short of a "characterization" of the dual $R^\infty(T)^\star$, since not much is known about singular functionals. In fact, one can only show their existence for certain classes of spaces and semigroups (such as those in **2.6**), but there is not a single "explicit construction" of a singular functional.

There is also evidence that singular functionals may be much more numerous than regular functionals; there are examples (see Miscellaneous Notes in **2.8**) where $\mathcal{R}(T)$ is separable while $R^\infty(T)^\star$ is not, so that regular functionals are a tiny island in a vast ocean of functionals either singular or having nonzero singular part.

**2.5. The maximum principle.** Let $\bar{u}(\sigma)$ be a norm optimal control in the interval $0 \leq t \leq T$ under the target condition

$$y(t, \zeta, u) = \bar{y}\,. \tag{2.5.1}$$

Then (by definition of norm optimality)

$$\bar{y} - S(T)\zeta = \int_0^T S(T-\sigma)\bar{u}(\sigma)d\sigma \tag{2.5.2}$$

is a boundary point of the ball $B^\infty_\rho(T)$ of center 0 and radius $\rho = \|\bar{u}(\cdot)\|_{L^\infty(0,T;E)}$, thus it can be separated from $B^\infty_\rho(T)$ by a nonzero functional $\xi \in R^\infty(T)^\star$. The separation inequality is

$$\langle\!\langle \xi, y\rangle\!\rangle \leq \langle\!\langle \xi, \bar{y} - S(T)\zeta\rangle\!\rangle \qquad (\|y\|_{R^\infty(T)} \leq \rho)\,, \tag{2.5.3}$$

and is equivalent to

$$\left\langle\!\!\left\langle \xi, \int_0^T S(T-\sigma)u(\sigma)d\sigma \right\rangle\!\!\right\rangle \leq \left\langle\!\!\left\langle \xi, \int_0^T S(T-\sigma)\bar{u}(\sigma)d\sigma \right\rangle\!\!\right\rangle$$
$$(u(\cdot) \in L^\infty(0,T;E)\,, \quad \|u(\cdot)\|_{L^\infty(0,T;E)} \leq \rho)\,. \tag{2.5.4}$$

In fact, the implication (2.5.3) $\Rightarrow$ (2.5.4) is obvious. For the opposite implication, let $y \in R^\infty(T)$, $\|y\|_{R^\infty(T)} \leq \rho$. By definition of the norm of $R^\infty(T)$, for every $\varepsilon > 0$ we can write

$$y = \int_0^T S(T-\sigma)u_\varepsilon(\sigma)d\sigma, \qquad \|u_\varepsilon(\cdot)\|_{L^\infty(0,T;E)} \leq \rho + \varepsilon\,.$$

Assume (2.5.4) holds. Then

$$\begin{aligned}\frac{\rho}{\rho+\varepsilon}\langle\langle \xi, y\rangle\rangle &= \left\langle\!\!\left\langle \xi\,, \int_0^T S(T-\sigma)\frac{\rho}{\rho+\varepsilon}u_\varepsilon(\sigma)d\sigma\right\rangle\!\!\right\rangle \\ &\le \left\langle\!\!\left\langle \xi\,, \int_0^T S(T-\sigma)\bar u(\sigma)d\sigma\right\rangle\!\!\right\rangle \le \langle\langle \xi, \bar y - S(T)\zeta\rangle\rangle\,,\end{aligned}$$

which, since $\varepsilon$ is arbitrary and $y$ does not depend on $\varepsilon$, implies (2.5.3).

If $\xi = \xi_z$ ($z \in Z_w(T)$) is a regular functional, (2.5.4) is

$$\int_0^T \langle S(T-\sigma)^* z, u(\sigma)\rangle d\sigma \le \int_0^T \langle S(T-\sigma)^* z, \bar u(\sigma)\rangle d\sigma \tag{2.5.5}$$

for $\|u(\cdot)\|_{L^\infty(0,T;E)} \le \rho$. We take the supremum of the left side over $u(\cdot) \in B_\rho^\infty(T)$ and use Lemma 2.2.1 (Lemma 2.2.10 if $E$ is not separable), obtaining

$$\rho\int_0^T \|S(T-\sigma)^* z\| d\sigma = \int_0^T \langle S(T-\sigma)^* z, \bar u(\sigma)\rangle d\sigma\,, \tag{2.5.6}$$

which, since $\langle S(T-\sigma)^* z, \bar u(\sigma)\rangle \le \rho\|S(T-\sigma)^* z\|$, implies

$$\langle S(T-\sigma)^* z, \bar u(\sigma)\rangle = \rho\|S(T-\sigma)^* z\| \quad \text{a. e.} \tag{2.5.7}$$

This is equivalent to the maximum principle

$$\langle S(T-t)^* z, \bar u(t)\rangle = \max_{\|u\|\le\rho}\langle S(T-t)^* z, u\rangle \quad \text{a. e. in } \ 0 \le t \le T\,.$$

**Theorem 2.5.1.** *Let $\bar u(t)$ drive $\zeta$ to $\bar y$ norm optimally, and*

$$\bar y - S(T)\zeta \in \overline{D(A)} \tag{2.5.8}$$

*(the bar indicates closure in $R^\infty(T)$). Then there exists a regular functional $\xi_z \ne 0$ separating $\bar y - S(T)\zeta$ from $B_\rho^\infty(T)$. If the drive is time optimal there exists a regular functional $\xi_z \ne 0$ separating $\bar y - S(T)\zeta$ from $B_1^\infty(T)$.*

We prove first an auxiliary result involving the decomposition $\xi = \xi_z + \xi_s$ in (2.4.1) of a functional in its regular and singular parts.

**Theorem 2.5.2.** *Let*

$$y = \int_0^T S(T-\sigma)\bar u(\sigma)d\sigma \in B_\rho^\infty(T)$$

*be a boundary point of $B_\rho^\infty(T)$. Assume $\xi = \xi_z + \xi_s \in R^\infty(T)^\star$ separates $y$ from $B_\rho^\infty(T)$. Then, if either (a) $y \in \overline{D(A)}$ or (b) $S(t)E \subseteq D(A)$ ($t > 0$), $\xi_z$ separates $y$ from $B_\rho^\infty(T)$.*

*Proof.* Let $u(\cdot)$ be an arbitrary element of $L^\infty(0,T;E)$ with $\|u(\cdot)\|_{L^\infty(0,T;E)} \le \rho$. Assume we can construct a sequence $\{u_n(\cdot)\} \subset L^\infty(0,T;E)$ such that

$$\begin{aligned} \|u_n(\cdot)\|_{L^\infty(0,T;E)} \le \rho\,, \quad u_n(\sigma) \to \bar{u}(\sigma) \quad \text{a. e. in } 0 \le \sigma \le T\,, \\ u_n(\sigma) \in D(A)\,, \quad \|Au_n(\sigma)\| \le C_n \quad \text{a. e. in } 0 \le \sigma \le T\,. \end{aligned} \tag{2.5.9}$$

The last line of (2.5.9) and Theorem 3.7.4 on "introduction of an operator under the integral sign" in Hille-Phillips [1957, p. 83] say that

$$\begin{aligned} &\int_0^T S(T-\sigma)u_n(\sigma)d\sigma \in D(A)\,, \\ A&\int_0^T S(T-\sigma)u_n(\sigma)d\sigma = \int_0^T S(T-\sigma)Au_n(\sigma)d\sigma\,. \end{aligned} \tag{2.5.10}$$

The statement that $\xi = \xi_z + \xi_s$ separates $y$ from $B_\rho^\infty(T)$ is inequality (2.5.4) and we use it for $u(\sigma) = u_n(\sigma)$. Under assumption $(a)$, $y \in \overline{D(A)}$. Due to this, the first line of (2.5.10) and the fact that $\xi_s$ vanishes in $\overline{D(A)}$, (2.5.4) becomes

$$\left\langle\!\!\left\langle \xi_z\,, \int_0^T S(T-\sigma)u_n(\sigma)d\sigma \right\rangle\!\!\right\rangle \le \left\langle\!\!\left\langle \xi_z\,, \int_0^T S(T-\sigma)\bar{u}(\sigma)d\sigma \right\rangle\!\!\right\rangle.$$

This is equivalent to

$$\int_0^T \langle S(T-\sigma)^* z, u_n(\sigma)\rangle d\sigma \le \int_0^T \langle S(T-\sigma)^* z, \bar{u}(\sigma)\rangle d\sigma\,. \tag{2.5.11}$$

We then use the first line of (2.5.9) and the dominated convergence theorem to take limits on the left side of (2.5.11). The result is

$$\left\langle\!\!\left\langle \xi_z\,, \int_0^T S(T-\sigma)u(\sigma)d\sigma \right\rangle\!\!\right\rangle \le \left\langle\!\!\left\langle \xi_z\,, \int_0^T S(T-\sigma)\bar{u}(\sigma)d\sigma \right\rangle\!\!\right\rangle, \tag{2.5.12}$$

which is the separation conclusion in Theorem 2.5.2.

It remains to construct a sequence[15] $\{u_n(\sigma)\}$ satisfying (2.5.9). Since the function $u(\cdot)$ is strongly measurable, for each $n$ there exists a countably valued function $w_n(\sigma) = \sum_{j=1}^\infty \chi_{nj}(\sigma)w_{nj}$ (the $\chi_{nj}(\cdot)$ characteristic functions of a finite or countable partition $\{e_{nj}\}$ of $[0,T]$ in measurable sets) with $\|w_n(\sigma) - u(\sigma)\| \le \rho/n$ in $0 \le \sigma \le T$. This implies $\|w_n(\sigma)\| \le \rho + \rho/n = \rho((n+1)/n)$. Accordingly, if we define

$$v_n(\sigma) = \frac{n}{n+2}\,w_n(\sigma) = \sum_{j=1}^\infty \chi_{nj}(\sigma)\frac{n}{n+2}\,w_{nj} = \sum_{j=1}^\infty \chi_{nj}(\sigma)v_{nj}\,,$$

---

[15] The obvious candidate for $u_n(\sigma)$ would be $u_n(\sigma) = \lambda_n R(\lambda_n; A)u(\sigma)$ for $\lambda_n \to \infty$, but $\|\lambda_n R(\lambda_n; A)\| \le 1$ only if $\|S(t)\| \le 1$.

we have

$$\begin{aligned}\|v_n(\sigma) - u(\sigma)\| &\le \|w_n(\sigma) - u(\sigma)\| + \|w_n(\sigma) - v_n(\sigma)\| \\ &\le \frac{\rho}{n} + \rho\Big(\frac{n+1}{n}\Big)\Big(1 - \frac{n}{n+2}\Big) < \frac{3\rho}{n}\,, \\ \|v_n(\sigma)\| &\le \frac{n+1}{n+2}\,\rho\,,\end{aligned}$$

in $0 \le \sigma \le T$ for $n \ge 1$. Finally, for each $n$ we select $k(n)$ such that

$$\Big|\bigcup_{j=1}^{k(n)} e_{nj}\Big| \ge T - \frac{1}{2^n}\,,$$

find a finite sequence $u_{n1}, \dots, u_{n,k(n)} \in D(A)$ such that $\|u_{nj} - v_{nj}\| \le \rho/(n+2)$ $(j = 1, \dots, k(n))$ and define

$$u_n(\sigma) = \sum_{j=1}^{k(n)} \chi_{nj}(\sigma) u_{nj}\,, \quad e_m = \bigcap_{n=m}^{\infty} \bigcup_{j=1}^{k(n)} e_{nj}\,, \quad e = \bigcup_{m=1}^{\infty} e_m\,.$$

We have

$$\|u_n(\sigma)\| \le \|v_n(\sigma)\| + \frac{\rho}{n+2} \le \frac{n+1}{n+2}\rho + \frac{1}{n+2}\rho = \rho \quad (n \ge 1)\,,$$

and the second line of (2.5.9) is obvious. On the other hand,

$$|e_m| = \Big|\bigcap_{n=m}^{\infty} \bigcup_{j=1}^{k(n)} e_{nj}\Big| \ge T - \frac{1}{2^m} - \frac{1}{2^{m+1}} \;\dots = T - \frac{1}{2^{m-1}}\,,$$

so that $e$ has full measure in $[0, T]$. If $\sigma \in e$ then $\sigma \in$ some $e_m$ and

$$\begin{aligned}\|u_n(\sigma) - u(\sigma)\| &\le \|v_n(\sigma) - u(\sigma)\| + \|u_n(\sigma) - v_n(\sigma)\| \\ &< \frac{3\rho}{n} + \frac{\rho}{n+2} < \frac{4\rho}{n} \quad (n \ge m)\,,\end{aligned}$$

which completes the construction of $\{u_n(\sigma)\}$ and thus the proof of Theorem 2.5.2 under hypothesis $(a)$.

Under $(b)$, we pick $u(\cdot) \in L^\infty(0, T; E)$ with $\|u(\cdot)\|_{L^\infty(0,T;E)} \le \rho$, $\delta > 0$ and define

$$v(\sigma) = \begin{cases} u(\sigma) & 0 \le \sigma < T - \delta \\ \bar{u}(\sigma) & T - \delta \le \sigma \le T \end{cases}.$$

We put $v(\cdot)$ in the separation inequality (2.5.4). Since the singular functional $\xi_s$ is localized at $T$ (Theorem 2.4.3) we have

$$\Big\langle\!\!\Big\langle \xi_s\,, \int_0^T S(T-\sigma)v(\sigma)d\sigma \Big\rangle\!\!\Big\rangle = \Big\langle\!\!\Big\langle \xi_s\,, \int_0^T S(T-\sigma)\bar{u}(\sigma)d\sigma \Big\rangle\!\!\Big\rangle\,,$$

and we can "cross out" $\xi_s$ from (2.5.4), that is, replace $\xi$ by $\xi_z$; on the other hand,

$$\int_{T-\delta}^{T} S(T-\sigma)u(\sigma)d\sigma = \int_{T-\delta}^{T} S(T-\sigma)\bar{u}(\sigma)d\sigma$$

so we can also cross out the portion of the integrals over $T-\delta \leq \sigma \leq T$ in (2.5.4). We end up with

$$\left\langle\!\!\left\langle \xi_z\,, \int_0^{T-\delta} S(T-\sigma)u(\sigma)d\sigma \right\rangle\!\!\right\rangle \leq \left\langle\!\!\left\langle \xi_z\,, \int_0^{T-\delta} S(T-\sigma)\bar{u}(\sigma)d\sigma \right\rangle\!\!\right\rangle,$$

or

$$\int_0^{T-\delta} \langle S(T-\sigma)^* z, u(\sigma)\rangle d\sigma \leq \int_0^{T-\delta} \langle S(T-\sigma)^* z, \bar{u}(\sigma)\rangle d\sigma\,.$$

In view of the integrability of $\|S(T-\sigma)^* z\|$ we can take limits as $\delta \to 0$ on the basis of the dominated convergence theorem and obtain (2.5.12). This completes the proof of Theorem 2.5.2.

*Proof of Theorem* 2.5.1.We separate $\bar{y} - S(T)\zeta$ from $B_\rho^\infty(T)$ with a functional $\xi \in R^\infty(T)^\star$, $\xi \neq 0$. The result is (2.5.4), but the separation theorem gives the right of fixing the value of the the right side, say $= 1$;

$$\left\langle\!\!\left\langle \xi\,, \int_0^{T} S(T-\sigma)u(\sigma)d\sigma \right\rangle\!\!\right\rangle \leq \left\langle\!\!\left\langle \xi\,, \int_0^{T} S(T-\sigma)\bar{u}(\sigma)d\sigma \right\rangle\!\!\right\rangle = 1\,. \tag{2.5.13}$$

Condition (2.5.8) is

$$\int_0^T S(T-\sigma)\bar{u}(\sigma)d\sigma \in \overline{D(A)}$$

and singular functionals $\xi_s$ are zero in $\overline{D(A)}$, hence $\xi = \xi_z + \xi_s$, $\xi_z \neq 0$. The result then follows from Theorem 2.5.2 $(a)$. To prove the statement for time optimal controls we use time optimality $\Rightarrow$ norm optimality (Theorem 2.1.2).

We summarize what has been proved so far in Theorem 2.5.3 below. In this result, "$\bar{u}(t)$ is norm optimal" means "$\bar{u}(t)$ drives some $\zeta$ to $\bar{y} = y(T, \zeta, \bar{u})$ norm optimally." Same for time optimal controls.

**Theorem 2.5.3** *Assume* (2.5.8) *holds. Then* $(a)$ *If* $\bar{u}(\cdot)$ *is norm optimal,*

$$\langle S(T-t)^* z, \bar{u}(t)\rangle = \max_{\|u\|\leq\rho} \langle S(T-t)^* z, u\rangle \quad a.\ e.\ in\ \ 0 \leq t \leq T \tag{2.5.14}$$

*with* $z \in Z_w(T)$, $z \neq 0$ *and* $\rho = \|\bar{u}(\cdot)\|_{L^\infty(0,T;E)}$. $(b)$ *If* $\bar{u}(\cdot)$ *is time optimal,*

$$\langle S(T-t)^* z, \bar{u}(t)\rangle = \max_{\|u\|\leq 1} \langle S(T-t)^* z, u\rangle \quad a.\ e.\ in\ \ 0 \leq t \leq T \tag{2.5.15}$$

*with* $z \in Z_w(T)$, $z \neq 0$.

Theorem 2.5.3 implies the bang-bang property

$$\|\bar{u}(t)\| = \rho \quad \text{a. e. in } 0 \le t \le T \tag{2.5.16}$$

for the norm optimal problem under (2.5.8)[16] if

$$S(T-t)^* z \neq 0 \quad (0 < t \le T)\,. \tag{2.5.17}$$

**Corollary 2.5.4.** *Let $E$ be strictly convex, and assume that* (2.5.8) *and* (2.5.17) *hold. Then norm optimal controls are unique.*

The proof is the same as that of Theorem 2.2.9 and is omitted.

The results below are on sufficient conditions. In them, "$\bar{u}(t)$ is norm optimal in $0 \le t \le T$" means $\bar{u}(t)$ drives *any* initial condition $\zeta$ to the target $\bar{y} = y(T, \zeta, \bar{u})$ norm optimally in $0 \le t \le T$. Similarly for time optimal controls, although some restrictions have to be placed on $\zeta$ and $\bar{y}$.

**Theorem 2.5.5.** *Assume that $\bar{u}(t)$ satisfies* (2.5.14) *with $z \in Z_w(T)$, $z \neq 0$ and $\rho = \|\bar{u}(\cdot)\|_{L^\infty(0,T;E)}$. Then $\bar{u}(t)$ is norm optimal in $0 \le t \le T$.*

*Proof.* It imitates that of the sufficiency part of Theorem 2.2.5.[17] We again take advantage of the fact that (2.5.14) is equivalent to

$$\int_0^T \langle S(T-\sigma)^* z, u(\sigma)\rangle d\sigma \le \int_0^T \langle S(T-\sigma)^* z, \bar{u}(\sigma)\rangle d\sigma$$
$$(u(\cdot) \in L^\infty(0,T;E)\,, \quad \|u(\cdot)\|_{L^\infty(0,T;E)} \le \rho)\,. \tag{2.5.18}$$

Arguing as we did after (2.5.5) we deduce that

$$\langle\!\langle \xi_z\,, \bar{y} - S(T)\zeta \rangle\!\rangle = \int_0^T \langle S(T-\sigma)^* z, \bar{u}(\sigma)\rangle d\sigma = \rho \int_0^T \|S(T-\sigma)^* z\| d\sigma\,. \tag{2.5.19}$$

If $\bar{u}(t)$ is not norm optimal then there exists a control $u(\cdot) \in L^\infty(0,T;E)$ with $\|u(\cdot)\|_{L^\infty(0,T;E)} = \rho' < \rho$ and $y(T,\zeta,u) = y(T,\zeta,\bar{u})$. This means

$$\langle\!\langle \xi_z\,, \bar{y} - S(T)\zeta \rangle\!\rangle = \int_0^T \langle S(T-\sigma)^* z, u(\sigma)\rangle d\sigma \le \rho' \int_0^T \|S(T-\sigma)^* z\| d\sigma\,,$$

[16] That time optimal controls satisfy $\|u(t)\| = 1$ a. e. (Theorem 2.1.3) is independent of the maximum principle. In particular, (2.5.8) is not needed. Same for uniqueness of time optimal controls in strictly convex spaces (Theorem 2.1.7).

[17] Unless in the reversible case in Theorem 2.2.5, we don't have a true "if-and-only-if" result; the necessity of the maximum principle (2.5.14) for norm optimality depends on (2.5.8). Without (2.5.8) there may be norm optimal controls not satisfying the maximum principle. Examples in **2.7** and **2.8**.

which contradicts (2.5.19). This completes the proof.

**Remark 2.5.6.** It is important to realize that the proof of Theorem 2.5.5 does *not* require $S(T-t)^*z \neq 0$ in the entire interval $0 \le t \le T$. Let $\delta > 0$ be such that $S(t)^*z \neq 0$ $(0 \le t < \delta)$, $S(t)^*z = 0$ $(t \ge \delta)$. Then

$$S(T-t)^*z \neq 0 \quad (T-\delta < \sigma \le T), \qquad S(T-\sigma)^*z = 0 \quad (0 \le \sigma \le T-\delta).$$

Any control $\bar{u}(t)$ that satisfies the maximum principle

$$\langle S(T-t)^*z, \bar{u}(t)\rangle = \max_{\|u\|\le\rho} \langle S(T-t)^*z, u\rangle \quad \text{a. e. in } T-\delta \le t \le T$$

$(\rho = \|\bar{u}(\cdot)\|_{L^\infty(T-\delta,T)})$ and is defined arbitrarily in $0 \le t \le T-\delta$, as long as

$$\|u(t)\| \le \rho \quad \text{a. e. in } 0 \le t \le T-\delta,$$

satisfies (2.5.14), thus is norm optimal. We will see examples in **2.7** and **2.8**.

**Theorem 2.5.7.** *Assume that $\bar{u}(t)$ satisfies* (2.5.15) *with $z \in Z_w(T)$, $z \neq 0$ and that either*

$$\begin{array}{ll} (a) & \bar{y} = y(T,\zeta,\bar{u}) \in D(A), \quad \|A\bar{y}\| < 1, \quad or \\ (b) & \qquad\qquad \zeta \in D(A), \quad \|A\zeta\| < 1, \quad S(t)^*z \neq 0 \ in \ 0 \le t < T. \end{array} \tag{2.5.20}$$

*Then $\bar{u}(t)$ drives $\zeta$ to $\bar{y}$ time optimally in $0 \le t \le T$.*

*Proof under* $(a)$. Assume $\bar{u}(\cdot)$ does not drive $\zeta$ to $\bar{y}$ time optimally. Then there exists $\delta > 0$ and a control $\tilde{u}(\cdot) \in L^\infty(0,T-\delta;E)$, $\|\tilde{u}(\cdot)\|_{L^\infty(0,T-\delta;E)} \le 1$, that drives $\zeta$ to $\bar{y}$ in time $T-\delta$. The control

$$v(\sigma) = \begin{cases} \tilde{u}(\sigma) & (0 \le \sigma \le T-\delta) \\ -A\bar{y} & (T-\delta \le \sigma \le T) \end{cases}$$

satisfies $\|v(\cdot)\|_{L^\infty(0,T;E)} \le 1$. We have

$$\begin{aligned} \bar{y} - S(t-(T-\delta))\bar{y} &= -\int_0^{t-(T-\delta)} S(\sigma)A\bar{y}\,d\sigma \\ &= -\int_0^{t-(T-\delta)} S(t-(T-\delta)-\sigma)A\bar{y}\,d\sigma \\ &= \int_{T-\delta}^{t} S(t-\sigma)(-A\bar{y})\,d\sigma \quad (T-\delta \le t \le T), \end{aligned} \tag{2.5.21}$$

hence the trajectory $y(t,\zeta,v)$ starts at $\zeta$, reaches $\bar{y}$ at time $T-\delta$ and stays at $\bar{y}$ for $T-\delta \le t \le T$;

$$y(t,\zeta,v) = \bar{y} \quad (T-\delta \le t \le T).$$

This can be also be seen noting that if $y(t) = \bar{y}$ then we have $y'(t) - Ay(t) = -A\bar{y}$ in $T - \delta \le t \le T$.

The maximum principle (2.5.15) is equivalent to (2.5.18), which is in turn equivalent to (2.5.4). We have

$$\int_0^T S(T-\sigma)v(\sigma)d\sigma = \int_0^T S(T-\sigma)\bar{u}(\sigma)d\sigma\,, \tag{2.5.22}$$

hence (2.5.4) gives

$$\left\langle\!\!\left\langle \xi_z\,, \int_0^T S(T-\sigma)u(\sigma)d\sigma \right\rangle\!\!\right\rangle \le \left\langle\!\!\left\langle \xi_z\,, \int_0^T S(T-\sigma)v(\sigma)d\sigma \right\rangle\!\!\right\rangle$$
$$(u(\cdot) \in L^\infty(0,T;E), \quad \|u(\cdot)\|_{L^\infty(0,T;E)} \le \rho)\,,$$

which is equivalent to

$$\langle S(T-\sigma)^* z, v(t)\rangle = \max_{\|u\|\le 1} \langle S(T-t)^* z, u\rangle \quad \text{a. e. in } 0 \le t \le T\,. \tag{2.5.23}$$

Since $S(T-\sigma)^* z \ne 0$ near $T$, the maximum principle (2.5.23) implies $\|v(\sigma)\| = 1$ near $T$. This is a contradiction since, by hypothesis, $\|A\bar{y}\| < 1$.

*Proof under* $(b)$. This time we define

$$v(\sigma) = \begin{cases} -A\zeta & (0 \le \sigma \le \delta) \\ \tilde{u}(\sigma - \delta) & (\delta \le \sigma \le T)\,. \end{cases}$$

As in (2.5.21) we have

$$\zeta - S(t)\zeta = \int_0^t S(t-\sigma)(-A\zeta)d\sigma\,,$$

hence the trajectory $y(t,\zeta,v)$ stays at $\zeta$ for $0 \le t \le \delta$,

$$y(t,\zeta,v) = \zeta \qquad (0 \le t \le \delta)\,,$$

and then starts for the target $\bar{y}$, which hits at time $T$. The proof ends in the same way as that of $(a)$ noting that $S(T-\sigma)^* z \ne 0$ in $0 \le \sigma \le T$ (in particular, in $0 \le \sigma \le \delta$) hence the maximum principle (2.5.15) implies $\|v(\sigma)\| = 1$ in $0 \le \sigma \le \delta$. This contradicts the fact that $\|v(\sigma)\| = \|A\zeta\| < 1$ in $0 \le \sigma \le \delta$, and completes the proof of Theorem 2.5.7.

**Remark 2.5.8.** The proof of Theorem 2.5.7 implies that of Theorem 2.2.6. Note that under the assumptions of Theorem 2.2.6 (semigroup satisfying $S(t)E = E$ for $t > 0$) we have $Z_w(T) = E^*$, so the multiplier $z$ in Theorem 2.5.7 is an element of $E^*$. Also, the condition that $S(t)^* z \ne 0$ in $0 \le t \le T$ is automatic (see (2.2.21)).

Another case where $z \neq 0 \Rightarrow S(t)^* z \neq 0$ for all $t$ is that where $S(t)$ is an analytic semigroup. This implies $S(t)^*$ is analytic in $t > 0$. If $S(t)^* z = 0$ for some $t > 0$ the semigroup equation implies $S(\sigma)^* z = S(\sigma - t)^* S(t)^* z = 0$ for $\sigma \geq t$, thus, by analyticity, $S(\sigma)^* z = 0$ for $\sigma > 0$, which is equivalent to $z = 0$.

Example 1.1.5 shows that assumption (2.5.20) on $\zeta$, $\bar{y}$ cannot be discarded, even in dimension 1. Since in this counterexample we have $\|A\zeta\| = \|A\bar{y}\| = 1$, it also shows that (2.5.20) is the best possible condition (of its kind).

For another (family of) counterexamples, let $S(t)$ be a strongly continuous semigroup such that $S(t)S(t)^* = I$. As shown in Remark 2.2.7, if $T > 0$ satisfies

$$T = \|S(T)\zeta - \bar{y}\|, \tag{2.5.24}$$

then the control

$$u(t) = S(T-t)^* z, \quad z = \frac{\bar{y} - S(T)\zeta}{\|\bar{y} - S(T)\zeta\|} \tag{2.5.25}$$

satisfies the maximum principle (2.5.15) and drives $\zeta$ to $\bar{y}$ in time $T$.

**Theorem 2.5.9.** *Let $E$ be reflexive and $\|S(t)\| = o(t)$ as $t \to \infty$. Assume that, either $(a)$ $\zeta \in D(A), \|A\zeta\| > 1$, or $(b)$ $\zeta \notin D(A)$. Then there exists a control of the form (2.5.25) that drives $\zeta$ to $\zeta$ in time $T > 0$, thus is not time optimal.*

*Proof.* We write (2.5.24) as

$$\frac{\|S(T)\zeta - \zeta\|}{T} = 1. \tag{2.5.26}$$

For every $\zeta \in E$,

$$\lim_{t \to \infty} \frac{\|S(t)\zeta - \zeta\|}{t} = 0.$$

In case $(a)$ we have

$$\lim_{t \to 0+} \frac{\|S(t)\zeta - \zeta\|}{t} = \lim_{t \to 0+} \left\| \frac{S(t)\zeta - \zeta}{t} \right\| \to \|A\zeta\| > 1,$$

while in case $(b)$,

$$\liminf_{t \to 0+} \frac{\|S(t)\zeta - \zeta\|}{t} = \liminf_{t \to 0+} \left\| \frac{S(t)\zeta - \zeta}{t} \right\| = \infty,$$

since[18] $\zeta \notin D(A)$. In either case $(a)$ or $(b)$, (2.5.26) has a solution $T > 0$, thus the control (2.5.25) satisfies the maximum principle in $0 \leq t \leq T$ and drives $\zeta$ to $\zeta$. This ends the proof.

---

[18] That the $\liminf$ is $= \infty$ follows from Theorem 2.4.2 with the roles of $E$, $E^*$ and of $S(t)$, $S(t)^*$ reversed; since $E$ is reflexive we have $E = (E^*)^*$ and $S(t) = (S(t)^*)^*$.

**Example 2.5.10.** Consider the unitary semigroup

$$S(t) = e^{At} = \begin{bmatrix} \cos t & \sin t \\ -\sin t & \cos t \end{bmatrix}, \quad A = \begin{bmatrix} 0 & 1 \\ -1 & 0 \end{bmatrix}$$

in $\mathbb{R}^2$. Writing $\zeta = (r\cos\theta,\, r\sin\theta)$, $\bar{y} = (s\cos\varphi,\, s\sin\varphi)$ in polar coordinates,

$$\begin{aligned}\|S(t)\zeta - \bar{y}\|^2 &= \left\| S(t)\begin{bmatrix} r\cos\theta \\ r\sin\theta \end{bmatrix} - \begin{bmatrix} s\cos\varphi \\ s\sin\varphi \end{bmatrix} \right\|^2 \\ &= (r\cos t\cos\theta - s\cos\varphi + r\sin t\sin\theta)^2 \\ &+ (-r\cos\theta\sin t + r\cos t\sin\theta - s\sin\varphi)^2 \\ &= r^2 + s^2 - 2rs\cos(t - \theta + \varphi)\,.\end{aligned}$$

For $\zeta = \bar{y} = (1.1, 0)$ ($\|A\zeta\| = \|A\bar{y}\| = 1.1$) we have $r = s = 1.1$, $\theta = \varphi = 0$. Equation (2.5.26) has a positive solution (Figure 2.5.1)

$$T = 1.49797\,, \tag{2.5.27}$$

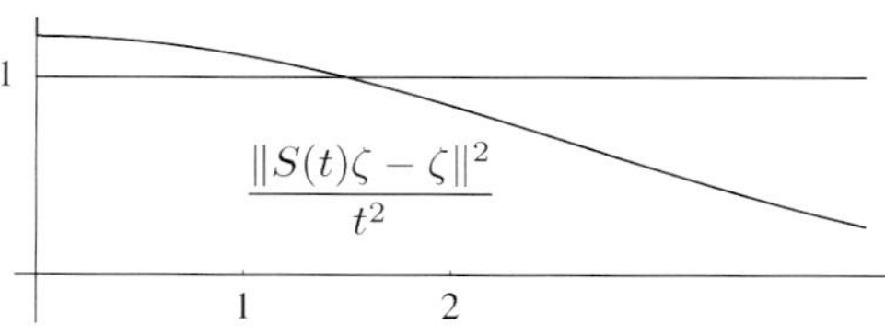


Figure 2.5.1.

thus we can drive from $\zeta = (1.1, 0)$ back to $\zeta$ in time $T$ with a control

$$\bar{u}(\sigma) = S(T-\sigma)^\star z\,, \quad z = \frac{\zeta - S(T)\zeta}{\|\zeta - S(T)\zeta\|} = (0.68090, 0.73238)\,. \tag{2.5.28}$$

Figure 2.5.2 below shows the drive (moving clockwise).

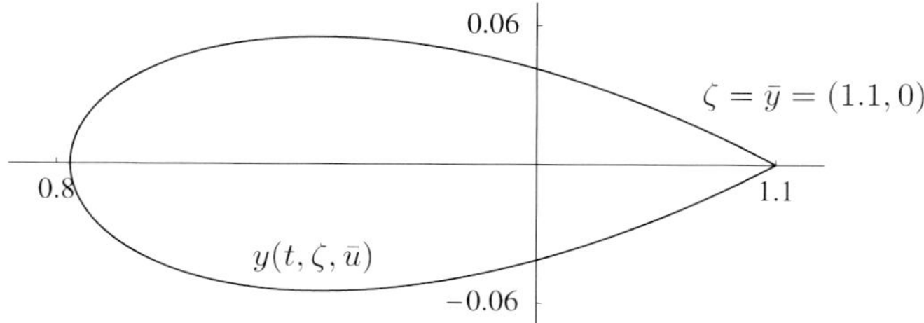


Figure 2.5.2.

For

$$\zeta = (4,0)\,, \quad \bar{y} = (-4,0)$$

we have $r = s = 4$, $\theta = 0$, $\varphi = \pi$ and the equation (2.5.26) has three solutions shown in Figure 2.5.3,

$$T_0 = 2.5047\,, \quad T_1 = 4.2667\,, \quad T_2 = 7.1906\,, \tag{2.5.29}$$

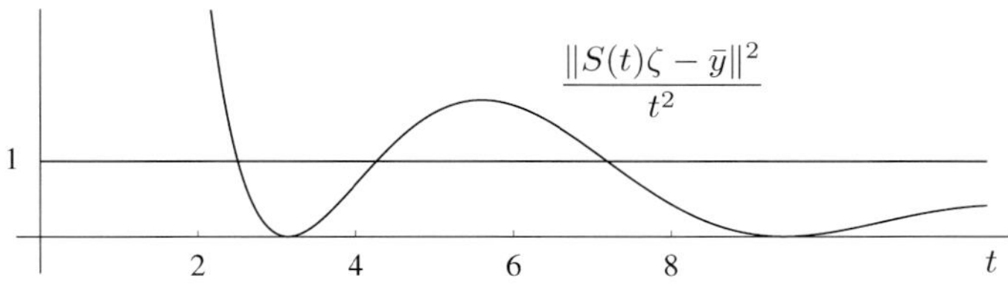


Figure 2.5.3.

thus we can drive from $(4,0)$ to $(-4,0)$ with three different controls of the form

$$\bar{u}_j(t) = S(T_j - \sigma)^* z_j\,, \quad z_j = \frac{\bar{y} - S(T_j)\zeta}{\|\bar{y} - S(T_j)\zeta\|} \quad (j = 0\,,1\,,2)\,, \tag{2.5.30}$$

$$z_0 = (-0.31309, 0.94972)\,,\ z_1 = (-0.53333, -0.84591)\,,\ z_2 = (-0.89883, 0.43831)\,.$$

Figure 2.5.4 shows the three trajectories, each in the interval $0 \le t \le T_j$; only the lower one (thicker curve) is time optimal.

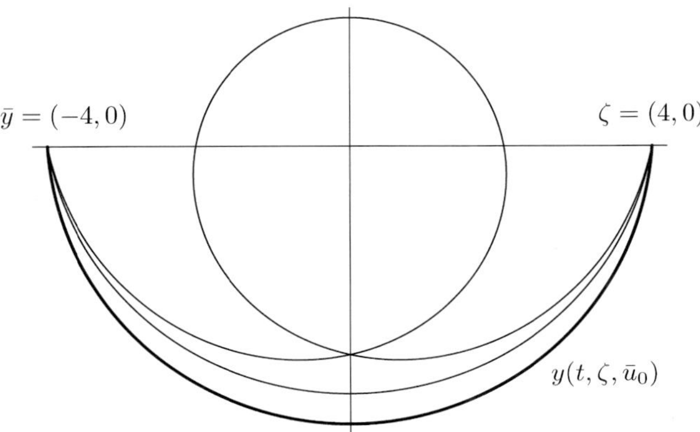


Figure 2.5.4.

**Remark 2.5.11.** The maximum principle (2.5.14) is a sufficient condition for norm optimality (Theorem 2.5.5) with no conditions on $\zeta$ or $\bar{y}$ so that each of the controls $\bar{u}_j(t)$, $j = 0, 1, 2$ in (2.5.30) is norm optimal in its own interval; if $\zeta = (4, 0)$ can be driven to $\bar{y} = (-4, 0)$ in the interval $0 \le t \le T_j$ by means of a control $u(t)$ then

$$\|u(\cdot)\|_{L^\infty(0,T_j;E)} \ge 1 = \|\bar{u}_j(\cdot)\|_{L^\infty(0,T_j;E)} .$$

The same observation applies to the control (2.5.28); it drives $\zeta$ back to $\zeta$ norm optimally in the interval $0 \le t \le T$.

**Miscellaneous notes.** Since Pontryagin *et al.* [1961] certain terminology has been associated with the maximum principle. The $E^*$-valued function

$$z(t) = S(T - t)^* z \tag{2.5.31}$$

in Theorem 2.5.3 is called the *costate* (associated with the optimal trajectory $y(t, \zeta, \bar{u})$), while $z$ is the *multiplier.* Formally, the costate satisfies the *reversed adjoint equation*

$$z'(t) = -A^* z(t) \ \ (0 \le t < T) , \quad z(T) = z , \tag{2.5.32}$$

and, just as for the trajectory $y(t, \zeta, \bar{u})$, it is possible to show that both the reverse adjoint equation and the assumption of the initial (or, rather, final) condition are satisfied in suitable weak senses (for the equation, in the sense of vector valued distributions). This is not particularly relevant in the linear theory.

In terms of the costate, the maximum principle (2.5.15) becomes

$$\langle z(t), \bar{u}(t)\rangle = \max_{\|u\| \le 1} \langle z(t), u\rangle \quad \text{a. e. in } 0 \le t \le T \tag{2.5.33}$$

and (2.5.14) can be rewritten in the same way.

Most of this section is based on the author [2001:3]. Some of the results appeared in a particular case (Hilbert space, self adjoint infinitesimal generator) in the author [2000]. Theorems 2.5.7, 2.5.9 and Example 2.5.10 (as well as Example 1.1.5) are taken from the author [T. A.:1]. Although it is hard to believe that examples like 2.5.10 and 1.1.5 are not somewhere in existing literature, we have been unable to find references.

The maximum principle (2.5.15) as necessary condition for time optimality was obtained in the author [1974/75] under the assumption $\zeta - S(T)\bar{y} \in D(A)$. The treatment there avoids (partly) the problem of figuring out the dual of $R^\infty(T)$ by doing the separation in $D(A)$. The maximum principle is obtained in the form (2.5.33) with costate $z(t) = A^* S(T - t)^* y^*$, where $y^* \in E^*$ is an element of $E^*$ such that $S(t)^* y^* \in D(A^*)$. However, not enough attention is paid to $E$-weak measurability, thus the treatment is only satisfactory in such particular situations as $E$ reflexive, $E^*$ separable, or $S(t)E \subseteq D(A)$.

**2.6. Vanishing of the costate and nonuniqueness in norm optimality.** Consider the space $E = L^2(0,\infty)$, all functions $y(x)$, $z(x)\dots$ defined in $x \geq 0$ and extended $= 0$ for $x < 0$. The (strongly continuous, isometric) *right translation* semigroup $S(t)$ in $L^2(0,\infty)$ is defined by[19]

$$S(t)y(x) = \begin{cases} y(x-t) & (x \geq t)\,, \\ 0 & (x < t)\,. \end{cases} \tag{2.6.1}$$

The adjoint semigroup is the *left translation* (and chop-off) semigroup

$$S(t)^* z(x) = \begin{cases} z(x+t) & (x \geq 0)\,, \\ 0 & (x < 0)\,. \end{cases} \tag{2.6.2}$$

The infinitesimal generator $A$ of $S(t)$ is

$$Ay(x) = -y'(x)\,, \tag{2.6.3}$$

with domain $D(A)$ = {all absolutely continuous $y(x)$ with derivative $y'(x)$ in $L^2(0,\infty)$ and $y(0) = 0$}. Equivalently, $D(A)$ consists of all $y(\cdot) \in L^2(0,\infty)$ such that $y'(\cdot)$, taken in the sense of distributions, belongs to $L^2(0,\infty)$ and $y(0) = 0$. The semigroup $S(t)$ is associated with the control system

$$\begin{aligned} \frac{\partial y(t,x)}{\partial t} &= -\frac{\partial y(t,x)}{\partial x} + u(t,x) \quad (0 \leq t,\, x < \infty)\,, \\ y(0,x) &= \zeta(x)\,, \qquad y(t,0) = 0\,, \end{aligned} \tag{2.6.4}$$

in the sense that $S(t)$ is the propagation semigroup of the homogeneous equation ($u(t,x) = 0$). The variation-of-constants formula corresponding to the control $u(t)(x) = u(t,x)$ is

$$\begin{aligned} y(t,x,\zeta,u) = y(t,\zeta,u)(x) &= \left( S(t)\zeta + \int_0^t S(t-\sigma)u(\sigma)d\sigma \right)(x) \\ &= \zeta(x-t) + \int_0^t u(\sigma, x-(t-\sigma))d\sigma \end{aligned} \tag{2.6.5}$$

thus the contribution of the control $u(\sigma,x)$ to $y(t,x,\zeta,u)$ is the integral of $u(\sigma,x)$ over the intersection with the positive quadrant of the line $(\sigma, x-(t-\sigma))$ joining $(0, x-t)$ with $(t,x)$ as seen in Figure 2.6.1.

---

[19] The second line of (2.6.1) is actually redundant (since $y(x) = 0$ for $x < 0$) and will be omitted in all subsequent formulas.

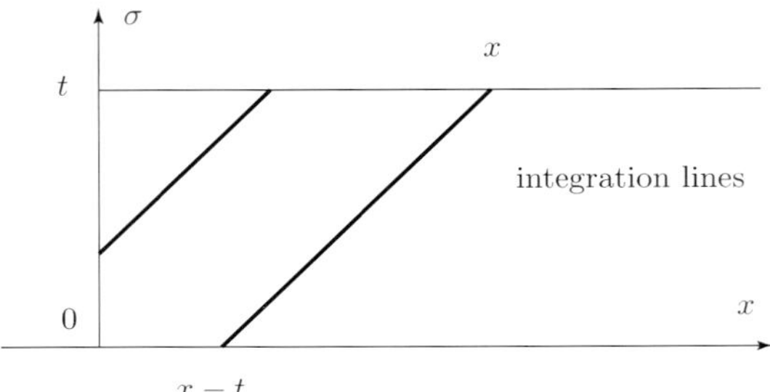


Figure 2.6.1

For this space and semigroup, we call $\mathcal{Z}$ the space of all measurable $z(x)$ defined a. e. in $x \geq 0$ and such that

$$\begin{aligned}\|S(\sigma)^* z(\cdot)\| &= \sqrt{\int_0^\infty z(x+\sigma)^2 dx} \\ &= \sqrt{\int_\sigma^\infty z(x)^2 dx} = \kappa(\sigma, z) < \infty \quad (\sigma > 0)\,. \end{aligned} \tag{2.6.6}$$

The space $Z(T)$ consists of all $z(\cdot) \in \mathcal{Z}$ such that[20]

$$\int_0^T \|S(\sigma)^* z(\cdot)\| d\sigma = \int_0^T \kappa(\sigma, z) d\sigma < \infty\,. \tag{2.6.7}$$

Since we are in a Hilbert space, any control that satisfies the maximum principle

$$\langle S(T-\sigma)^* z, \bar{u}(\sigma)\rangle = \max_{\|u\| \leq \rho} \langle S(T-\sigma)^* z, u\rangle \quad \text{a. e. in } 0 \leq \sigma \leq T \tag{2.6.8}$$

is given a. e. by

$$\bar{u}(\sigma, x) = \rho \frac{S(T-\sigma)^* z(x)}{\|S(T-\sigma)^* z(\cdot)\|} = \rho\chi(x) \frac{z(x + (T-\sigma))}{\kappa(T-\sigma, z)} \quad (T - \delta < \sigma \leq T)\,, \tag{2.6.9}$$

where $\chi(x)$ is the characteristic function of $x \geq 0$ and $0 \leq t < \delta$ is the maximal interval where $S(t)^* z \neq 0$; if $\delta \geq T$ the interval in (2.6.9) is $0 < \sigma \leq T$.

We use Theorem 2.5.5 and Remark 2.5.6 to construct norm optimal controls that are undetermined (by the maximum principle) in part of the control interval. We take $\rho = 1$ for simplicity. For $\delta > 0$ let $z(\cdot) \in Z(T)$ have support $= [0, \delta]$. According to Theorem 2.5.5, the control

[20] The space $Z(T)$ is the same space defined in Sections **2.2** and **2.3**. See Remark 2.8.3.

$$\bar{u}(\sigma, x) = \begin{cases} \dfrac{S(T-\sigma)^* z(x)}{\|S(T-\sigma)^* z(\cdot)\|} & (T - \delta < \sigma \le T)\,, \\ \bar{u}(\sigma, x) & (0 \le \sigma \le T - \delta)\,, \end{cases} \tag{2.6.10}$$

where $\bar{u}(\sigma, x)$ is measurable and satisfies

$$\int_0^\infty |\bar{u}(\sigma, x)|^2 dx \le 1 \quad \text{a. e. in } 0 \le \sigma \le T - \delta\,, \tag{2.6.11}$$

drives norm optimally an arbitrary initial condition $\zeta(x)$ to the target

$$\bar{y}(x) = y(T, x, \zeta, \bar{u}) = \zeta(x - T) + \int_0^T \bar{u}(\sigma, x - (T - \sigma))d\sigma\,.$$

**Theorem 2.6.1.** *Let $T > 0$. There exist targets $\bar{y}(\cdot)$ such that (a) the norm optimal control driving $0$ to $\bar{y}(\cdot)$ in $0 \le t \le T$ is not unique, (b) the costate $S(T-\sigma)^* z$ is zero in an interval $0 \le \sigma \le T - \delta < T$ with $\delta$ as small as desired.*

*Proof.* Choose $z(x)$ with support $= [0, \delta]$, $\delta < T$, define $\bar{u}(\sigma, x)$ by (2.6.10) and then select $\bar{u}(\sigma, x)$ with support in the strip $0 \le \sigma \le T - \delta$, $0 \le x < \infty$ satisfying (2.6.11) and

$$\int_0^T \bar{u}(\sigma, x - (T - \sigma))d\sigma = 0 \quad (\delta < x < \infty)\,. \tag{2.6.12}$$

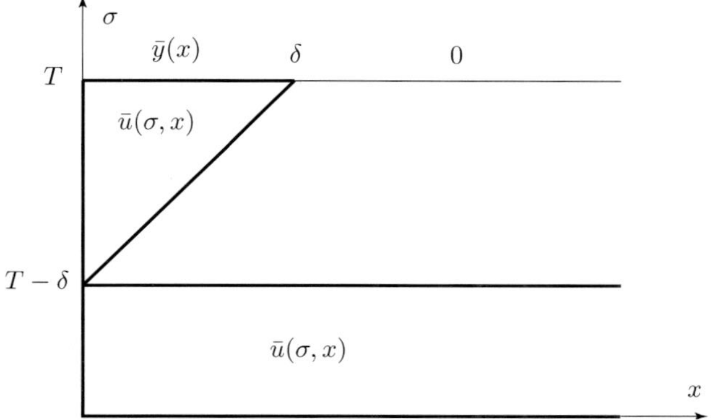


Figure 2.6.2

Clearly, we can achieve (2.6.12) and (2.6.11) with a nonzero control (even with compact support). The control $\bar{u}(\sigma, x)$ so constructed and the control $\bar{u}_0(\sigma, x)$ defined by (2.6.10) with $\bar{u}_0(\sigma, x) = 0$ in $0 \le \sigma \le T - \delta$ both drive $\zeta(\cdot) = 0$ to

$$\bar{y}(x) = \int_0^T \bar{u}(\sigma, x - (T - \sigma))d\sigma \quad (0 \le x \le \delta)\,, \qquad \bar{y}(x) = 0 \quad (x \ge \delta)$$

norm optimally in $0 \le t \le T$. This proves $(a)$. As for $(b)$, $\delta > 0$ is arbitrary.

**Corollary 2.6.2.** *There exist norm optimal controls that are not time optimal.*[21]

*Proof.* The control $\bar{u}_0(\sigma, x)$ above does the job: it does not satisfy the mandatory condition $\|u(\sigma)\| = 1$ a. e. for time optimal controls (Theorem 2.1.3) in the interval $0 \le \sigma \le T - \delta$.

It is clear that the example in Theorem 2.6.1 and Corollary 2.6.2 cannot be made to work if we require the target to have support in $x \ge T$ (which is equivalent to $\bar{y}(\cdot) = S(T)\eta(\cdot), \eta(\cdot) \in E$). The next result, which is valid for an arbitrary Banach space $E$ and an arbitrary strongly continuous semigroup $S(t)$, greatly generalizes this observation.

**Theorem 2.6.3.** *Assume $\bar{u}(\cdot) \in L^\infty(0, T; E)$ satisfies*

$$\langle S(T-\sigma)^* z. \bar{u}(\sigma)\rangle = \max_{\|u\| \le \rho} \langle S(T-\sigma)^* z, u\rangle \quad a.\ e.\ in\ 0 \le \sigma \le T \tag{2.6.13}$$

*with $z \in Z(T), z \ne 0$. Then, if $\bar{u}(\cdot)$ drives $\zeta$ to $\bar{y} = S(T)\eta$, that is, if*

$$\bar{y} = y(T, \zeta, \bar{u}) = S(T)\zeta + \int_0^T S(T-\sigma)\bar{u}(\sigma)d\sigma = S(T)\eta\,, \tag{2.6.14}$$

*we have*

$$\langle S(T)^* z, \eta - \zeta\rangle \ne 0\,. \tag{2.6.15}$$

*In particular, $S(T)^* z \ne 0$, thus*

$$z(\sigma) = S(T-\sigma)^* z \ne 0 \quad (0 \le \sigma \le T)\,. \tag{2.6.16}$$

*Proof.* (2.6.13) is equivalent to

$$\left\langle\!\!\left\langle \xi_z, \int_0^T S(T-\sigma)u(\sigma)d\sigma \right\rangle\!\!\right\rangle \le \left\langle\!\!\left\langle \xi_z, \int_0^T S(T-\sigma)\bar{u}(\sigma)d\sigma \right\rangle\!\!\right\rangle \tag{2.6.17}$$

for all $u(\cdot) \in L^\infty(0, T; E)$, $\|u(\cdot)\|_{L^\infty(0,T;E)} \le \rho$, where $\xi_z \in R^\infty(T)^\star$ is the regular functional

$$\left\langle\!\!\left\langle \xi_z, \int_0^T S(T-\sigma)u(\sigma)d\sigma \right\rangle\!\!\right\rangle = \int_0^T \langle S(T-\sigma)^* z, u(\sigma)\rangle d\sigma\,. \tag{2.6.18}$$

On the other hand, (2.6.14) is

$$S(T)(\eta - \zeta) = \int_0^T S(T-\sigma)\bar{u}(\sigma)d\sigma\,, \tag{2.6.19}$$

[21] That is, the converse of Theorem 1.1.2 is not true.

and (from the semigroup equation) we also have

$$S(T)(\eta - \zeta) = \int_0^T S(T-\sigma)\Big(\frac{S(\sigma)(\eta-\zeta)}{T}\Big)d\sigma\,. \tag{2.6.20}$$

Accordingly, application of the functional $\xi_z$ works as follows:[22]

$$\begin{aligned}&\Big\langle\!\Big\langle \xi_z, \int_0^T S(T-\sigma)\bar{u}(\sigma)d\sigma \Big\rangle\!\Big\rangle = \Big\langle\!\Big\langle \xi_z, \int_0^T S(T-\sigma)\Big(\frac{S(\sigma)(\eta-\zeta)}{T}\Big)d\sigma \Big\rangle\!\Big\rangle \\ &= \frac{1}{T}\int_0^T \langle S(T-\sigma)^* z, S(\sigma)(\eta-\zeta)\rangle d\sigma = \frac{1}{T}\int_0^T \langle S(\sigma)^* S(T-\sigma)^* z, \eta-\zeta\rangle d\sigma \\ &= \langle S(T)^* z, \eta-\zeta\rangle\,. \end{aligned} \tag{2.6.21}$$

If $\langle S(T)^* z, \eta-\zeta\rangle = 0$ then (2.6.17) is not possible, since the right side is $= 0$ and the left side can take both positive and negative values (unless $\xi_z = 0$, a contradiction). This ends the proof.

**Remark 2.6.4.** That the drive in Theorem 2.6.3. is norm optimal follows from (2.6.13) and Theorem 2.5.5. According to Theorem 2.5.7, it is also time optimal if $\rho = 1$ and, either if $\zeta \in D(A)$ and $\|A\zeta\| < 1$ or if $\bar{y} = S(T)\eta \in D(A)$ with $\|A\bar{y}\| = \|AS(T)\eta\| < 1$.

**Corollary 2.6.5.** *Assume that* $\bar{u}(\cdot) \in L^\infty(0,T;E)$, $\|\bar{u}(\cdot)\|_{L^\infty(0,T;E)} \le 1$ *drives* $\zeta$ *norm or time optimally to* $\bar{y} = S(T)\eta$ *in time* $T$ *and that* $S(T)(\eta-\zeta) \in D(A)$. *Then* $\bar{u}(\sigma)$ *satisfies* (2.6.13), *where* $z \in Z(T)$ *is such that condition* (2.6.15) (*in particular condition* (2.6.16)) *holds.*

*Proof.* The assumptions give (2.5.8), which guarantees the maximum principle (2.6.13) for the time optimal problem (Theorem 2.5.5) and for the time optimal problem (Theorem 2.5.7). The result then follows from Theorem 2.6.3.

Theorem 2.6.3 and Corollary 2.6.5 guarantee the complete nontriviality of the costate (here, "completely nontrivial" means "not zero in the whole interval $0 \le t \le T$" under the condition $\bar{y} \in S(T)E$. It is then natural to ask:

*Is the costate completely nontrivial under any other target restriction, for instance the smoothness conditions*[23] $\bar{y} \in D(A)$, $\bar{y} \in D(A^n)$ *for sufficiently high* $n$ *or* $\bar{y} \in D(A^\infty) = \cap_{n\ge 1} D(A^n)$?

---

[22] The value of the regular functional defined by (2.6.18) pays heed to the equivalence relation in the definition of $R^\infty(T)$, that is, is the same for any $u(\cdot)$, $v(\cdot)$ giving $y(T,0,u) = y(T,0,v)$. In particular, it has the same value for the two controls in (2.6.19) and (2.6.20).

[23] For an analytic semigroup, $\bar{y} \in S(T)E$ is a smoothness condition far stronger that $\bar{y} \in D(A^\infty)$. For a semigroup like (2.6.1) the two conditions are unrelated and the first does not imply smoothness. The second does; $\bar{y}(\cdot) \in D(A^\infty)$ is equivalent to infinite differentiability of $\bar{y}(x)$ in $x \ge 0$ and $\bar{y}^{(n)}(0) = 0$ for all $n$.

To deal with this question, we take a second look at the variation-of-constants formula (2.6.5), with a control given by (2.6.9) in $0 \le \sigma \le T$; this means, we assume $\kappa(\sigma, z) \neq 0$ in $0 \le \sigma < T$, so that

$$\bar{u}(\sigma, x) = \rho \frac{S(T-\sigma)^* z(x)}{\|S(T-\sigma)^* z(\cdot)\|} = \rho \chi(x) \frac{z(x + (T-\sigma))}{\kappa(T-\sigma, z)} \quad (0 \le \sigma < T). \tag{2.6.22}$$

To simplify, we assume $\rho = 1$. Using (2.6.22) we have

$$\begin{aligned} &S(T-\sigma)\bar{u}(\sigma) = \bar{u}(\sigma, x - (T-\sigma)) \\ &= \chi(x - (T-\sigma)) \frac{z((x + (T-\sigma)) - (T-\sigma))}{\kappa(T-\sigma, z)} = \frac{\chi_{T-\sigma}(x) z(x)}{\kappa(T-\sigma, z)} \quad (0 \le \sigma < T), \end{aligned}$$

where $\chi_s(x)$ is the characteristic function of $[s, \infty)$. Formula (2.6.5) for $t = T$ becomes

$$\begin{aligned} &y(T, x, \zeta, \bar{u})(x) = S(T)\zeta(x) + \int_0^T S(T-\sigma)\bar{u}(\sigma, x) d\sigma \\ &= \zeta(x - T) + \int_0^T \frac{\chi_{T-\sigma}(x) z(x)}{\kappa(T-\sigma, z)} d\sigma = \zeta(x - T) + z(x) \int_0^T \frac{\chi_{T-\sigma}(x)}{\kappa(T-\sigma, z)} d\sigma \\ &= \zeta(x - T) + z(x) \int_0^T \frac{\chi_\sigma(x)}{\kappa(\sigma, z)} d\sigma = \zeta(x - T) + z(x)\omega(T, x, z), \end{aligned} \tag{2.6.23}$$

where

$$\omega(T, x, z) = \int_0^T \frac{\chi_\sigma(x)}{\kappa(\sigma, z)} d\sigma = \int_0^{\min(x,T)} \frac{d\sigma}{\kappa(\sigma, z)}. \tag{2.6.24}$$

If we drive from 0 in time $T$, the target $\bar{y}(x)$ and the costate $z(x)$ are related by

$$\bar{y}(x) = y(T, x, 0, \bar{u}) = z(x)\omega(T, x, z). \tag{2.6.25}$$

The objective is to make $\bar{y}(x)$ as smooth as possible by suitable choice of $z(x)$. Formula (2.6.25) says that it's not enough to make $z(x)$ smooth. In fact, (2.6.25) gives $\omega(T, x, z) = \omega(T, T, z) = \text{const.}$ for $x \ge T$, thus $\omega'(T, x, z) = 0$ in $x \ge T$; on the other hand,

$$\omega'(T, x, z) = \frac{1}{\kappa(x, z)} \quad (0 \le x < T)$$

so that $\omega'(T, x, z)$ has a jump[24] as $x \to T$ (in fact, it blows up as $x \to T$ if $z(x)$ has support in $[0, T]$). Using again (2.6.25),

$$\bar{y}'(x) = z'(x)\omega(T, x, z) + z(x)\omega'(T, x, z),$$

[24] This jump in the derivative corresponds to the integration lines in Figure 2.6.1 hitting the corner (0, 0) of the strip $0 \le x < \infty$, $0 \le \sigma \le T$; this happens when $x = T$.

thus the discontinuity in $\omega'(T, x, z)$ will communicate to $\bar{y}'(x)$ unless $z(T) = 0$. A similar argument shows that for continuity of higher derivatives we need $z'(T) = 0$, $z''(T) = 0, \ldots$ and so on, thus $z(x)$ must "flatten out to zero" in a hyperpower fashion at $x = T$. One such function (with $T = 1$) is

$$z(x) = \phi(x)\frac{\sqrt{2\alpha}\, e^{-1/(1-x)^\alpha}}{(1-x)^{(\alpha+1)/2}} \quad (0 \le x \le 1), \quad z(x) = 0 \quad (x \ge 1), \tag{2.6.26}$$

with $\alpha > 0$, $\phi(x)$ a $C^\infty$ function with $\phi(x) = 0$ in $0 \le x \le a < 1/2$, $\phi(x) > 0$ in $x > a$ and $\phi(x) = 1$ in $x \ge 1/2$. This function appears as a likely candidate to produce a smooth $\bar{y}(x)$. This suspicion is justified (but only up to a finite degree of smoothness) by Theorem 2.6.6. below.

**Theorem 2.6.6.** *If $z(x)$ is given by (2.6.26), the function*

$$\bar{y}(x) = y(1, x, 0, \bar{u}) = z(x)\omega(1, x, z) \tag{2.6.27}$$

*belongs to $D(A^n)$ if and only if*

$$n < 1 + \frac{\alpha}{2}. \tag{2.6.28}$$

*Proof.* The function $\kappa(\sigma, z)$ defined from $z(x)$ by (2.6.6) is $C^\infty$ and positive in $0 \le x < 1$, so that $\omega(1, x, z)$ in (2.6.24) is $C^\infty$ in $0 \le x < 1$. Accordingly, (2.6.27) defines a $\bar{y}(\cdot) \in C^\infty$ in $0 \le x < 1$ with $\bar{y}(x) = 0$ in $x \ge 1$ and $\bar{y}(x) = 0$ in $0 \le x \le a$ (the latter takes care of the boundary condition $y(0) = 0$ for $\bar{y}(x)$ and all its derivatives). It follows from these observations that to check that $\bar{y}(\cdot)$ belongs (or doesn't belong) to the domain of a power of $A$ we only need to examine its behavior for $x < 1$ near 1. We have

$$\begin{aligned}\kappa(\sigma, z)^2 &= \int_\sigma^\infty z(x)^2 dx = \int_\sigma^1 e^{-2/(1-x)^\alpha} \frac{2\alpha}{(1-x)^{\alpha+1}} dx \\ &= -e^{-2/(1-x)^\alpha}\Big|_{x=\sigma}^{x=1} = e^{-2/(1-\sigma)^\alpha} \quad (1/2 \le \sigma < 1),\end{aligned}$$

hence

$$\begin{aligned}\omega(1, x, z) &= \int_0^{1/2} \frac{d\sigma}{\kappa(\sigma, z)} + \int_{1/2}^x e^{1/(1-\sigma)^\alpha} d\sigma = C + \int_0^x e^{1/(1-\sigma)^\alpha} d\sigma \\ &= C + \frac{1}{\alpha}\int_1^{1/(1-x)^\alpha} \frac{e^\tau}{\tau^{1/\alpha+1}} d\tau = C + \frac{1}{\alpha} F\left(\frac{1}{(1-x)^\alpha}\right),\end{aligned} \tag{2.6.29}$$

with

$$F(z) = \int_1^z \frac{e^\tau}{\tau^{1/\alpha+1}} d\tau$$

through the change of variables $1/(1-\sigma)^\alpha = \tau \Leftrightarrow \sigma = 1-\tau^{-1/\alpha}$, $d\sigma = \tau^{-1/\alpha-1}/\alpha$. To deal with the asymptotics we note that if $\rho(z)$ is a sufficiently differentiable function we obtain after repeated integration by parts,

$$\begin{aligned}
&\int_c^z \rho(\tau)e^\tau d\tau = \rho(z)e^z - \rho(c)e^c - \int_c^z \rho'(\tau)e^\tau d\tau \\
&= (\rho(z)e^z - \rho(c)e^c) - (\rho'(z)e^z - \rho'(c)e^c) + \dots \\
&\quad + (-1)^n(\rho^{(n)}(z)e^z - \rho^{(n)}(c)e^c) + (-1)^{n+1}\int_c^z \rho^{(n+1)}(\tau)e^\tau d\tau\,. \qquad (2.6.30)
\end{aligned}$$

Applying this to $c = 1$, $\rho(\tau) = \tau^{-\theta}$ and multiplying both sides by $e^{-z}$ we obtain (here and below $C, C_1, C_2, \dots$ are generic constants)

$$\begin{aligned}
&e^{-z}\int_1^z \tau^{-\theta}e^\tau d\tau = z^{-\theta} + \theta z^{-\theta-1} + \theta(\theta+1)z^{-\theta-2} + \dots \\
&+ \theta(\theta+1)\dots(\theta+n-1)z^{-\theta-n} + C_1e^{-z} + C_2R_n(z)\,,
\end{aligned}$$

where

$$\begin{aligned}
R_n(z) &= e^{-z}\int_1^z \tau^{-\theta-n-1}e^\tau d\tau = \int_1^z \tau^{-\theta-n-1}e^{\tau-z}d\tau \\
&= \int_1^{z/2} \tau^{-\theta-n-1}e^{\tau-z}d\tau + \int_{z/2}^z \tau^{-\theta-n-1}e^{\tau-z}d\tau \\
&\le e^{-z/2}\int_0^{z/2} d\tau + \Big(\frac{z}{2}\Big)^{-\theta-n-1} e^{-z}\int_{z/2}^z e^\tau d\tau = O(z^{-\theta-n-1})\,.
\end{aligned}$$

Using this for $\theta = 1/\alpha + 1$ we get the asymptotic estimation

$$F(z) = \int_1^z \frac{e^\tau}{\tau^{1/\alpha+1}}d\tau = e^z\Big(\sum_{j=1}^n \frac{a_{\alpha j}}{z^{1/\alpha+j}} + O\Big(\frac{1}{z^{1/\alpha+n+1}}\Big)\Big) \qquad (2.6.31)$$

where $a_{\alpha j} \ne 0$. Putting $z = 1/(1-x)^\alpha$ the estimation becomes

$$F\Big(\frac{1}{(1-x)^\alpha}\Big) = e^{1/(1-x)^\alpha}\Big(\sum_{j=1}^n a_{\alpha j}(1-x)^{1+\alpha j} + O\big((1-x)^{1+\alpha(n+1)}\big)\Big)\,,$$

thus (2.6.29) and (2.6.27) give

$$\begin{aligned}
\bar y(x) &= z(x)\omega(1,x,z) = C\phi(x)\frac{e^{-1/(1-x)^\alpha}}{(1-x)^{(\alpha+1)/2}} \\
&\quad + \phi(x)\frac{\sqrt{2\alpha}e^{-1/(1-x)^\alpha}}{\alpha(1-x)^{(\alpha+1)/2}}F\Big(\frac{1}{(1-x)^\alpha}\Big) \\
&= C\phi(x)\frac{e^{-1/(1-x)^\alpha}}{(1-x)^{(\alpha+1)/2}} \\
&\quad + C_1\sum_{j=1}^n a_{\alpha j}(1-x)^{((2j-1)\alpha+1)/2} + O\big((1-x)^{((2n+1)\alpha+1)/2}\big) \qquad (2.6.32)
\end{aligned}$$

for $x < 1$; $\bar{y}(x) = 0$ for $x \geq 1$. In order to estimate derivatives, we need to differentiate repeatedly both sides of the asymptotic expansion (2.6.32), which reduces to the same operation for (2.6.31). For this, we differentiate formally (2.6.31) any number of times and then justify by applying (2.6.30) to the resulting integral terms. This granted and translated to (2.6.32), we note that the first term on the right side is $C^\infty$ on $0 \leq x \leq 1$. The less differentiable term on the right side is the first under the summation sign,

$$\begin{cases} C_1 a_{\alpha 1}(1-x)^{(\alpha+1)/2} & (x < 1)\,, \\ 0 & (x \geq 1)\,. \end{cases}$$

The $n$-th derivative of the function on the first line is a nonzero constant times $(1-x)^{(\alpha+1-2n)/2}$, which is in $L^2$ if and only if $\alpha + 1 - 2n > -1$. This inequality is equivalent to (2.6.28), thus the proof of Theorem 2.6.6 is complete.

To "make $D(A^\infty)$" we use the even flatter function

$$z(x) = \phi(x)\frac{2e^{1/(1-x)}e^{-e^{2/(1-x)}}}{1-x} \quad (0 \leq x \leq 1), \quad z(x) = 0 \quad (x \geq 1)\,. \tag{2.6.33}$$

**Theorem 2.6.7.** *If $z(x)$ is the function* (2.6.33), *the target $\bar{y}(\cdot)$ given by* (2.6.27) *belongs to $D(A^\infty)$.*

*Proof.* We have

$$\begin{aligned} \kappa(\sigma, z)^2 &= \int_\sigma^1 z(x)^2 dx = \int_\sigma^1 e^{-2e^{2/(1-x)}} 2e^{2/(1-x)} \frac{2}{(1-x)^2} dx \\ &= -e^{-2e^{2/(1-x)}}\Big|_{x=\sigma}^{x=1} = e^{-2e^{2/(1-\sigma)}}\,, \\ \omega(x, z) &= C + \int_0^x e^{e^{2/(1-\sigma)}} d\sigma \\ &= C + 2\int_{e^2}^{e^{2/(1-x)}} \frac{e^\tau}{\tau(\log\tau)^2} d\tau = C + 2F\big(e^{2/(1-x)}\big)\,, \end{aligned} \tag{2.6.34}$$

with

$$F(z) = \int_{e^2}^z \frac{e^\tau}{\tau(\log\tau)^2} d\tau$$

through the change of variables $e^{2/(1-\sigma)} = \tau \Leftrightarrow \sigma = 1 - 2/\log\tau$, $d\sigma = 2/\tau(\log\tau)^2$. This time we apply (2.6.30) with $c = e^2$, $\rho(\tau) = 1/\tau(\log\tau)^2$. We have

$$\rho'(\tau) = -\frac{2 + \log\tau}{\tau^2(\log\tau)^3}\,, \quad \rho''(\tau) = \frac{6 + 6\log\tau + 2(\log\tau)^2}{\tau^3(\log\tau)^4}\,, \ldots$$

and, in general,

$$\rho^{(n)}(\tau) = \frac{a_{n0} + a_{n1}\log\tau + a_{n2}(\log\tau)^2 + \ldots + a_{nn}(\log\tau)^n}{\tau^{n+1}(\log\tau)^{n+2}}\,. \tag{2.6.35}$$

Estimation of the integral remainder in (2.6.30) is base on the obvious inequality

$$\frac{1}{\tau^a(\log\tau)^b} \le \frac{1}{c^a(\log c)^b} \quad (\tau \ge c > 1)\,. \tag{2.6.36}$$

Formula (2.6.35) for the derivatives $\rho^{(n)}(\tau)$ and (2.6.36) for $a = 0$ give

$$|\rho^{(n+1)}(\tau)| \le \frac{C}{\tau^{n+2}(\log\tau)^2} \quad (\tau \ge e^2)\,,$$

and we use this inequality plus (2.6.36) again to bound the integral:

$$\begin{aligned}
&\left| e^{-z}\int_{e^2}^{z} \rho^{(n+1)}(\tau)e^{\tau}\,d\tau \right| \le e^{-z}\int_{e^2}^{z} \left|\rho^{(n+1)}(\tau)\right| e^{\tau}\,d\tau \\
&\le Ce^{-z}\int_{e^2}^{z} \frac{e^{\tau}}{\tau^{n+2}(\log\tau)^2}\,d\tau = C\int_{e^2}^{z} \frac{e^{\tau-z}}{\tau^{n+2}(\log\tau)^2}\,d\tau \\
&= C\int_{e^2}^{z/2} \frac{e^{\tau-z}}{\tau^{n+2}(\log\tau)^2}\,d\tau + C\int_{z/2}^{z} \frac{e^{\tau-z}}{\tau^{n+2}(\log\tau)^2}\,d\tau \\
&\le C_1 e^{-z/2}\int_{e^2}^{z/2} d\tau + \frac{C}{(z/2)^{n+2}(\log(z/2))^2}\int_{z/2}^{z} e^{\tau-z}\,d\tau \\
&= O\Big(\frac{1}{z^{n+2}(\log z)^2}\Big)\,,
\end{aligned}$$

since $\log(z/2)^2 = (\log z - \log 2)^2 \ge (\log z)^2/2$ for $z \ge 11$. We obtain

$$\begin{aligned}
F(z) &= \int_{e^2}^{z} \frac{e^{\tau}}{\tau(\log\tau)^2}\,d\tau \\
&= e^z\Big(\sum_{j=0}^{n} \frac{1}{z^{j+1}(\log z)^{j+2}}\sum_{k=0}^{j} a_{jk}(\log z)^k + O\Big(\frac{1}{z^{n+2}(\log z)^2}\Big)\Big)\,.
\end{aligned} \tag{2.6.37}$$

Setting $z = e^{2/(1-x)}$ we obtain an expansion whose generic term is

$$a_{jk}\frac{e^z}{z^{j+1}(\log z)^{j-k+2}} = b_{jk}\, e^{e^{2/(1-x)}} e^{-2(j+1)/(1-x)}(1-x)^{j-k+2}\,,$$

where $j, k \ge 0$, $j - k \ge 0$; the remainder is

$$O\big(e^{e^{2/(1-x)}} e^{-2(n+2)/(1-x)}(1-x)^{n+2}\big)\,.$$

Now, according to (2.6.27) and (2.6.34) we have

$$\begin{aligned}
\bar{y}(x) = z(x)\omega(1,x,z) &= C\phi(x)\frac{e^{1/(1-x)}e^{-e^{2/(1-x)}}}{1-x} \\
&\quad + \phi(x)\frac{2e^{1/(1-x)}e^{-e^{2/(1-x)}}}{1-x} 2F\big(e^{2/(1-x)}\big)
\end{aligned} \tag{2.6.38}$$

for $x < 1$; $\bar{y}(x) = 0$ for $x \geq 1$. Term-by-term differentiation of this expansion is justified in the same way as for (2.6.32). The first term of (2.6.38) is $C^\infty$ in $x \geq 0$. In the second we use the expansion (2.6.37) Accordingly, the generic term of the corresponding sum in the second term on the right of (2.6.38) is

$$\begin{cases} Ce^{-(2j+1)/(1-x)}(1-x)^{j-k+1} & (x < 1)\,, \\ 0 & (x \geq 1)\,, \end{cases}$$

which is $C^\infty$ in $x \geq 0$. This shows that $\bar{y}(x)$ is infinitely differentiable. Since $\phi(x) = 0$ near zero we have $\bar{y}^{(k)}(0) = 0$ for all $k$, which shows that $\bar{y}(\cdot) \in D(A^\infty)$.

It is obvious that the constructs in Theorem 2.6.6 and Theorem 2.6.7 can be extended to any $T > 0$ by rescaling the case $T = 1$, so that Theorem 2.6.1 remains true even if the target $\bar{y}(\cdot)$ is required to belong to $D(A^\infty)$. This in turn shows that Theorem 2.6.3 does not allow the replacement of the hypothesis that $\bar{y} \in S(T)\eta$ by the hypothesis that $\bar{y} \in D(A^\infty)$. For completeness we state the result:

**Corollary 2.6.8.** *Let $T > 0$. There exist targets $\bar{y}(\cdot) \in D(A^\infty)$ such that (a) the norm optimal control driving 0 to $\bar{y}(\cdot)$ in $0 \leq t \leq T$ is not unique, (b) the costate $S(T-\sigma)^*z$ is zero in an interval $0 \leq \sigma \leq T - \delta < T$ with $\delta$ as small as desired.*

**Miscellaneous notes.** This section is based on the author [2004].

**2.7. Vanishing of the costate for time optimal controls.** We use again the right translation semigroup (2.6.1) associated with the control system (2.6.4).

**Theorem 2.7.1.** *Given $T > 0$ and $0 < \delta < T$ there exists $\bar{y}(\cdot) \in L^2(0, \infty)$ such that the control $\bar{u}(\sigma, x)$ driving 0 to $\bar{y}(x)$ in optimal time $T$ satisfies*

$$\bar{u}(\sigma, x) = \frac{S(T-\sigma)^* z(x)}{\|S(T-\sigma)^* z(\cdot)\|} \qquad (T - \delta < \sigma \leq T) \tag{2.7.1}$$

*with $z(\cdot) \in L^2(0, \infty)$ such that*

$$\begin{aligned} z(\sigma, \cdot) &= S(T-\sigma)^* z(\cdot) \neq 0 \quad (T - \delta < \sigma \leq T)\,, \\ z(\sigma, \cdot) &= S(T-\sigma)^* z(\cdot) = 0 \quad (0 \leq \sigma \leq T - \delta)\,. \end{aligned} \tag{2.7.2}$$

*Proof.* We pick $z_1(\cdot) \in L^2(0, \infty)$ with support $= [a, T - \delta]$ $(0 \leq a < T - \delta)$ and define

$$\bar{u}_1(\sigma, x) = \frac{S(T-\delta-\sigma)^* z_1(x)}{\|S(T-\delta-\sigma)^* z_1(\cdot)\|}\,.$$

The support condition guarantees that

$$S(T-\delta-\sigma)^* z_1(\cdot) = \chi(\cdot) z(\cdot + T - \delta - \sigma) \neq 0 \quad (0 < \sigma \leq T - \delta)\,,$$

where $\chi(x)$ is the characteristic function of $x \geq 0$. According to Theorem 2.5.7, the drive from 0 to

$$\bar{y}_1(x) = y(T-\delta, x, 0, \bar{u}_1) = \int_0^{T-\delta} S(T-\delta-\sigma)\bar{u}_1(\sigma, x)d\sigma$$

is time optimal. We repeat *verbatim* this construct replacing $T-\delta$ by $\delta$; now, $z_2(\cdot)$ has support $= [b, \delta]$ with $0 \leq b < \delta$, so that

$$S(\delta-\sigma)^* z_2(\cdot) \neq 0 \quad (0 < \sigma \leq \delta)\,,$$

and

$$\bar{u}_2(\sigma, x) = \frac{S(\delta-\sigma)^* z_2(x)}{\|S(\delta-\sigma)^* z_2(\cdot)\|}$$

drives 0 to

$$\bar{y}_2(x) = y(\delta, x, 0, \bar{u}_2) = \int_0^{\delta} S(\delta-\sigma)\bar{u}_2(\sigma, x)d\sigma$$

in optimal time $\delta$. We then define

$$\bar{y}(x) = \begin{cases} \bar{y}_2(x) & (0 \leq x < \delta)\,, \\ \bar{y}_1(x-\delta) & (\delta \leq x \leq T)\,, \\ 0 & (x > T)\,. \end{cases}$$

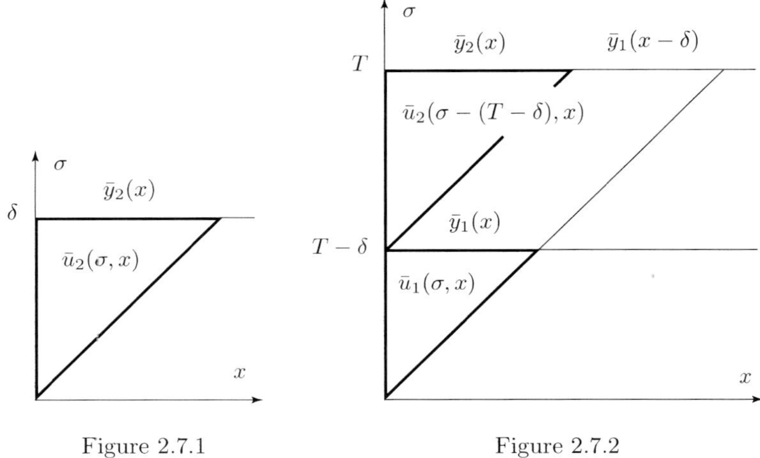


Figure 2.7.1 Figure 2.7.2

We can drive 0 to $\bar{y}(x)$ in time $T$ by means of the control

$$\bar{u}(\sigma, x) = \begin{cases} \bar{u}_1(\sigma, x) & (0 \leq \sigma < T-\delta)\,, \\ \bar{u}_2(\sigma-(T-\delta), x) & (T-\delta \leq \sigma \leq T)\,. \end{cases} \tag{2.7.3}$$

In fact,

$$\begin{aligned} y(T,x,0,\bar u) &= \int_0^T S(T-\sigma)\bar u(\sigma,x)d\sigma \\ &= \int_0^{T-\delta} S(T-\sigma)\bar u_1(\sigma,x)d\sigma + \int_{T-\delta}^{T} S(T-\sigma)\bar u_2(\sigma-(T-\delta),x)d\sigma \\ &= S(\delta)\int_0^{T-\delta} S(T-\delta-\sigma)\bar u_1(\sigma,x)d\sigma + \int_0^{\delta} S(\delta-\sigma)\bar u_2(\sigma,x)d\sigma \\ &= S(\delta)\bar y_1(x) + \bar y_2(x) = \bar y(x)\,. \end{aligned}$$

It follows that the time $\mathcal{T}$ to drive optimally from $0$ to $\bar y(x)$ satisfies the second inequality

$$\delta \le \mathcal{T} \le T\,. \tag{2.7.4}$$

For the first, note that if we can reach $\bar y(x)$ from $0$ in time $\mathcal{T} < \delta$ driving with a control $u(\sigma,x)$, then we can reach $\bar y_2(\cdot)$ from zero in the same time $\mathcal{T}$ using the restriction of $u(\sigma,x)$ to the trapezoid of vertices $(0,\mathcal{T})$, $(\delta,\mathcal{T})$, $(\delta-\mathcal{T},0)$, $(0,0)$ described by

$$0 \le x \le \delta - \mathcal{T} + \sigma\,, \quad 0 \le \sigma \le \mathcal{T},$$

which contradicts the fact that $\delta$ is the optimal driving time from $0$ to $\bar y_2(\cdot)$.

Let $\bar v(\sigma,x)$ be the admissible control[25] that drives $0$ to $\bar y(x)$ in optimal time $\mathcal{T} \le T$. Define

$$v(\sigma,x) = \begin{cases} 0 & (0 \le \sigma < T-\mathcal{T})\,, \\ \bar v(\sigma-(T-\mathcal{T}),x) & (T-\mathcal{T} \le \sigma \le T)\,. \end{cases} \tag{2.7.5}$$

The solution $y(t,x,0,v)$ corresponding to this control satisfies

$$y(T-\mathcal{T},x,0,v) = y(0,x,0,\bar v) = 0\,, \quad y(T,x,0,v) = y(\mathcal{T},x,0,\bar v) = \bar y(x)\,. \tag{2.7.6}$$

Due to the way the variation-of-constants formula (2.6.5) works (see Figure 2.6.1) if $v_\delta(\sigma,x)$ is the restriction of $v(\sigma,x)$ to the triangle of vertices $(0,T)$, $(\delta,T)$, $(0,T-\delta)$ described by[26]

$$0 \le x \le \delta - T + \sigma, \quad T-\delta \le \sigma \le T \tag{2.7.7}$$

(the upper triangle in Figure 2.7.2) we have $y(T-\delta,x,0,v_\delta) = 0$, $y(T,x,0,v_\delta) = \bar y_2(x)$. Hence, if

$$w(\sigma,x) = v_\delta(\sigma+T-\delta,x)$$

---

[25] We are using an existence result. Existence of optimal controls will be treated in Chapter 3, but we can also deduce existence from Lemma 2.7.3 below.

[26] Here we use $T-\mathcal{T} \le T-\delta$ (see Figure 2.7.2). This is a consequence of the first inequality (2.7.4).

we have

$$y(0, x, 0, w) = 0\,, \qquad y(\delta, x, 0, w) = \bar{y}_2(x)\,,$$

so that $w(\sigma, x)$ drives 0 to $\bar{y}_2(x)$ in optimal time $\delta$. In the Hilbert space $L^2(0, \infty)$ we have uniqueness of time optimal controls (Theorem 2.1.7), thus[27] $w(\sigma, x) = \bar{u}_2(\sigma, x)$; then

$$v_\delta(\sigma, x) = w(\sigma - (T - \delta), x) = \bar{u}_2(\sigma - (T - \delta), x) = \bar{u}(\sigma, x) \tag{2.7.8}$$

in the triangle of vertices $(0, T)$, $(\delta, T)$, $(0, T - \delta)$ described by (2.7.7).

Both $\bar{u}_2(\sigma, x)$ and $\bar{v}(\sigma, x)$ are time optimal controls (the first in the interval $0 \le \sigma \le \delta$, the second in the interval $0 \le \sigma \le \mathcal{T}$) thus they must satisfy the bang-bang property (2.1.19). This means

$$\int_0^\sigma \bar{u}_2(\sigma, x)^2 dx = 1 \quad (0 \le \sigma \le \delta)\,, \qquad \int_0^\sigma \bar{v}(\sigma, x)^2 dx = 1 \quad (0 \le \sigma \le \mathcal{T})\,.$$

The second equality implies

$$\int_0^\sigma v(\sigma, x)^2 dx = 1 \quad (T - \mathcal{T} \le \sigma \le T)\,, \tag{2.7.9}$$

while the first, combined with (2.7.8) gives

$$\int_0^\sigma v_\delta(\sigma, x)^2 dx = 1 \quad (T - \delta \le \sigma \le T)\,. \tag{2.7.10}$$

Taking into account that $T - \mathcal{T} \le T - \delta$ (consequence of the first inequality (2.7.4)) we deduce from (2.7.9) and (2.7.10) that $v(\sigma, x) = 0$ outside of the triangle (2.7.7):

$$v(\sigma, x) = 0 \quad (\delta - T + \sigma \le x < \infty\,,\ T - \delta \le \sigma \le T)\,.$$

Combined with (2.7.8), this gives

$$v(\sigma, x) = \bar{u}_2(\sigma - (T - \delta), x) = \bar{u}(\sigma, x) \quad (0 \le x < \infty,\ T - \delta \le \sigma \le T)\,. \tag{2.7.11}$$

Due to (2.7.11) and the variation-of-constants formula (2.6.5) (see again Figure 2.7.2) we must have

$$y(T - \delta, x, 0, v) = \bar{y}_1(x)$$

which, combined with the first equality (2.7.6) says that the drive performed by $v(\sigma, x)$ starting at 0 at time $T - \mathcal{T}$ reaches $\bar{y}_1(x)$ at time $T - \delta$. Equivalently, after a translation by $T - \mathcal{T}$,

$$y(0, x, 0, \bar{v}) = 0\,, \quad y(T - \delta - (T - \mathcal{T}), x, 0, \bar{v}) = \bar{y}_1(x)\,.$$

[27] This equality (and others below) should be qualified by "a. e.". We gloss over this to simplify some statements.

If $\mathcal{T} < T$ we have a contradiction with the fact that the optimal driving time from 0 to $\bar{y}_1(x)$ is $T - \delta$. It then follows that $\mathcal{T} = T$, thus the control (2.7.3) is time optimal. The control $\bar{u}(\sigma, x)$ does all that is required in Theorem 2.7.1 plus an additional trick. In fact, it "switches costates" at $\sigma = T - \delta$. At the top of the control interval we have

$$\bar{u}(\sigma, x) = \bar{u}_2(\sigma - (T - \delta), x) = \frac{S(T-\sigma)^* z_2(x)}{\|S(T-\sigma)^* z_2(\cdot)\|} \quad (T - \delta \leq \sigma \leq T),$$

the costate $z_2(\sigma, x) = S(T-\sigma)^* z_2(x)$ doing both lines of of (2.7.2). At the top of the lower control interval $0 \leq \sigma < T - \delta$ the costate $z_2(\sigma, x) = S(T-\sigma)^* z_2(x)$ gives up (becomes zero) and the "second costate" $z_1(\sigma, x) = S(T - \delta - \sigma)^* z_1(x)$ takes over:

$$\bar{u}(\sigma, x) = \bar{u}_1(\sigma, x) = \frac{S(T-\delta-\sigma)^* z_1(x)}{\|S(T-\delta-\sigma)^* z_1(\cdot)\|} \quad (0 \leq \sigma < T - \delta).$$

The next result produces time optimal controls with an arbitrary (finite) number of costate switches.

**Theorem 2.7.2.** *Given $T > 0$, $0 = T_0 < T_1 < \ldots < T_n = T$ there exists $\bar{y}(\cdot) \in L^2(0, \infty)$ such that the admissible control $\bar{u}(\sigma, x)$ driving 0 to $\bar{y}(x)$ in optimal time $T$ satisfies*

$$\begin{aligned} \bar{u}(\sigma, x) &= \frac{S(T_j - \sigma)^* z_j(x)}{\|S(T_j - \sigma)^* z_j(\cdot)\|} && (T_{j-1} < \sigma \leq T_j\,,\ z_j(\cdot) \in L^2(0, \infty))\,, \\ z_j(\sigma, \cdot) &= S(T_j - \sigma)^* z_j(\cdot) \neq 0 && (T_{j-1} < \sigma \leq T_j)\,, \\ z_j(T_{j-1}, \cdot) &= S(T_j - T_{j-1})^* z_j(\cdot) = 0 && (j = 1, \ldots, n)\,. \end{aligned}$$

The proof is an obvious extension of that of Theorem 2.7.1. We pick $z_j(x)$ with support $= [a_j, T_j - T_{j-1}]$ with $0 \leq a_j < T_j - T_{j-1}$ $(j = 1, 2, \ldots, n)$. The control

$$\bar{u}_j(\sigma, x) = \frac{S(T_j - T_{j-1} - \sigma) z_j(x)}{\|S(T_j - T_{j-1} - \sigma) z_j(\cdot)\|}$$

drives 0 to

$$\bar{y}_j(x) = \int_0^{T_j - T_{j-1}} S(T_j - T_{j-1} - \sigma) \bar{u}_j(\sigma, x) d\sigma$$

time optimally; $\bar{y}_j(x)$ has support $\subseteq [a_j, T_j - T_{j-1}]$. The control $\bar{u}(\sigma, x)$ is defined overall by

$$\bar{u}(\sigma, x) = \bar{u}_j(\sigma - T_{j-1}, x) \qquad (T_{j-1} < \sigma \leq T_j\,,\ j = 1, \ldots, n)\,,$$

and the target is made up from all the $\bar{y}_j(x)$ by

$$\bar{y}(x) = \bar{y}_j(x - (T - T_j)) \quad (T - T_j \leq x \leq T - T_{j-1}\,,\ j = 1, 2, \ldots n)\,,$$

with $\bar{y}(x) = 0$ elsewhere. To show that $\bar{u}(\sigma, x)$ drives 0 to $\bar{y}(x)$ time optimally we do induction. For $n = 2$ we have Theorem 2.7.1. For $n > 2$ assume Theorem 2.7.2 is true for $n - 1$; this means that the control $\bar{u}(\sigma, x)$ drives 0 in optimal time $T_{n-1}$ to the target

$$\bar{y}^1(x) = \bar{y}(x + (T - T_{n-1})) .$$

Let $\mathcal{T} \le T$ be the optimal time for driving from 0 to $\bar{y}(x)$. Then we prove as in Theorem 2.7.1 that $T - T_{n-1} \le \mathcal{T} \le T$. If $\bar{v}(\sigma, x)$ is the control that drives 0 to $\bar{y}(x)$ in optimal time $\mathcal{T}$ we define $v(\sigma, x)$ by (2.7.5) and prove, again as in Theorem 2.7.1 that

$$v(\sigma, x) = \bar{u}(\sigma, x) \qquad (T_{n-1} \le \sigma \le T_n = T) .$$

It then follows that

$$y(T - \mathcal{T}, x, 0, v) = 0 , \quad y(T_{n-1}, x, 0, v) = \bar{y}(x + (T_n - T_{n-1})) = \bar{y}^1(x) ,$$

hence

$$y(0, x, 0, \bar{v}) = 0 , \quad y(T_{n-1} - (T - \mathcal{T}), x, 0, \bar{v}) = \bar{y}^1(x) ,$$

which contradicts time optimality of $\bar{u}(\sigma, x)$ in the interval $0 \le \sigma \le T_{n-1}$. This completes the proof.

It is not difficult to pass from a finite to an infinite number of switchings of the costate, which we do below after some preparatory work. Consider the control system

$$y'(t) = Ay(t) + u(t) \tag{2.7.12}$$

in a Hilbert space $E$, $A$ the generator of an arbitrary strongly continuous semigroup $S(t)$. The *value function* $\mathcal{T} : E \to \mathbb{R}_+ \cup \{+\infty\}$ of the time optimal problem is defined thus: $\mathcal{T}(y)$ is the optimal driving time from[28] 0 to $y$. It follows from the coming Theorem 3.1.2 that if we can drive from 0 to $y$ in any time by means of a control satisfying

$$\|u(t)\| \le 1 \quad \text{a. e.} \tag{2.7.13}$$

then a time optimal control exists, hence $\mathcal{T}(y) < \infty$; if it's not possible to drive from 0 to $y$ under the constraint (2.7.13) we set $\mathcal{T}(y) = +\infty$.

**Lemma 2.7.3.** *Let $E$ be a Hilbert space, and let $\{y_n\} \subset E$ be a sequence with $y_n \to y \in E$. Then*

$$\mathcal{T}(y) \le \liminf_{n \to \infty} \mathcal{T}(y_n) . \tag{2.7.14}$$

*Proof.* If the lim inf is $= +\infty$ there is nothing to prove; assume then it's finite. This implies that there exists a subsequence (equally named) with $\mathcal{T}(y_n) < +\infty$

[28] The value function is usually defined in relation to driving from $y$ to 0; it's different from this one.

and $\lim_{n\to\infty} \mathcal{T}(y_n)$ equal to the lim inf on the right of (2.7.14). Accordingly, if (2.7.14) fails,

$$T = \lim_{n\to\infty} \mathcal{T}(y_n) < \mathcal{T}(y) . \tag{2.7.15}$$

Let $\bar{u}_n(\cdot)$ be the control that drives 0 to $y_n$ in optimal time $T_n = \mathcal{T}(y_n)$. Then

$$\int_0^{T_n} S(T_n - \sigma)u_n(\sigma)d\sigma = y_n . \tag{2.7.16}$$

Select $b \geq$ all $T_n$ and a subsequence of $\{u_n(\cdot)\}$ (equally named) with $\bar{u}_n(\cdot) \to \bar{u}(\cdot)$ weakly in $L^2(0, b; E)$ (we set $\bar{u}_n(\sigma) = 0$ for $\sigma \geq T_n$). Taking the scalar product of both sides of (2.7.16) with an element $z \in E$,

$$\int_0^{T_n} \langle S(T_n - \sigma)^* z, u_n(\sigma)\rangle d\sigma = \langle z, y_n\rangle ,$$

and taking limits using weak convergence

$$\int_0^{T} \langle S(T - \sigma)^* z, \bar{u}(\sigma)\rangle d\sigma = \langle z, y\rangle .$$

This, since $z$ is arbitrary, shows that $\bar{u}(\cdot)$ drives 0 to $y$ in time $T = \lim_{n\to\infty} \mathcal{T}(y_n)$, hence $\mathcal{T}(y) \leq T$, a contradiction with (2.7.15) that completes the proof.

**Theorem 2.7.4.** *Let $E = L^2(0, \infty)$, $S(t)$ the left translation semigroup* (2.6.1). *Then there exists $T > 0$, a sequence $0 = T_0 < T_1 < \ldots < T_n < \ldots < T$ with $T_n \to T$ and $\bar{y}(\cdot) \in L^2(0, \infty)$ such that the control $\bar{u}(\cdot)$ driving 0 to $\bar{y}$ in optimal time $T$ satisfies*

$$\begin{aligned} \bar{u}(\sigma, x) &= \frac{S(T_j - \sigma)^* z_j(x)}{\|S(T_j - \sigma)^* z_j(\cdot)\|} && (T_{j-1} < \sigma \leq T_j\,,\ z_j \in L^2(0, \infty))\,, \\ z_j(\sigma, \cdot) &= S(T_j - \sigma)^* z_j(\cdot) \neq 0 && (T_{j-1} < \sigma \leq T_j)\,, \\ z_j(T_{j-1}, \cdot) &= S(T_j - T_{j-1})^* z_j(\cdot) = 0 && (j = 1, 2, \ldots)\,. \end{aligned}$$

*Proof.* We assume for the moment that the (increasing, bounded) sequence $\{T_n\}$ has been selected, and define

$$T = \sup_{n\geq 1} T_n = \lim_{n\to\infty} T_n . \tag{2.7.17}$$

This done, we pick $z_j(x)$ with support $= [a_j, T_j - T_{j-1}]$, $0 \leq a_j < T_j - T_{j-1}$ $(j = 1, 2, \ldots)$. The control

$$\bar{u}_j(\sigma, x) = \frac{S(T_j - T_{j-1} - \sigma)z_j(x)}{\|S(T_j - T_{j-1} - \sigma)z_j(\cdot)\|}$$

drives 0 to

$$\bar{y}_j(x) = \int_0^{T_j - T_{j-1}} S(T_j - T_{j-1} - \sigma)\bar{u}_j(\sigma, x)d\sigma$$

time optimally. The control $\bar{u}(\sigma, x)$ is defined piece-by-piece by

$$\bar{u}(\sigma, x) = \bar{u}_j(\sigma - T_{j-1}, x) \quad (T_{j-1} \le \sigma < T_j\,,\ j = 1, 2, \ldots)\,, \tag{2.7.18}$$

and the target is

$$\bar{y}(x) = \bar{y}_j(x - (T - T_j)) \quad (T - T_j \le x < T - T_{j-1}\,,\ j = 1, 2, \ldots)\,. \tag{2.7.19}$$

It is clear from the definition (2.7.18) that $\bar{u}(\sigma, x)$ drives 0 to $\bar{y}(x)$. The control (2.7.18) is strongly continuous in each interval $T_{j-1} \le t < T$ and

$$\|\bar{u}(\sigma, \cdot)\| = 1 \quad (0 \le \sigma \le T)\,,$$

so that $\bar{u}(\cdot, \cdot) \in L^\infty(0, T; L^2(0, \infty))$ and $\bar{y}(\cdot) \in L^2(0, \infty)$. Define next

$$\begin{aligned} \bar{y}^n(x) = \bar{y}_j(x - (T_n - T_j)) &= \bar{y}(x + (T - T_n)) \\ (T_n - T_j \le x \le T_n - T_{j-1}\,,&\ j = 1, \ldots, n)\,. \end{aligned} \tag{2.7.20}$$

We check easily that the control $\bar{u}(\sigma, x)$ drives 0 to $\bar{y}_n(x)$ in time $T_n$ (see Figure 2.7.3 below) and we know from Theorem 2.5.7 that this drive is time optimal. We show below that $\bar{u}(\sigma)$ is also time optimal in the entire interval $0 \le t \le T$ but this requires a specific choice of the $T_n$, which proceeds in an inductive way and involves another sequence $\{T^n\}$ such that

$$T^0 > T^1 > \ldots > T^n > \ldots, \quad T_n < T^n\,, \quad T^n - T_n \to 0\,. \tag{2.7.21}$$

The starting elements are $T_0 = 0$ and $T^0 > 0$ arbitrary. Assume that $T_n$ and $T^n$ have been chosen; this means the control $\bar{u}(\sigma, x)$ in (2.7.18) has already been constructed in $0 \le \sigma \le T_n$ so that the definition

$$\bar{u}_n(\sigma, x) = \begin{cases} \bar{u}_j(\sigma - T_{j-1}, x) & (T_{j-1} \le \sigma < T_j\,,\ j = 1, 2, \ldots, n) \\ 0 & (\sigma \ge T_n) \end{cases}$$

makes sense. So does the definition of $\bar{y}^n(x)$ in (2.7.20) by $\bar{y}^n(x) = \bar{y}_j(x - (T_n - T_j))$, since we don't have to use $\bar{y}(x)$ at all, just $\bar{y}_j(x)$, whose definition doesn't depend on $T$. We have

$$y(\tau, x, 0, \bar{u}_n) = \bar{y}^n(x - (\tau - T_n)) \quad (\tau > T_n)\,,$$

and, since $\bar{y}^n(\cdot - (\tau - T_n)) \to \bar{y}^n(\cdot)$ in $L^2(0, \infty)$ as $\tau \to 0$, given $\epsilon > 0$ arbitrary we can choose $T^{n+1} > T_n$ such that

$$T_n \le \tau \le T^{n+1} \implies \|\bar{y}^n(\cdot - (\tau - T_n)) - \bar{y}^n(\cdot)\|_{L^2(0,\infty)} \le \epsilon\,, \tag{2.7.22}$$

and then, using Lemma 2.7.3 make $\epsilon$ so small that (2.7.22) implies

$$T_n \le \tau \le T^{n+1} \implies \mathcal{T}(\bar{y}^n(\cdot - (\tau - T_n)) \ge \mathcal{T}(\bar{y}^n(\cdot)) - \frac{1}{n} = T_n - \frac{1}{n}. \quad (2.7.23)$$

Once $T^{n+1}$ has been chosen, we pick $T_{n+1} > T_n$ in the open interval $(T_n, T^{n+1})$ under the additional condition $T^{n+1} - T_{n+1} \le 1/n$ in order to insure that, as the inductive construction proceeds, the condition $T^n - T_n \to 0$ demanded by (2.7.21) will hold.

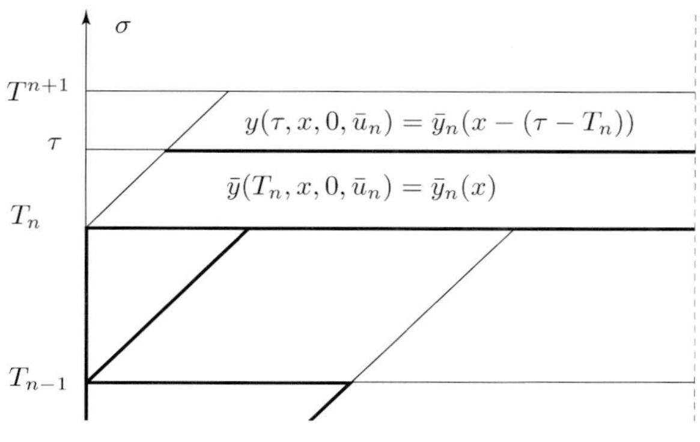


Figure 2.7.3

We show that the control $\bar{u}(\sigma, x)$ defined in (2.7.18) drives 0 to $\bar{y}(x)$ in optimal time $T$ defined by (2.7.17). Note first that

$$\bar{y}^n(x - (T - T_n)) = \chi_{T-T_n}(x)\bar{y}(x) \quad (0 \le x < \infty) \quad (2.7.24)$$

($\chi_s(x)$ the characteristic function of $[s, \infty)$), and let $\bar{v}(\sigma, x)$ be the control that does the drive from 0 to $\bar{y}(\cdot)$ in optimal time $\mathcal{T} = \mathcal{T}(\bar{y}(\cdot))$ (since we can drive from 0 to $\bar{y}(\cdot)$ with an admissible control, this optimal control exists). Take $n$ so large that

$$T - T_n < \mathcal{T},$$

and let $\mathcal{R}_n$ be the triangle of vertices $(0, \mathcal{T} - (T - T_n))$, $(0, \mathcal{T})$, $(T - T_n, \mathcal{T})$, described by

$$0 \le x \le \sigma + T - T_n - \mathcal{T}, \quad T - T_n - \mathcal{T} \le \sigma \le \mathcal{T}.$$

Define

$$\bar{v}_n(\sigma, x) = \begin{cases} \bar{v}(\sigma, x) & (\sigma, x) \notin \mathcal{R}_n, \\ 0 & (\sigma, x) \in \mathcal{R}_n. \end{cases}$$

We have

$$y(\mathcal{T}, x, 0, \bar{v}_n) = \chi_{T-T_n}(x)\bar{y}(x) = \bar{y}^n(x - (T - T_n))$$

(the second equality because of (2.7.24)) thus we have, using (2.7.23) and the fact that $T \in [T_n, T^n]$ for all $n$,

$$\mathcal{T} = \mathcal{T}(\bar{y}(\cdot)) \geq \mathcal{T}(\bar{y}^n(\cdot - (T - T_n))) \geq T_n - \frac{1}{n}\,,$$

showing that $\mathcal{T} \geq T$. This ends the proof.

**Miscellaneous notes.** This section is based on the author [2004].

**2.8. Singular norm optimal controls.** We return to the generality of the initial sections of this chapter: $E$ is an arbitrary Banach space and $S(t)$ is an arbitrary strongly continuous semigroup. A *multiplier space* $\mathcal{K}$ is any space such that $(a)$ there exists an imbedding $E^* \hookrightarrow \mathcal{K}$, $(b)$ $S(t)^*$ is defined in $\mathcal{K}$ and

$$S(t)^*\mathcal{K} \subseteq E^* \quad (t > 0)\,. \tag{2.8.1}$$

The elements of $\mathcal{K}$ are called *multipliers.*[29] As an example, the spaces $\mathcal{Z}$ and $Z_w(T) \subseteq \mathcal{Z}$ defined in **2.3** are multiplier spaces; on the other hand, the space $E^*_{-1}$ is not necessarily a multiplier space since $S(t)^*$, extended to $E^*_{-1}$ by $S(t)^\bullet$ may not take values in $E^*$. Note that (2.8.1) bears only on the behavior of $S(t)^*$ for small $t$, as the semigroup equation in the form

$$S(t)^* = S(t-h)^*S(h)^* \tag{2.8.2}$$

shows. It also follows from (2.8.2) that $S(t)^*\psi$ is bounded in $0 < h \leq t \leq T$ and

$$\langle S(t)^*\psi, y\rangle = \langle S(h)^*\psi, S(t-h)y\rangle \quad (y \in E)\,,$$

so that $S(t)^*\psi$ is $E$-weakly continuous in $t \geq h > 0$, thus in $t > 0$. The maximum principle

$$\langle S(T-t)^*\psi, \bar{u}(t)\rangle = \max_{\|u\|\leq\rho} \langle S(T-t)^*\psi, u\rangle \quad \text{a. e. in } 0 \leq t \leq T \tag{2.8.3}$$

makes sense for $\psi$ in any multiplier space; $t = T$ is the only point where the costate $S(T-t)^*z$ may not belong to $E^*$, and "a. e." doesn't mind the loss of one point.

We call $\mathcal{M}$ the space of all multipliers, or rather its quotient by the equivalence relation

$$\psi_1 = \psi_2 \iff S(t)^*\psi_1 = S(t)^*\psi_2 \quad (t > 0)\,.$$

The following result is a characterization of $Z_w(T)$ among all multiplier spaces, or, equivalently, as a subspace of $\mathcal{M}$.

[29] We do not require $\mathcal{K}$ to be a normed space, in fact not even a linear space; the imbedding $E^* \hookrightarrow \mathcal{K}$ is simply a 1-1 map from $E^*$ into $\mathcal{K}$. We shall see, however (Corollary 2.8.2 below) that multipliers have a linear structure.

**Lemma 2.8.1.** *Let* $\psi$ *be a multiplier such that*

$$\int_0^T \|S(t)^*\psi\|_{E^*}dt < \infty\,. \tag{2.8.4}$$

*Then* $\psi \in Z_w(T)$, *that is, there exists* $z \in Z_w(T)$ *with*

$$S(t)^*\psi = S(t)^*z \quad (t > 0)\,.$$

*Proof.* Let $\mu > \omega$ ($\omega$ the constant in (2.3.1)) and let $0 < s < t$. Using the fact that $A - \mu I$ is the infinitesimal generator of $e^{-\mu t}S(t)$ we obtain

$$\begin{aligned}
&-\langle (R(\mu;A)^*e^{-\mu t}S(t)^*\psi - R(\mu;A)^*e^{-\mu s}S(s)^*\psi\,,\ y\rangle \\
&\quad = \langle (A-\mu I)^*(e^{-\mu(t-s)}S(t-s)^* - I)e^{-\mu s}S(s)^*\psi\,,\ y\rangle \\
&\quad = \langle e^{-\mu s}S(s)^*\psi\,,\ (A-\mu I)^{-1}(e^{-\mu(t-s)}S(t-s) - I)y\rangle \\
&\quad = \left\langle e^{-\mu s}S(s)^*\psi\,,\ \int_0^{t-s} e^{-\mu\sigma}S(\sigma)y d\sigma \right\rangle \\
&\quad = \int_0^{t-s} \langle e^{-\mu s}S(s)^*\psi,\ e^{-\mu\sigma}S(\sigma)y\rangle d\sigma \\
&\quad = \int_0^{t-s} \langle e^{-\mu\sigma}e^{-\mu s}S(\sigma)^*S(s)^*\psi,\ y\rangle d\sigma \\
&\quad = \int_0^{t-s} \langle e^{-\mu(\sigma+s)}S(\sigma+s)^*\psi,\ y\rangle d\sigma \\
&\quad = \int_s^t \langle e^{-\mu\sigma}S(\sigma)^*\psi,\ y\rangle d\sigma \quad (y \in E)\,.
\end{aligned} \tag{2.8.5}$$

Let $\{t_n\}$ be a decreasing sequence of positive numbers with $t_n \to 0$. Using (2.8.5) for $s = t_n, t = t_m$ $(m < n)$ we obtain

$$\begin{aligned}
&|\langle R(\mu;A)^*e^{-\mu t_m}S(t_m)^*\psi - R(\mu;A)^*e^{-\mu t_n}S(t_n)^*\psi,\ y\rangle| \\
&\qquad \le \|y\| \int_{t_n}^{t_m} e^{-\mu\sigma}\|S(\sigma)^*\psi\|_{E^*}d\sigma \quad (y \in E)\,,
\end{aligned}$$

hence

$$\begin{aligned}
&\|R(\mu;A)^*e^{-\mu t_m}S(t_m)^*\psi - R(\mu;A)^*e^{-\mu t_n}S(t_n)^*\psi\|_{E^*} \\
&\qquad \le \int_{t_n}^{t_m} e^{\mu\sigma}\|S(\sigma)^*\psi\|_{E^*}d\sigma\,.
\end{aligned}$$

This estimation says that $\{R(\mu;A)^*e^{-\mu t_n}S(t_n)^*\psi\}$ is a Cauchy sequence, hence

$$R(\mu;A)^*e^{-\mu t_n}S(t_n)^*\psi \to y^* \in \overline{D(A^*)} \subseteq E^*.$$

On the other hand, heeding the definition of $E^*_{-1}$ in **2.3**,

$$e^{-\mu t_n} S(t_n)^* \psi \to z \in E^*_{-1}$$

and

$$R(\mu; A)^\bullet z = y^* \iff z = (\mu I^* - A^\bullet) y^*$$

(see (2.3.12) and (2.3.14)). The first convergence relation gives

$$\lim_{n\to\infty} S(t - t_n)^* R(\mu; A)^* e^{-\mu t_n} S(t_n)^* \psi = S(t)^* y^* ,$$

and the second

$$\lim_{n\to\infty} (\mu I^* - A^*) S(t - t_n)^* R(\mu; A)^* e^{-\mu t_n} S(t_n)^* \psi$$
$$= \lim_{n\to\infty} S(t - t_n)^* e^{-\mu t_n} S(t_n)^* \psi = \lim_{n\to\infty} S(t)^* \psi = S(t)^* \psi ,$$

so that $S(t)^* y^* \in D(A^*)$ and

$$S(t)^* \psi = (\mu I^* - A^*) S(t)^* y^* = S(t)^* (\mu I^* - A^\bullet) y^* = S(t)^* z \quad (t > 0) .$$

This ends the proof (the fact that $z \in Z_w(T)$ follows from (2.8.4)).

Let $\eta^*(t)$ $(t > 0)$ be an $E^*$-valued function satisfying

$$\eta^*(s + t) = S(t)^* \eta^*(s) \quad (s, t > 0) . \tag{2.8.6}$$

Equation (2.8.6) in the form $\eta^*(t) = S^*(t - h)\eta^*(h)$ $(t \geq h > 0)$ implies that $\eta^*(t)$ is bounded in $0 < h \leq t \leq T$ and $E$-weakly measurable in $t > 0$. We call $\mathcal{N}$ the (obviously linear) space of all $E^*$- valued functions satisfying (2.8.6) and denote by $\mathcal{N}_1(T)$ the subspace of $\mathcal{N}$ consisting of all $\eta^*(\cdot)$ satisfying

$$\|\eta^*(\cdot)\|_{\mathcal{N}_1(T)} = \int_0^T \|\eta^*(t)\|_{E^*} d\sigma < \infty . \tag{2.8.7}$$

Obviously the $\mathcal{N}_1(T)$ are the same (algebraically) for all $T > 0$. Not only this: the norms $\| \cdot \|_{\mathcal{N}_1(T)}$ are all equivalent. We don't prove this last statement, since Corollary 2.8.2 below (combined with the comments after (2.3.23)) obviates the need for a proof.

A multiplier $\psi \in \mathcal{M}$ originates a $\eta^*(\cdot) \in \mathcal{N}$ through

$$\eta^*(t) = S(t)^* \psi \quad (t > 0) ,$$

and, conversely, any $\eta^*(\cdot) \in \mathcal{N}$ defines a multiplier $\psi \in \mathcal{M}$ through

$$S(t)^* \psi = \eta^*(t) \quad (t > 0) .$$

These two observations prove $(a)$ of the result below.

**Corollary 2.8.2.** $(a)$ *The correspondence*

$$\mathcal{M} \ni \psi \longleftrightarrow \eta^*(\cdot) = S(t)^*\psi \in \mathcal{N}$$

*is one-to-one and onto, thus it makes* $\mathcal{M}$ *a linear space.* $(b)$ *The correspondence*

$$Z_w(T) \ni z \longleftrightarrow \eta^*(\cdot) = S(t)^*z \in \mathcal{N}_1(T)$$

*is a linear and isometric isomorphism.*

*Proof.* Only $(b)$ remains, and it follows directly from Lemma 2.8.1.

**Remark 2.8.3.** Corollary 2.8.2 doesn't say much over Lemma 2.8.1, but gives a way to identify the space $Z_w(T)$ (via its isomorph $\mathcal{N}_1(T)$) for specific spaces and semigroups without "going the $E_{-1}^*$ route" as in **2.3**. We illustrate this with the space $E = L^2(0,\infty)$ and the right translation semigroup (2.6.1),

$$S(t)y(x) = \begin{cases} y(x-t) & (x \ge t)\,, \\ 0 & (x \le t)\,, \end{cases}$$

whose adjoint is is the left translation (and chop-off) semigroup

$$S(t)^*z(x) = \begin{cases} z(x+t) & (x \ge 0)\,, \\ 0 & (x \le 0)\,. \end{cases}$$

If $\eta^*(\cdot) \in \mathcal{N}_1(T)$ then $\eta^*(t)(\cdot) \in E^* = L^2(0,\infty)$. If $t \ge h > 0$,

$$\eta^*(t)(x) = S(t-h)^*\eta^*(h)(x) = \eta^*(h)(x+t-h)\,.$$

The function on the right side does not depend on $h$ (and is defined for $t, x > 0$) thus we may call it $\phi(t,x)$. We have

$$\phi(t-k, x+k) = \eta^*(h)((x+k)+(t-k)-h) = \eta^*(h)(x+t-h) = \phi(t,x)\,,$$

so that

$$\eta^*(t)(x) = \phi(t,x) = \psi(t+x)$$

and we can identify the space $\mathcal{M}$ of all multipliers with the space $L^2(0+,\infty)$ of all functions defined in $(0,\infty)$ which belong to $L^2(h,\infty)$ for $h > 0$. The subspace $Z_w(T) = Z(T) \approx \mathcal{N}_1(T)$ can then be identified with the subspace of $L^2(0+,\infty)$ consisting of all functions $z(\cdot)$ that satisfy

$$\int_0^T \|\eta^*(t)\|_{L^2(0,\infty)}dt = \int_0^T \sqrt{\int_0^\infty |z(t+x)|^2 dx}\; dt < \infty\,.$$

Below, an *extremal* is a norm optimal or time optimal control.

An extremal is *regular* if it satisfies (2.8.3) with $\psi = z \in Z_w(T)$, $z \neq 0$; in view of Lemma 2.8.1, the requirement that $\psi \in Z_w(T)$ is the summability assumption (2.8.4). The fact that $z \neq 0$ does not necessarily imply that $S(T-t)^*z \neq 0$ in the entire interval $0 \leq t \leq T$ so that (2.8.3) may fail to give complete information on the regular extremal $\bar{u}(\cdot)$. The norm optimal controls constructed in Theorem 2.6.1 and the time optimal controls in Theorem 2.7.1 are examples; these controls qualify as regular although $S(T-t)^*z$ vanishes in part of the interval $0 \leq t \leq T$.

An extremal is *weakly singular* if it satisfies the maximum principle (2.8.3) with $\psi \in \mathcal{M}$ such that the costate $S(T-t)^*\psi$ is not identically zero in $0 \leq t < T$, but it does *not* satisfy (2.8.3) with $\psi \in Z_w(T)$; this means that for *every* $\psi \in \mathcal{M}$ enabling the maximum principle (2.8.3) with costate not identically zero we have

$$\int_0^T \|S(t)^*\psi\|_{E^*} = \infty\,. \tag{2.8.8}$$

The requirement that $S(T-t)^*\psi$ be not identically zero in $0 \leq t < T$, and the semigroup equation for $S(t)^*$ imply there exists $\delta > 0$ such that

$$S(T-t)^*\psi \neq 0 \quad (T-\delta < t < T)\,. \tag{2.8.9}$$

In Hilbert spaces, the maximum principle (2.8.3) and condition (2.8.9) imply

$$\bar{u}(t) = \rho\frac{S(T-t)^*\psi}{\|S(T-t)^*\psi\|} \qquad \text{a. e. in } T-\delta < t < T\,. \tag{2.8.10}$$

Finally, an extremal is *strongly singular* (or simply *singular*) if it does *not* satisfy (2.8.3)-(2.8.9) with $\psi$ in *any* multiplier space. Since (2.8.10) for arbitrary $\psi \in \mathcal{M}$ implies smoothness of $\bar{u}(t)$ in $T-\delta < t < T$ (after possible modification in a null set), one method to construct strongly singular controls is to produce optimal controls that are "essentially discontinuous" (that is, not improvable to continuous by modification in a null set) in any interval $T-\delta < t < T$. There are other ways to foil (2.8.3), however; see Theorem 2.8.6 below.

**Theorem 2.8.4.** *Strongly singular norm and time optimal controls exist.*

*Proof.* We use $E = L^\infty(0,\infty)$ and the left translation semigroup (2.6.1). The control $\bar{u}(\cdot)$ constructed in Theorem 2.7.4 is time optimal (thus also norm optimal). If $T_{j-1} < \sigma \leq T_j$ it follows from the definition (2.7.18) of $\bar{u}_j(\sigma, x)$ (see also Figure 2.7.3) that the support of $\bar{u}(\sigma,\cdot)$ is contained in $[0, \sigma - T_{j-1}]$; since, on the other hand, $\|\bar{u}(\sigma,\cdot)\|_{L^2(0,\infty)} = 1$, the limit

$$\lim_{\sigma\to T_{j-1}+} u(\sigma,\cdot)$$

does not exist in $L^2(0,\infty)$, and nothing is gained by modification in a null set. All the $T_j$ are "essential" discontinuity points of $\bar{u}(\sigma,\cdot)$, and the $T_j$ accumulate at the endpoint $T$. This ends the proof.

**Corollary 2.8.5.** *Assume a weakly singular or strongly singular norm optimal control exists. Then singular functionals in* $R^\infty(T)$ *exist, that is* $\mathcal{S}(T) \neq \{0\}$*; equivalently,* $\mathcal{R}(T) \neq R^\infty(T)^\star$.

*Proof.* If $\bar{u}(t)$ is norm optimal then

$$\int_0^T S(T-\sigma)\bar{u}(\sigma)d\sigma \in B_\rho^\infty(T)$$

is a boundary point of the ball $B_\rho^\infty(T)$, thus it can be separated from the ball by a nonzero functional $\xi \in R^\infty(T)^\star$. If all functionals are regular then $\xi = \xi_z$ is regular, and the separation inequality is

$$\int_0^T \langle S(T-\sigma)^* z, u(\sigma)\rangle d\sigma \leq \int_0^T \langle S(T-\sigma)^* z, \bar{u}(\sigma)\rangle d\sigma$$
$$u(\cdot) \in L^\infty(0,T;E)\,, \quad \|u(\cdot)\|_{L^\infty(0,T;E)} \leq \rho\,,$$

for some $z \in Z_w(T)$, $z \neq 0$ which is equivalent to (2.8.3); this means *all extremals are regular,* and we have a contradiction.

Corollary 2.8.5 plus the strongly singular extremals identified in Theorem 2.8.4 provide the first evidence that *singular functionals actually exist.* Weakly singular extremals will be constructed in next chapter. In this section and the next we show that strongly singular *norm optimal* controls can be constructed not only for specific examples of spaces and semigroups (such as $L^2(0,\infty)$ and the translation semigroup in Theorem 2.8.4) but for arbitrary spaces and a numerous family of semigroups including analytic semigroups with unbounded infinitesimal generators. However, the argument does not cover time optimal controls.

Below we set $\mu > \omega$, where $\omega$ is the constant in (2.3.1). Since $A - \mu I$ is the infinitesimal generator of the strongly continuous semigroup $e^{-\mu t}S(t)$, we have

$$-y = \lim_{t\to\infty} e^{-\mu t}S(t)y - S(0)y = \int_0^\infty e^{-\mu\sigma}(A-\mu I)S(\sigma)y d\sigma \tag{2.8.11}$$

for $y \in D(A-\mu I) = D(A)$. Accordingly,

$$\|y\| \leq \int_0^\infty e^{-\mu\sigma}\|(\mu I - A)S(\sigma)y\|d\sigma \quad (y \in D(A))\,. \tag{2.8.12}$$

We say that $S(t)$ *satisfies* $\mathcal{P}(a,b,\mu)$ $(a \geq 0, b > 0, \mu > \omega)$ if

$$b\|y\| \leq \int_a^\infty e^{-\mu\sigma}\|(\mu I - A)S(\sigma)y\|d\sigma \quad (y \in D(A))\,. \tag{2.8.13}$$

By (2.8.12), every semigroup $S(t)$ satisfies $\mathcal{P}(0,1,\mu)$ for any $\mu > \omega$. We say that $S(t)^*$ *satisfies* $\mathcal{P}(a,b,\mu)$ $(a \geq 0, b > 0, \mu > \omega)$ if

$$b\|y^*\|_{E^*} \leq \int_a^\infty e^{-\mu\sigma}\|(\mu I^* - A^*)S(\sigma)^* y^*\|_{E^*}d\sigma \quad (y^* \in D(A^*)) \tag{2.8.14}$$

but the integral must be justified since $S(t)^*$ may not be strongly continuous. To this end, note that $\langle A^* S(\sigma)^* y^*, y\rangle = \langle S(\sigma)^* A^* y^*, y\rangle = \langle A^* y^*, S(\sigma) y\rangle$ so that $A^* S(\sigma)^* y^*$ is $E$-weakly continuous, hence $\|A^* S(\sigma)^* y^*\|_{E^*}$ is lower semicontinuous (Lemma 2.3.4). To check that any adjoint semigroup $S(t)^*$ satisfies $\mathcal{P}(0, 1, \mu)$, take $y \in D(A)$, $y^* \in D(A^*)$; (2.8.11) gives

$$
\begin{aligned}
-\langle y^*, y\rangle &= \left\langle y^*, \int_0^\infty e^{-\mu\sigma}(A - \mu I)S(\sigma) y d\sigma \right\rangle \\
&= \int_0^\infty \langle y^*, e^{-\mu\sigma}(A - \mu I)S(\sigma) y\rangle d\sigma \\
&= \int_0^\infty \langle e^{-\mu\sigma}(A^* - \mu I^*)S(\sigma)^* y^*, y\rangle d\sigma\,,
\end{aligned}
$$

so that

$$
|\langle y^*, y\rangle| \le \|y\| \int_0^\infty e^{-\mu\sigma}\|(\mu I^* - A^*)S(\sigma)^* y^*\|_{E^*} d\sigma \quad (y \in D(A)\,, y^* \in D(A^*))
$$

and, since $D(A)$ is dense in $E$,

$$
\|y^*\| \le \int_0^\infty e^{-\mu\sigma}\|(\mu I^* - A^*)S(\sigma)^* y^*\|_{E^*} d\sigma \quad (y^* \in D(A^*))\,. \tag{2.8.15}
$$

**Theorem 2.8.6.** *Assume there exists $\mu > \omega$ such that $S(t)^*$ does not satisfy $\mathcal{P}(a, b, \mu)$ for any $a > 0, b > 0$. Then there exists a (strongly) singular norm optimal control, that is, a control $\bar{u}(t)$ that does not satisfy (2.8.3) for any $\psi$ in a multiplier space $\mathcal{M}$ unless $S(t)^*\psi = 0$ $(t > 0)$.*

*Proof.* We select at will a positive sequence $\{\epsilon_n\}$ with $\epsilon_n \to 0$ and then pick another positive sequence $\{\delta_n\}$ beginning with $\delta_1 < \min(1, T)$. Since $S(t)^*$ does not satisfy $\mathcal{P}(\delta_1, \epsilon_1 e^{-\mu T}, \mu)$, there exists an element $y_1^* \in D(A^*)$ with norm $\|y_1^*\| = 1$ and

$$
\int_{\delta_1}^\infty e^{-\mu\sigma}\|(\mu I^* - A^*)S(\sigma)^* y_1^*\|_{E^*} d\sigma < \epsilon_1 e^{-\mu T}\|y_1^*\|_{E^*} = \epsilon_1 e^{-\mu T}\,.
$$

We have

$$
\begin{aligned}
\int_{\delta_1}^T \|(\mu I^* - A^*)S(\sigma)^* y_1^*\|_{E^*} d\sigma &\le e^{\mu T} \int_{\delta_1}^T e^{-\mu\sigma}\|(\mu I^* - A^*)S(\sigma)^* y_1^*\|_{E^*} \\
&\le e^{\mu T} \int_{\delta_1}^\infty e^{-\mu\sigma}\|(\mu I^* - A^*)S(\sigma)^* y_1^*\|_{E^*}\,,
\end{aligned}
$$

thus

$$
\int_{\delta_1}^T \|(\mu I^* - A^*)S(\sigma)^* y_1^*\|_{E^*} d\sigma \le \epsilon_1\,.
$$

We then choose $0 < \delta_2 < \min(\delta_1, T/2)$ so small that

$$\int_0^{\delta_2} \|(\mu I^* - A^*)S(\sigma)^* y_1^*\|_{E^*} d\sigma + \int_{\delta_1}^{T} \|(\mu I^* - A^*)S(\sigma)^* y_1^*\|_{E^*} d\sigma \le 2\epsilon_1 .$$

Since $S(t)^*$ does not satisfy $\mathcal{P}(\delta_2, \epsilon_2 e^{-\mu T}, \mu)$, we construct $y_2^*$ and $\delta_3$ similarly, this time with $\delta_3 < \min(\delta_2, T/3)$. Continuing in the same way, we end up with a sequence $\{\delta_n\}$ such that $T \ge \delta_1 > \delta_2 > \ldots > \delta_n > 0$, $\delta_n \to 0$, and a sequence $\{y_n^*\} \subseteq D(A^*)$, $\|y_n^*\| = 1$ such that

$$\int_{\delta_n}^{\infty} e^{-\mu\sigma} \|(\mu I^* - A^*)S(\sigma)^* y_n^*\|_{E^*} d\sigma \le \epsilon_n e^{-\mu T} < \epsilon_n , \tag{2.8.16}$$

$$\int_0^{\delta_{n+1}} \|(\mu I^* - A^*)S(\sigma)^* y_n^*\|_{E^*} d\sigma + \int_{\delta_n}^{T} \|(\mu I^* - A^*)S(\sigma)^* y_n^*\|_{E^*} d\sigma \le 2\epsilon_n \quad (n = 1, 2, \ldots) . \tag{2.8.17}$$

On the other hand, (2.8.15) gives

$$\int_0^{\infty} e^{-\mu\sigma} \|(\mu I^* - A^*)S(\sigma)^* y_n^*\|_{E^*} d\sigma \ge \|y^*\| = 1 , \tag{2.8.18}$$

so that

$$\begin{aligned} \int_0^{T} \|(\mu I^* - A^*)S(\sigma)^* y_n^*\|_{E^*} d\sigma &\ge \int_0^{T} e^{-\mu\sigma} \|(\mu I^* - A^*)S(\sigma)^* y_n^*\|_{E^*} d\sigma \\ &= \int_0^{\infty} e^{-\mu\sigma} \|(\mu I^* - A^*)S(\sigma)^* y_n^*\|_{E^*} d\sigma \\ &\quad - \int_T^{\infty} e^{-\mu\sigma} \|(\mu I^* - A^*)S(\sigma)^* y_n^*\|_{E^*} d\sigma \ge 1 - \epsilon_n , \end{aligned} \tag{2.8.19}$$

where we have used (2.8.16), (2.8.17) and (2.8.18). We obtain from (2.8.17) that

$$\begin{aligned} &\int_{T-\delta_n}^{T-\delta_{n+1}} \|(\mu I^* - A^*)S(T - \sigma)^* y_n^*\|_{E^*} d\sigma \\ &= \int_{\delta_{n+1}}^{\delta_n} \|(\mu I^* - A^*)S(\sigma)^* y_n^*\|_{E^*} d\sigma \\ &\ge \int_0^{T} \|(\mu I^* - A^*)S(\sigma)^* y_n^*\|_{E^*} - 2\epsilon_n . \end{aligned} \tag{2.8.20}$$

writing the integral in $[\delta_{n+1}, \delta_n]$ as the difference of the integrals in $[0, T]$ and in $[0, \delta_{n+1}] \cup [\delta_n, T]$.

Now, (2.8.20) and Lemma 2.2.1 (or Lemma 2.2.10 when $E$ is not separable)[30] produce a function

$$u_n(\cdot) \in L^\infty(T-\delta_n, T-\delta_{n+1}; E)\,, \quad \|u_n(\cdot)\|_{L^\infty(T-\delta_n, T-\delta_{n+1}; E)} = 1\,,$$

with

$$\begin{aligned}&\int_{T-\delta_n}^{T-\delta_{n+1}} \langle (\mu I^* - A^*)S(T-\sigma)^* y_n^*, u_n(\sigma)\rangle d\sigma \\ &\geq \int_{T-\delta_n}^{T-\delta_{n+1}} \|(\mu I^* - A^*)S(T-\sigma)^* y_n^*\|_{E^*} d\sigma - \epsilon_n\,. \end{aligned} \tag{2.8.21}$$

We define

$$\tilde{u}(\sigma) = u_n(\sigma) \quad (T-\delta_n \leq t < T-\delta_{n+1})\,, \tag{2.8.22}$$

and show that

$$y(T, 0, \tilde{u}) = \int_0^T S(T-\sigma)\tilde{u}(\sigma) d\sigma$$

is a boundary point of the unit ball $B_1^\infty(T)$ of $R^\infty(T)$. If this is not the case, there exists $v(\cdot) \in L^\infty(0, T; E)$, $\|v(\cdot)\|_{L^\infty(0,T;E)} = \rho < 1$ such that

$$\int_0^T S(T-\sigma)\tilde{u}(\sigma) d\sigma = \int_0^T S(T-\sigma)v(\sigma) d\sigma\,. \tag{2.8.23}$$

We have

$$\begin{aligned}&\left\langle (\mu I^* - A^*)y_n^*, \int_0^T S(T-\sigma)\tilde{u}(\sigma) d\sigma \right\rangle \\ &= \int_0^T \langle (\mu I^* - A^*)S(T-\sigma)^* y_n^*,\ \tilde{u}(\sigma)\rangle d\sigma \\ &= \int_{T-\delta_n}^{T-\delta_{n+1}} \langle (\mu I^* - A^*)S(T-\sigma)y_n^*,\ u_n(\sigma)\rangle d\sigma \\ &+ \left(\int_0^{T-\delta_n} + \int_{T-\delta_{n+1}}^T\right) \langle (\mu I^* - A^*)S(T-\sigma)y_n^*,\ u_n(\sigma)\rangle d\sigma \\ &\geq \int_{T-\delta_n}^{T-\delta_{n+1}} \|(\mu I^* - A^*)S(T-\sigma)y_n^*\|_{E^*} d\sigma - \epsilon_n \\ &- \left(\int_0^{T-\delta_n} + \int_{T-\delta_{n+1}}^T\right) \|(\mu I^* - A^*)S(T-\sigma)y_n^*\| d\sigma\,, \end{aligned} \tag{2.8.24}$$

where we have employed (2.8.21) to bound from below the integral in the interval $[T-\delta_n, T-\delta_{n+1}]$. To bound from above the integral in $[0, T-\delta_n] \cup [T-\delta_{n+1}]$ we use (2.8.17), obtaining

[30] That $(\mu I^* - A^*)S(\sigma)^* y^* = S(\sigma)^*(\mu I^* - A^*)y^*$ is $E-$weakly measurable is obvious.

$$\begin{aligned}&\left\langle (\mu I^* - A^*)y_n^*, \int_0^T S(T-\sigma)\tilde{u}(\sigma)d\sigma \right\rangle\\&\qquad \geq \int_{T-\delta_n}^{T-\delta_{n+1}} \|(\mu I^* - A^*)S(T-\sigma)y_n^*\|_{E^*} d\sigma - 3\epsilon_n\,.\end{aligned}$$

Finally, using (2.8.20),

$$\begin{aligned}&\left\langle (\mu I^* - A^*)y_n^*, \int_0^T S(T-\sigma)\tilde{u}(\sigma)d\sigma \right\rangle\\&\qquad \geq \int_0^T \|(\mu I^* - A^*)S(T-\sigma)y_n^*\|_{E^*} d\sigma - 5\epsilon_n\,. \qquad (2.8.25)\end{aligned}$$

We go back to (2.8.23). We have

$$\begin{aligned}&\left\langle (\mu I^* - A^*)y_n^*, \int_0^T S(T-\sigma)v(\sigma)d\sigma \right\rangle\\&\qquad = \int_0^T \langle (\mu I^* - A^*)S(T-\sigma)^* y_n^*, v(\sigma)\rangle d\sigma\\&\qquad \leq \rho \int_0^T \|(\mu I^* - A^*)S(T-\sigma)^* y_n^*\|_{E^*} d\sigma\,, \qquad (2.8.26)\end{aligned}$$

hence, combining (2.8.23), (2.8.25) and (2.8.26),

$$\begin{aligned}&\int_0^T \|(\mu I^* - A^*)S(T-\sigma)^* y_n^*\|_{E^*} d\sigma - 5\epsilon_n\\&\qquad \leq \rho \int_0^T \|(\mu I^* - A^*)S(T-\sigma)^* y_n^*\|_{E^*} d\sigma\,,\end{aligned}$$

or

$$(1-\rho)\int_0^T \|(\mu I^* - A^*)S(T-\sigma)^* y_n^*\|_{E^*} d\sigma \leq 5\epsilon_n\,.$$

With the lower bound (2.8.19) on the integral, we obtain

$$(1-\rho)(1-\epsilon_n) \leq 5\epsilon_n\,, \qquad (2.8.27)$$

absurd for large $n$; this contradicts (2.8.23) and hence shows that $\tilde{u}(\sigma)$ is norm optimal. However, the candidate to singular control is not $\tilde{u}(\sigma)$ but

$$\bar{u}(\sigma) = \begin{cases} \tilde{u}(\sigma) = u_{2n-1}(\sigma) & (T-\delta_{2n-1} \leq \sigma < T-\delta_{2n})\,,\\ 0 & (T-\delta_{2n} \leq \sigma < T-\delta_{2n+1})\,. \end{cases} \qquad (2.8.28)$$

We prove exactly in the same way as for $\tilde{u}(\cdot)$ that

$$y(T,0,\bar{u}) = \int_0^T S(T-\sigma)\bar{u}(\sigma)d\sigma \qquad (2.8.29)$$

is norm optimal. To justify this claim, note that the *only* place where information on $\tilde{u}_n(\cdot)$ outside of the interval $T - \delta_n \le \sigma \le T - \delta_{n+1}$ is actually used is in the last inequality (2.8.24), and this information reduces to $\|\tilde{u}(\sigma)\| \le 1$. This time, (2.8.27) is only obtained for $n$ odd, but the contradiction is produced all the same.

That $\bar{u}(\cdot)$ does not satisfy the maximum principle (2.8.3) with the qualification (2.8.9) is shown as follows. If (2.8.3) holds, we must have $S(T-\sigma_n)^*\psi = 0$ in all intervals $T - \delta_{2n} \le \sigma_n < T - \delta_{2n+1}$ thus by the semigroup equation for $S(t)^*$,

$$S(T-\sigma)^*\psi = 0 \quad (0 \le \sigma < T).$$

This ends the proof of Theorem 2.8.6.

**Theorem 2.8.7.** *Let $S(t)$ be a strongly continuous semigroup in a Banach space $E$. Assume that*

$$\overline{D(A)} = R^\infty(T). \tag{2.8.30}$$

*Then, for any $\mu > \omega$, $S(t)^*$ satisfies $\mathcal{P}(a, b, \mu)$ for some $a > 0, b > 0$.*

*Proof.* Assume there exists $\mu > \omega$ such that $S(t)^*$ does not satisfy $\mathcal{P}(a, b, \mu)$ for any $a > 0, b > 0$. Theorem 2.8.6 says we have a strongly singular norm optimal control $\bar{u}(\cdot)$ of norm 1 in $L^\infty(0, T; E)$ (recall that each $u_n(\cdot)$ has norm $= 1$). Then $y(T, 0, \bar{u})$, given by (2.8.29), can be separated from $B_1^\infty(T)$ only by a singular functional $\xi_s \ne 0$, hence $\mathcal{S}(T) \ne \{0\}$. The proof concludes observing that singular functionals vanish in $D(A)$, thus if nontrivial singular functionals exist, $D(A)$ cannot be dense in $R^\infty(T)$.

**Theorem 2.8.8.** *Assume that there exists $\mu > \omega$ such that $S(t)^*$ does not satisfy $\mathcal{P}(a, b, \mu)$ for any $a > 0, b > 0$. Then $R^\infty(T)$ is not separable.*

*Proof.* Let $\mathcal{N}$ be the space of all sequences

$$\xi = \{\xi_0, \xi_1, \ldots\} \quad \text{with } \xi_n = 0 \text{ or } \xi_n = 1,$$

and let $\mathcal{N}_0 \subseteq \mathcal{N}$ be the subspace consisting of all sequences $\xi$ with $\xi_n = 0$ for $n$ large enough (possibly depending on $\xi$). Define an equivalence relation $\approx$ in $\mathcal{N}$ by:

$$\xi \approx \eta \quad \text{if } \xi - \eta \in \mathcal{N}_0 .$$

Since $\mathcal{N}_0$ is countable, the equivalence class $\xi + \mathcal{N}_0$ of each $\xi \in \mathcal{N}$ is countable. The space $\mathcal{N}$ itself is uncountable, thus the number of disjoint equivalence classes must be uncountable. Pick one element out of every equivalence class and put all these chosen elements together in a set $\mathcal{N}_e \subseteq \mathcal{N}$; clearly, $\mathcal{N}_e$ must be uncountable. For every $\xi = \{\xi_0, \xi_1, \ldots\} \in \mathcal{N}_e$ define a control $\bar{u}_\xi(\cdot)$ modifying (2.8.28) as follows:

$$\bar{u}_\xi(\sigma) = \begin{cases} u_n(\sigma) & \text{in } T - \delta_n \le \sigma < T - \delta_{n+1} \text{ if } \xi_n = 1, \\ 0 & \text{in } T - \delta_n \le \sigma < T - \delta_{n+1} \text{ if } \xi_n = 0. \end{cases} \tag{2.8.31}$$

If $\xi, \eta \in \mathcal{N}_e$ then

$$\bar{u}_\xi(\sigma) - \bar{u}_\eta(\sigma) = \begin{cases} \pm u_n(\sigma) & \text{in } T - \delta_n \le \sigma < T - \delta_{n+1} \text{ if } \xi_n \ne \eta_n \,, \\ 0 & \text{in } T - \delta_n \le \sigma < T - \delta_{n+1} \text{ if } \xi_n = \eta_n \,. \end{cases} \tag{2.8.32}$$

If $\xi$, $\eta$ are distinct, $\xi - \eta \notin \mathcal{N}_0$ (otherwise $\xi, \eta$ would belong to the same equivalence class). Accordingly, we have $\xi_n \ne \eta_n$ for infinitely many indices $n$ and the first line of (2.8.32) will occur infinitely many times. We then show that

$$y(T, 0, \bar{u}_\xi) - y(T, 0, \bar{u}_\eta) = \int_0^T S(T - \sigma)(\bar{u}_\xi(\sigma) - \bar{u}_\eta(\sigma) d\sigma$$

is an extremal point of $B_1^\infty(T)$ arguing as in Theorem 2.8.6. This statement means

$$\|y(T, 0; \bar{u}_\xi) - y(T, 0; \bar{u}_\eta)\|_{R^\infty(T)} = 1 \,,$$

so we have an uncountable set $\{y(T, 0, \bar{u}_\xi);\ \xi \in \mathcal{N}_e\} \subseteq R^\infty(T)$ where any two different elements lie at a distance $= 1$. This ends the proof.

Under the assumptions of Theorem 2.8.6, nonseparability of $R^\infty(T)$ means the construction of the dual $R^\infty(T)^\star$ involves the Hahn-Banach theorem "in full force", that is, it makes use of the unrestricted axiom of choice. This makes unlikely that any constructive characterization of particular singular functionals exist. We don't even know of any example assembled with the help of some nonconstructive but familiar process (say, a Banach limit).

As a first example of a semigroup $S(t)$ such that $S(t)^*$ does not satisfy $\mathcal{P}(a, b, \mu)$ for any $a, b > 0$, $\mu > \omega$ we have the right translation semigroup (2.6.1), the adjoint semigroup given by (2.6.2). The adjoint of the operator $A$ in (2.6.3) is

$$A^* y(x) = y'(x)$$

with domain $D(A^*) = \{$all absolutely continuous $y(x)$ with derivative $y'(x)$ in $L^2(0, \infty)$, no boundary condition$\}$. Inequality (2.8.14) for $y(\cdot) \in D(A^*) \subset E^* = E = L^2(0, \infty)$ is

$$\begin{aligned} b\|y\|_{L^2(0,\infty)} &\le \int_a^\infty e^{-\mu\sigma} \|(\mu I - A^*) S(\sigma)^* y\|_{L^2(0,\infty)} d\sigma \\ &= \int_a^\infty e^{-\mu\sigma} \|\mu y(\cdot + \sigma) - y'(\cdot + \sigma)\|_{L^2(0,\infty)} d\sigma \,, \end{aligned}$$

thus it will be thwarted by any nonzero $y(x)$ with support $\subseteq [0, a]$. The results in this section add something to the theory of the control system (2.6.4); a large family of singular norm optimal controls is added to the example in **2.7** and we learn that $R^\infty(T)$ is not separable.

**Miscellaneous notes.** The first examples of singular norm optimal controls were given in the author [2000] in a particular setting (Hilbert space, analytic semigroups). The material in this section is taken from the author [2001:3] with some corrections and improvements; in particular, the requirement of negative exponential decay on the semigroup has been eliminated.

The fact that $R^\infty(T)$ is not separable implies that its dual $R^\infty(T)^\star$ is not separable (Dunford - Schwartz [1958, p. 65, Lemma 16]) thus for the operators in Theorem 2.8.8, $R^\infty(T)^\star$ is not separable. On the other hand, if $E$ is a Banach space such that $E^*$ is separable and $S(t)$ a strongly continuous semigroup, we have $E^\odot = E^*$ since $S(t)^*$ is a strongly continuous semigroup in $E^*$. This is condition $(a)$ in (2.4.19) and guarantees that $Z_w(T) = Z(T)$. The definition (2.3.22) of the norm of $Z(T)$ says that we have the isometric imbedding $Z_w(T) \overset{i}{\hookrightarrow} L^1(0, T; E^*)$ where, since $E^*$ is separable $L^1(0, T; E^*)$ is separable, thus $Z(T)$ is itself separable. Now, (2.3.34) says that $Z(T)$ and the space $\mathcal{R}(T) \subseteq R^\infty(T)^\star$ of all regular functionals are isometric, hence we have the proof of

**Theorem 2.8.9.** *Let $S(t)$ be a strongly continuous semigroup in a Banach space $E$ with separable dual. Assume there exists $\mu > \omega$ such that $S(t)^*$ does not satisfy $\mathcal{P}(a, b, \mu)$ for any $a > 0, b > 0$. Then $(a)$ The space $R^\infty(T)^\star$ is not separable, $(b)$ The subspace $\mathcal{R}(T) \subset R^\infty(T)^\star$ is separable.*

Of course, we have so far one solitary example of space and semigroup where all this happens, namely $E = L^2(0, \infty)$ and the right translation semigroup. Many other semigroups will be identified in next section.

**2.9. Singular norm optimal controls and singular functionals.** The next result simplifies the checking of $\mathcal{P}(a, b, \mu)$. $C$ and $\omega$ are the constants in (2.3.1).

**Theorem 2.9.1.** *Assume*

$$\|(\mu I^* - A^*)S(r)^* y^*\| \geq m\|y^*\| \quad (y^* \in D(A^*)) \tag{2.9.1}$$

*for some $r, m > 0, \mu > \omega$. Then $S(t)^*$ satisfies $\mathcal{P}(a, b, \mu)$ for*

$$0 < a < r\,, \qquad b = \frac{(r-a)m}{C} e^{-\mu r}\,. \tag{2.9.2}$$

*Conversely, assume $S(t)^*$ satisfies $\mathcal{P}(a, b, \mu)$ for some $a > 0$, $b > 0$, $\mu > \omega$. Then $S(t)^*$ satisfies* (2.9.1) *for*

$$r = a\,, \qquad m = \frac{b(\mu - \omega)}{C} e^{\mu a}\,. \tag{2.9.3}$$

*Proof.* Assuming (2.9.1) and using (2.3.1),

$$\begin{aligned} &Ce^{\mu(r-\sigma)}\|(\mu I^* - A^*)S^*(\sigma)y^*\| \geq e^{\omega(r-\sigma)}\|(\mu I^* - A^*)S^*(\sigma)y^*\| \\ &\geq \|S(r-\sigma)^*\|\|(\mu I^* - A^*)S(\sigma)y^*\| \geq \|S(r-\sigma)^*(\mu I^* - A^*)S(\sigma)^* y^*\| \\ &= \|(\mu I^* - A^*)S(r)^* y^*\| \geq m\|y^*\| \quad (0 \leq \sigma \leq r)\,. \end{aligned}$$

If $a < r$ we integrate both sides in $a \le \sigma \le r$ and obtain

$$\begin{aligned}(r-a)m\|y^*\| &\le Ce^{\mu r}\int_a^r e^{-\mu\sigma}\|(\mu I^* - A^*)S(\sigma)^*y^*\|d\sigma \\ &\le Ce^{\mu r}\int_a^\infty e^{-\mu\sigma}\|(\mu I^* - A^*)S(\sigma)^*y^*\|d\sigma\,,\end{aligned}$$

which is $\mathcal{P}(a, b, \mu)$ with $b$ given by (2.9.2). Conversely, assume that $S(t)^*$ satisfies $\mathcal{P}(a, b, \mu)$ with $a > 0, b > 0, \mu > \omega$. This means

$$b\|y^*\| \le \int_a^\infty e^{-\mu\sigma}\|(\mu I^* - A^*)S(\sigma)^*y^*\|d\sigma \quad (y^* \in D(A^*))\,.$$

Given $\epsilon > 0$ and $y^* \in D(A^*)$, Lemma 2.2.1 (Lemma 2.2.11 for $E$ nonseparable) produces

$$u_{\epsilon,n}(\cdot) \in L^\infty(a+n-1, a+n; E)\,, \qquad \|u_{\epsilon,n}(\cdot)\|_{L^\infty(a+n-1,a+n;E)} = 1\,,$$

such that

$$\begin{aligned}&\int_{a+n-1}^{a+n} \langle e^{-\mu\sigma}(\mu I^* - A^*)S(\sigma)^*y^*, u_{\epsilon,n}(\sigma)\rangle d\sigma \\ &\ge \int_{a+n-1}^{a+n} e^{-\mu\sigma}\|(\mu I^* - A^*)S(\sigma)^*y^*\|d\sigma - \frac{\epsilon}{2^n}\end{aligned}$$

for $n = 1, 2, \ldots$. Define

$$u_\epsilon(\sigma) = u_{\epsilon,n}(\sigma) \qquad (a+n-1 \le \sigma \le a+n)\,.$$

Then

$$\begin{aligned}b\|y^*\| - \epsilon &\le \int_a^\infty e^{-\mu\sigma}\|(\mu I^* - A^*)S(\sigma)^*y^*\|d\sigma - \epsilon \\ &= \sum_{n=1}^\infty \left(\int_{a+n-1}^{a+n} e^{-\mu\sigma}\|(\mu I^* - A^*)S(\sigma)^*y^*\|d\sigma - \frac{\epsilon}{2^n}\right) \\ &\le \sum_{n=1}^\infty \int_{a+n-1}^{a+n} \langle e^{-\mu\sigma}(\mu I^* - A^*)S(\sigma)^*y^*,\, u_{\epsilon,n}(\sigma)\rangle d\sigma \\ &= \int_a^\infty \langle e^{-\mu\sigma}(\mu I^* - A^*)S(\sigma)^*y^*,\, u_\epsilon(\sigma)\rangle d\sigma \\ &\le \int_a^\infty \langle(\mu I^* - A^*)S(a)^*y^*,\, e^{-\mu\sigma}S(\sigma - a)u_\epsilon(\sigma)\rangle d\sigma \\ &= \left\langle(\mu I^* - A^*)S(a)^*y^*,\, e^{-\mu a}\int_a^\infty e^{-\mu(\sigma-a)}S(\sigma - a)u_\epsilon(\sigma)d\sigma\right\rangle \\ &= \left\langle(\mu I^* - A^*)S(a)^*y^*,\, e^{-\mu a}\int_0^\infty e^{-\mu\sigma}S(\sigma)u_\epsilon(\sigma + a)d\sigma\right\rangle \\ &\le \frac{Ce^{-\mu a}}{\mu - \omega}\|(\mu I^* - A^*)S(a)^*y^*\| \quad (y^* \in D(A^*))\,.\end{aligned}$$

Letting $\epsilon \to 0$, (2.9.1) follows with $m$ given by (2.9.3). This ends the proof.

**Theorem 2.9.2.** *Let $S(t)^*$ satisfy* (2.9.1) *with $r, m > 0$. Assume*

$$S(a)E \subseteq D(A) \tag{2.9.4}$$

*for some $a < r$. Then $A$ is bounded.*

*Proof.* Since $S(r) = S(r-a)S(a)$, (2.9.4) implies that $S(r)E \subseteq D(A)$ for $r \geq a$, so that, by the closed graph theorem $(\mu I - A)S(r)$ is everywhere defined and bounded. If $A$ is unbounded, (2.9.4) also implies

$$S(t)E \subset E \quad \text{(strict inclusion)} \quad (t > 0)\,. \tag{2.9.5}$$

To see this, assume $S(t)E = E$. Then $S(t)E = S(\sigma)S(t-\sigma)E$, so that $S(\sigma)E = E$ for $0 \leq \sigma \leq t$ and then, writing $S(\sigma) = S(\sigma/n)^n$ for large enough $n$, for all $\sigma \geq 0$. Taking $\sigma = a$, (2.9.4) says that $D(A) = E$, which means $A$ is bounded.

With (2.9.5) granted, we write $(\mu I - A)S(r)E = S(r-a)(\mu I - A)S(a)E$ for $r > a$ and deduce that the bounded operator $(\mu I - A)S(r)$ satisfies

$$(\mu I - A)S(r)E \subset E \quad \text{(strict inclusion)}\,. \tag{2.9.6}$$

On the other hand,

$$((\mu I - A)S(r))^* = (S(r)(\mu I - A))^* = (\mu I^* - A^*)S(r)^*\,,$$

so that $S(t)^*E^* \subseteq D(A^*)$ and $(\mu I^* - A^*)S(r)^*$ is bounded. Using (2.9.1) combined with Theorem 2.2.4 we obtain

$$(\mu I - A)S(r)E = E\,,$$

which contradicts (2.9.6) and ends the proof of Theorem 2.9.2. The fact that $a < r$ in (2.9.4) is essential, as Example 2.9.3 below shows: (2.9.4) for $a = r$ does not imply $A$ is bounded.

The singular norm optimal controls constructed in **2.8** (these controls imply the existence of nontrivial singular functionals in $R^\infty(T)^\star$) need a semigroup such that $S(t)^*$ does *not* satisfy $\mathcal{P}(a, b, \mu)$ for some $\mu > \omega$ and every $a, b > 0$. Theorem 2.9.2 shows that the construction will work for an arbitrary analytic semigroup with unbounded infinitesimal generator, or more in general, for any semigroup satisfying $S(t)E \subseteq D(A)$ $(t > 0)$ having unbounded infinitesimal generator. For all these semigroups, $\mathcal{S}(T) \neq \{0\}$ or, equivalently, $\overline{D(A)} \neq R^\infty(T)$.

On the other hand, there exist semigroups $S(t)$ with unbounded infinitesimal generators that satisfy (2.9.1). If

$$S(t)E = E \quad (t \geq 0) \tag{2.9.7}$$

then it follows from Theorem 2.2.4 that $\|S(t)^*y^*\| \geq m\|y^*\|$ ($y^* \in E^*$), so that

$$m\|R(\mu;A)y^*\| \leq \|R(\mu;A)^*S(t)^*y^*\| \leq \frac{C}{\mu-\omega}\|S(t)^*y^*\| \quad (y^* \in E^*),$$

which is the same as (2.9.1). That (2.9.1) does not imply (2.9.7) is shown below.

**Example 2.9.3.** Let $E = L^2(0,\infty)$ (complex) and $A$ the unbounded normal operator

$$Ay(\lambda) = (i\lambda - \phi(\lambda))y(\lambda) \tag{2.9.8}$$

with $\phi(\lambda) \geq 0$ (maximal domain). The adjoint is $A^*y(\lambda) = (-i\lambda - \phi(\lambda))y(\lambda)$ with the same domain. The operator $A$ generates the strongly continuous semigroup

$$S(t)y(\lambda) = e^{i\lambda t}e^{-\phi(\lambda)t}y(\lambda) \quad (t \geq 0), \tag{2.9.9}$$

with adjoint $S(t)^*y(\lambda) = e^{-i\lambda t}e^{-\phi(\lambda)t}y(\lambda)$. We have

$$(\mu I^* - A^*)S(t)^*y(\lambda) = (\mu + i\lambda + \phi(\lambda))e^{-i\lambda t}e^{-\phi(\lambda)t}y(\lambda) = \omega^*(\mu,t,\lambda)y(\lambda).$$

Set $\phi(\lambda) = \log\max(1,\lambda)$, and take $\mu \geq 1$. Then

$$|\omega^*(\mu,1,\lambda)| = \begin{cases} \sqrt{\lambda^2+\mu^2} & (0 \leq \lambda \leq 1), \\ \dfrac{\sqrt{\lambda^2+(\phi(\lambda)+\mu)^2}}{\lambda^t} & (\lambda \geq 1), \end{cases}$$

so that

$$|\omega^*(\mu,t,\lambda)| \geq 1 \quad (-\infty < \lambda < \infty,\ 0 \leq t \leq 1),$$

and we obtain (2.9.1) for $0 \leq r \leq 1$,

$$\|(\mu I^* - A^*)S(r)^*y^*\| \geq \|y^*\| \quad (y^* \in D(A^*)).$$

On the other hand,

$$AS(t)y(\lambda) = (i\lambda - \phi(\lambda))e^{i\lambda t}e^{-\phi(\lambda)t}y(\lambda) = \omega(t,\lambda)y(\lambda)$$

with

$$|\omega(t,\lambda)| = \begin{cases} |\lambda| & (0 \leq \lambda \leq 1), \\ \dfrac{\sqrt{\lambda^2+\phi(\lambda)^2}}{\lambda^t} & (\lambda \geq 1), \end{cases}$$

so that

$$|\omega(t,\lambda)| \leq C \quad (-\infty < \lambda < \infty,\ t \geq 1).$$

In particular $AS(1)$ is bounded, or equivalently $S(1)E \subseteq D(A)$.

Other sorts of behavior of $A$, $S(t)$ are produced via a change of $\phi(\lambda)$. For instance, for $\phi(\lambda) = \log\log(\max(e, \lambda))$, (2.9.9) gives a semigroup satisfying (2.9.1) but such that $S(t)E \not\subseteq D(A)$ for all $t > 0$.

**Problem 2.9.4.** *Is the converse of Theorem* 2.8.7 *true, that is, if $A$ and $S(t)$ satisfy* (2.9.1) *do we have*

$$\overline{D(A)} = R^\infty(T)\,? \tag{2.9.10}$$

**Problem 2.9.5.** *Is there any semigroup $S(t)$ not satisfying* (2.9.7) *and such that its infinitesimal generator $A$ satisfies* (2.9.10)?

Obviously, Problem 2.9.5 is less ambitious than Problem 2.9.4; if the answer to the first is "yes" then so is the answer to the second. The reason for the qualification "not satisfying (2.9.7)" in Problem 2.9.5 is, if (2.9.7) is satisfied then $R^\infty(T) = E$ with equivalent norms (see (2.2.2) and following comments) so that in this case all that (2.9.10) says is: $D(A)$ is dense in $E$ in the norm of $E$. This is not news: it's a standing assumption.

If (2.9.10) in Problem 2.9.5 is strengthened to

$$D(A) = R^\infty(T)\,, \tag{2.9.11}$$

the the answer is an unequivocal "no". Precisely,

**Theorem 2.9.6.** *Let $A$ be an infinitesimal generator such that* (2.9.11) *holds. Then $A$ is bounded.*

The result follows from Theorem 2.9.7 below, whose proof uses the theory of the Phillips adjoint (see **2.4** and Hille-Phillips [1957, Chapter 14]). We need the imbedding

$$E \hookrightarrow (E^\odot)^*\,. \tag{2.9.12}$$

To prove it, it suffices to note that if we renorm $E$ with

$$\|y\|_\odot = \sup_{\|y^*\|_{E^\odot} \le 1} |\langle y^*, y\rangle|\,, \tag{2.9.13}$$

we have (Hille-Phillips [1957, Theorem 14.2.1, p. 422])

$$\|y\|_\odot \le \|y\| \le C\|y\|_\odot\,. \tag{2.9.14}$$

**Theorem 2.9.7.** *The following statements are equivalent for a strongly continuous semigroup $S(t)$ and its infinitesimal generator $A$; (a) $R^\infty(T) \subseteq D((A^\odot)^*)$, (b) $S(t)$ is analytic and*[31]

$$\int_0^T \|A^\odot S(\sigma)^\odot y^*\| d\sigma \le C\|y^*\| \quad (y^* \in D(A^\odot))\,. \tag{2.9.15}$$

[31] Below, $C$ indicates a generic constant.

*Proof.* $(b) \Rightarrow (a)$. If $y^* \in D(A^{\odot})$ and $u(\cdot) \in L^\infty(0,T;E)$ we have

$$\left\langle A^{\odot}y^*, \int_0^T S(T-\sigma)u(\sigma)d\sigma \right\rangle = \int_0^T \langle A^{\odot}S^{\odot}(T-\sigma)y^*, u(\sigma)\rangle d\sigma\,. \qquad (2.9.16)$$

Using (2.9.15),

$$\left|\left\langle A^{\odot}y^*, \int_0^T S(T-\sigma)u(\sigma)d\sigma \right\rangle\right| \le \int_0^T |\langle A^{\odot}S^{\odot}(T-\sigma)y^*, u(\sigma)\rangle| d\sigma$$
$$\le \|u(\cdot)\|_{L^\infty(0,T;E)} \int_0^T \|A^{\odot}S^{\odot}(T-\sigma)y^*\| d\sigma \le C\|u(\cdot)\|_{L^\infty(0,T;E)}\|y^*\|\,,$$

hence the left side of (2.9.16) is a bounded operator of $y^*$. It follows that

$$\int_0^T S(T-\sigma)u(\sigma)d\sigma \in D((A^{\odot})^*) \quad (u(\cdot) \in L^\infty(0,T;E))\,,$$

which is $(a)$. To show the implication $(a) \Rightarrow (b)$, assume

$$\int_0^T S(T-\sigma)u(\sigma)d\sigma \in D((A^{\odot})^*)$$

for all $u(\cdot) \in L^\infty(0,T;E)$. Then the operator $\mathcal{A} : L^\infty(0,T;E) \to (E^{\odot})^*$ defined by

$$\mathcal{A}u(\cdot) = (A^{\odot})^* \int_0^T S(T-\sigma)u(\sigma)d\sigma$$

is everywhere defined. It is also closed; in fact, assume

$$u_n(\cdot) \to u(\cdot) \text{ in } L^\infty(0,T;E), \qquad \mathcal{A}u_n(\cdot) \to y \text{ in } (E^{\odot})^*\,. \qquad (2.9.17)$$

It follows from the first convergence statement (2.9.17) that

$$\int_0^T S(T-\sigma)u_n(\sigma)d\sigma \to \int_0^T S(T-\sigma)u(\sigma)d\sigma \text{ in } E$$

and $(A^{\odot})^* : D((A^{\odot})^*) \to (E^{\odot})^*$ is closed, thus the second statement (2.9.17) gives

$$\int_0^T S(T-\sigma)u(\sigma)d\sigma \in D(A^{\odot})^*\,, \qquad (A^{\odot})^* \int_0^T S(T-\sigma)u(\sigma)d\sigma = y\,.$$

The closed graph theorem says that the closed, everywhere defined operator $\mathcal{A}$ is bounded. Accordingly, if $y^* \in D(A^{\odot})$,

$$\left|\int_0^T \langle A^{\odot}S^{\odot}(T-\sigma)y^*, u(\sigma)\rangle d\sigma\right| = \left|\int_0^T \langle A^{\odot}y^*, S(T-\sigma)u(\sigma)\rangle d\sigma\right|$$
$$= \left|\left\langle y^*, (A^{\odot})^* \int_0^T S(T-\sigma)u(\sigma)d\sigma \right\rangle\right| \le C\|y^*\|\|u(\cdot)\|_{L^\infty(0,T;E)}\,,$$

and, applying Lemma 2.2.1 (Lemma 2.2.10 for $E$ nonseparable) we obtain (2.9.15). Next, we extend (2.9.15) to $y^* \in E^\odot$, that is, we prove

$$\int_0^T \|A^\odot S(t)^\odot y^*\| dt \le C\|y^*\| \quad (y^* \in E^\odot) \tag{2.9.18}$$

as follows. Take a sequence $\{y_n^*\} \in D(A^\odot)$ with $y_n^* \to y^*$. Using inequality (2.9.15) for $y_n^* - y_m^*$ we obtain

$$\int_0^T \|A^\odot S^\odot(\sigma) y_n^* - A^\odot S^\odot(\sigma) y_m^*\| \le C\|y_n^* - y_m^*\| ,$$

so that the sequence $\{A^\odot S^\odot(\cdot) y_n^*\}$ is Cauchy in $L^1(0,T;E^\odot)$, hence convergent to $\zeta(\cdot) \in L^1(0,T;E^\odot)$. Accordingly, there exists a set $e$ of full measure in $[0,T]$ such that, if necessary passing to a subsequence, we have

$$S^\odot(\sigma) y_n^* \to S^\odot(\sigma) y^*, \quad A^\odot S^\odot(\sigma) y_n^* \to \zeta(\sigma) \quad (\sigma \in e) .$$

Since $A^\odot$ is closed,

$$S^\odot(\sigma) y^* \in D(A^\odot) , \quad A^\odot S^\odot(\sigma) y^* = \zeta(\sigma) \quad (\sigma \in e) .$$

Using the semigroup equation, $S^\odot(\sigma) y^* \in D(A^\odot)$ $(\sigma > 0)$ and we have

$$\begin{aligned} \int_0^T \|A^\odot S^\odot(\sigma) y^*\| d\sigma &= \lim_{n\to\infty} \int_0^T \|A^\odot S^\odot(\sigma) y_n^*\| d\sigma \\ &\le \lim_{n\to\infty} C\|y_n^*\| = C\|y^*\| \end{aligned}$$

which is (2.9.18). We prove finally that $S(t)^\odot$ is analytic showing

$$\|A^\odot S^\odot(t)\| \le \frac{C}{t} \qquad (0 < t \le T) \tag{2.9.19}$$

(see Pazy [1983, Theorem 5.2, p. 61] or the author [1983, Theorem 4.1.5, p. 176]). By the uniform boundedness theorem it is enough to show that $\|tA^\odot S^\odot(t) y^*\|$ is bounded for all $y^* \in E^\odot$. We prove more:

$$\lim_{t\to 0+} t\|A^\odot S^\odot(t) y^*\| = 0 \quad (y^* \in E^\odot) . \tag{2.9.20}$$

Assume (2.9.20) fails for some $y^* \in E^\odot$. Then there exists $\delta > 0$ and a sequence $\{t_n\}$, $T = t_1 > t_2 > \ldots > 0$, $t_n \to 0$ such that

$$\|A^\odot S^\odot(t_n) y^*\| \ge \frac{\delta}{t_n} . \tag{2.9.21}$$

We may obviously assume that $t_{n+1}/t_n \to 0$. If $t_{n+1} \le \sigma \le t_n$ then we have $S^\odot(t_n - \sigma)S^\odot(\sigma) = S^\odot(t_n)$; accordingly,

$$\|S^\odot(t_n - \sigma)\|\|A^\odot S^\odot(\sigma)y^*\| \ge \|A^\odot S^\odot(t_n)y^*\|\,,$$

and it follows from this inequality and (2.9.21) that

$$\|A^\odot S(\sigma)^\odot y^*\| \ge \frac{\|A^\odot S^\odot(t_n)y^*\|}{\|S^\odot(t_n - \sigma)\|} \ge \frac{\delta}{Mt_n} \quad (t_{n+1} \le \sigma \le t_n)\,,$$

where $M$ is the maximum of $\|S^\odot(t)\|$ in $0 \le t \le T$. Integrating,

$$\begin{aligned}\int_{t_{n+1}}^{t_n} \|A^\odot S^\odot(\sigma)y^*\|d\sigma &\ge \frac{\delta(t_n - t_{n+1})}{Mt_n} \\ &= \frac{\delta}{M}\Big(1 - \frac{t_{n+1}}{t_n}\Big) \to \frac{\delta}{M} \quad (n \to \infty) \qquad (2.9.22)\end{aligned}$$

which contradicts (2.9.18); in fact, (2.9.22) implies that the integral in $0 \le t \le T$ is infinite. This completes the proof that $S^\odot(t)$ is an analytic semigroup.

We prove next that $S(t)$ itself is an analytic semigroup. We have

$$\langle y^*, S(t)y\rangle = \langle S^\odot(t)y^*, y\rangle \qquad (y \in E, y^* \in E^\odot)$$

hence, by (2.9.12) $S(t)$ is $E^\odot$-weakly analytic, thus analytic in the same open sector as $S^\odot(t)$. This implies $S(t)E \in D(A)$ for $t > 0$. Finally, in view of (2.9.19),

$$\begin{aligned}|\langle y^*, tAS(t)y\rangle| &= |\langle tA^\odot S^\odot(t)y^*, y\rangle| \\ &\le C\|y^*\|\|y\| \quad (0 \le t \le T,\ y \in E,\ y^* \in E^\odot)\,,\end{aligned}$$

so that by (2.9.13) and (2.9.14) we have

$$\|tAS(t)y\| \le C\|y\| \quad (0 \le t \le T,\ y \in E)\,.$$

The uniform boundedness principle says that

$$\|AS(t)y\| \le \frac{C}{t} \quad (0 < t \le T)\,,$$

thus $S(t)$ is analytic as claimed.

*Proof of Theorem* 2.9.6. (2.9.11) implies statement $(a)$ of Theorem 2.9.7, thus by its equivalent $(b)$ $S(t)$ is an analytic semigroup. We have seen in Theorem 2.9.2 that if $A$ is unbounded and $S(t)$ is an analytic semigroup $S(t)^*$ cannot satisfy (2.9.1). Accordingly, Theorem 2.8.6 guarantees the existence of singular extremals, hence the existence of singular functionals. This implies that $\overline{D(A)} \ne R^\infty(T)$, a contradiction.

**Example 2.9.8.** The assumptions of Theorem 2.9.7 do *not* imply that $A$ is bounded, as this example (Baillon [1980]) shows. It uses the space $\ell^\infty$ of all sequences $y = \{y_1, y_2, \ldots\}$ with $\|y\|_{\ell^\infty} = \sup |y_n| < \infty$ equipped with the norm $\|y\|_{\ell^\infty}$. We set $E = \ell^0 \subseteq \ell^\infty$, where $\ell_0$ is the (closed) subspace of all sequences with $\lim_{n\to\infty} |y_n| = 0$. The operator and semigroup are

$$A\{y_n\} = \{-ny_n\}\,, \qquad S(t)\{y_n\} = \{e^{-nt}y_n\}\,,$$

$D(A)$ the set of all $\{y_n\} \in \ell^0$ with $\{ny_n\} \in \ell^0$. The dual of $E$ is the space $E^* = \ell^1$ of all sequences $y^* = \{y_n\}$ with

$$\|y^*\|_{\ell^1} = \sum_{n=1}^{\infty} |y_n| < \infty\,.$$

The adjoints $A^*$ and $S(t)^*$ are given by the same formulas,

$$A\{y_n\} = \{-ny_n\}\,, \qquad S(t)\{y_n\} = \{e^{-nt}y_n\}\,,$$

$D(A^*)$ the set of all $\{y_n\} \in \ell^1$ with $\{ny_n\} \in \ell^1$. Since $D(A^*)$ is dense in $E^*$, we have $E^\odot = E^*$, $A^\odot = A^*$, $S^\odot(t) = S(t)^*$. If $y^* = \{y_n\} \in \ell_1$ we have

$$\int_0^\infty \|A^*S(t)^*y^*\|_{\ell^1} dt = \sum_{n=1}^{\infty} \left( n \int_0^\infty e^{-nt} dt \right) |y_n| = \sum_{n=1}^{\infty} |y_n| = \|y^*\|_{\ell^1}\,.$$

Note that

$$t\|A^*S(t)^*y^*\|_{\ell_1} = \sum_{n=1}^{\infty} nte^{-nt}|y_n| \to 0$$

by the discrete version of the dominated convergence theorem, confirming (2.9.20).

**Miscellaneous notes.** The material in this section is mostly taken from the author [2001:3]. When $E$ is reflexive, Theorem 2.9.7 says that any of the two equivalent conditions $(a)$ $R^\infty(T) = D(A)$, $(b)$

$$\int_0^T \|A^*S(\sigma)^*y^*\|d\sigma \le C\|y^*\| \quad (y^* \in D(A^*))$$

forces $A$ to be bounded. This result is due to Baillon [1980]. Example 2.9.8 shows that reflexiveness of $E$ cannot be dropped.

CHAPTER 3

# SYSTEMS WITH STRONGLY MEASURABLE CONTROLS, II

**3.1. Existence and uniqueness of optimal controls.** Existence theory is what is generalized the easiest from the finite dimensional case. We do this for a slightly more general equation than (2.1.1), namely

$$y'(t) = Ay(t) + Bu(t)\,, \quad y(0) = \zeta \tag{3.1.1}$$

with $A$ the infinitesimal generator of a strongly continuous semigroup $S(t)$ in the Banach space $E$. The (strongly measurable) controls $u(t)$ take values in a second Banach space $F$ and $B : F \to E$ is a linear bounded operator. As in **2.1**,

$$y(t, \zeta, u) = S(t)\zeta + \int_0^t S(t-\sigma)Bu(\sigma)d\sigma \tag{3.1.2}$$

denotes the solution of (3.1.1).

Existence is handled taking (weak) limits of *minimizing sequences.* For the norm optimal problem in an interval $0 \le t \le T$, a minimizing sequence is any sequence $\{u_n(\cdot)\} \subset L^\infty(0,T;F)$ satisfying

$$y(T, \zeta, u_n) = y_n \to \bar{y}\,, \quad \|u_n(\cdot)\|_{L^\infty(0,T;F)} \to \rho\,, \tag{3.1.3}$$

where

$$\rho = \inf\{\,\|u(\cdot)\|_{L^\infty(0,T;F)};\, y(T,\zeta,u) = \bar{y}\,\}\,. \tag{3.1.4}$$

For the time optimal problem, a minimizing sequence $\{u_n(\cdot)\}$, $u_n(\cdot) \in L^\infty(0,t_n;F)$ satisfies

$$y(t_n, \zeta, u_n) = y_n \to \bar{y}\,, \quad \|u_n(\cdot)\|_{L^\infty(0,t_n;F)} \le 1\,, \quad t_n \to T \tag{3.1.5}$$

where $T$ is the infimum of all driving times from $\zeta$ to $\bar{y}$. In the norm optimal problem, if $y(T,\zeta,u_n) = \bar{y}$ we may assume the sequence $\{\|u_n(\cdot)\|_{L^\infty(0,T;F)}\}$ is decreasing; likewise, in the time optimal problem, if $y(t_n,\zeta,u_n) = \bar{y}$ we may assume the sequence $\{t_n\}$ is decreasing.

**Lemma 3.1.1.** *(a) Assume there exists $u(\cdot) \in L^\infty(0,T;F)$ driving $\zeta$ to $\bar{y}$ in time $T$. Then there exists a minimizing sequence for the norm optimal problem in $0 \le t \le T$. (b) Assume there exists $u(\cdot) \in L^\infty(0,t;F)$, $\|u(\cdot)\|_{L^\infty(0,t;F)} \le 1$ driving $\zeta$ to $\bar{y}$ in any time $t$. Then there exists a minimizing sequence for the time optimal problem.*

The proof is obvious from the definitions. To upgrade Lemma 3.1.1 to an existence theorem, we need to take limits of minimizing sequences $\{u_n(\cdot)\}$. This can be done at least in two cases:

(*i*) $F$ is reflexive,

(*ii*) $F$ is separable and $F = X^*$ ($X$ another Banach space).

In fact, if either (*i*) or (*ii*) hold we have (see Ionescu Tulcea [1969] or the author [1999, Example 12.9.6, p. 673])

$$L^\infty(0,T;F) = L^1(0,T;X)^* \tag{3.1.6}$$

(if (*i*) holds $X = E^*$). Under (3.1.6) we can take weak limits of (subsequences of) minimizing sequences.[1] A technical point concerns the adjoint semigroup: in the present level of generality, $S(t)^*$ may not be strongly continuous. This is easily bypassed by using the Phillips dual $E^\odot$ and the Phillips adjoint semigroup $S^\odot(t)$, which is strongly continuous in $E^\odot$. (see **2.4**).

**Theorem 3.1.2.** *Assume that (i) or (ii) hold. Then, if minimizing sequences exist, optimal controls exist.*

*Proof.* If $u_n(\cdot) \subset L^\infty(0,T;X^*)$ is a minimizing sequence for the norm optimal problem we have

$$y_n - S(T)\zeta = \int_0^T S(T-\sigma)Bu_n(\sigma)d\sigma\,.$$

Applying a functional $y^* \in E^\odot$ to both sides,

$$\begin{aligned}\langle y^*, y_n - S(T)\zeta\rangle &= \left\langle y^*, \int_0^T S(T-\sigma)Bu_n(\sigma)d\sigma\right\rangle \\ &= \int_0^T \langle B^*S^\odot(T-\sigma)y^*, u_n(\sigma)\rangle d\sigma\,. \end{aligned} \tag{3.1.7}$$

[1] If $X$ is separable, then $L^1(0,T;X)$ is separable. Accordingly, the $L^1(0,T;X)$-weak topology of the unit ball of $L^\infty(0,T;X^*)$ is defined by a metric (Dunford - Schwartz [1958, Theorem 3, p. 434]) and we only need to use sequences and subsequences; generalized sequences are not needed. Under (*ii*) $X$ is separable (separability of $X^*$ implies separability of $X$). Under (*i*) $X$ is not necessarily separable, thus we may have to take generalized subsequences instead of subsequences. This does not modify much the arguments. Besides, applications don't seem to call for models where $X$ is not separable, thus, the reader only needs to think of subsequences.

Since the sequence $\{u_n(\cdot)\}$ is bounded in $L^\infty(0,T;X^*)$ we can select a subsequence (named in the same way) such that $\{u_n(\cdot)\}$ is $L^1(0,T;X)$-weakly convergent to an element $\bar{u}(\cdot) \in L^\infty(0,T;X^*)$. Taking limits[2] in (3.1.7),

$$\begin{aligned}\langle y^*, \bar{y} - S(T)\zeta\rangle &= \int_0^T \langle B^* S^\odot(T-\sigma)y^*, \bar{u}(\sigma)\rangle d\sigma \\ &= \left\langle y^*, \int_0^T S(T-\sigma)B\bar{u}(\sigma)d\sigma\right\rangle,\end{aligned}$$

which, due to the fact that $y^* \in E^\odot$ is arbitrary (Hille-Phillips [1957, Theorem 14.2.1, p. 422]) implies

$$\bar{y} - S(T)\zeta = \int_0^T S(T-\sigma)B\bar{u}(\sigma)d\sigma\,,$$

so that $\bar{u}(\cdot)$ drives $\zeta$ to $\bar{y}$. According to (3.1.3) the subsequence $\{u_n(\cdot)\}$ is eventually contained in the ball of center 0 and radius $\rho + \epsilon$ ($\epsilon > 0$ arbitrary) in the space $L^\infty(0,T;X^*)$. It follows that $\|\bar{u}(\cdot)\|_{L^\infty(0,T;X^*)} \le \rho + \epsilon$ for all $\epsilon$, thus we have $\|\bar{u}(\cdot)\|_{L^\infty(0,T;X^*)} \le \rho$ and $\bar{u}(\cdot)$ is norm optimal.

If $\{u_n(\cdot)\}$ is a minimizing sequence for the time optimal problem we have

$$y_n - S(t_n)\zeta = \int_0^{t_n} S(t_n-\sigma)Bu_n(\sigma)d\sigma$$

and, applying a functional $y^* \in E^\odot$ to both sides,

$$\begin{aligned}\langle y^*, y_n - S(t_n)\zeta\rangle &= \int_0^{t_n} \langle y^*, S(t_n-\sigma)Bu_n(\sigma)\rangle d\sigma \\ &= \int_0^{t_n} \langle B^* S^\odot(t_n-\sigma)y^*, u_n(\sigma)\rangle d\sigma\,. \end{aligned} \tag{3.1.8}$$

Take $\bar{t} \ge$ all $t_n$ and extend each $u_n(\cdot)$ to $t_n \le \sigma \le \bar{t}$ setting $u_n(\sigma) = 0$ there. Then select a subsequence of the extended $\{u_n(\cdot)\}$ (denoted with the same name) such that $\{u_n(\cdot)\}$ is $L^1(0,\bar{t};X)$-weakly convergent to an element $\bar{u}(\cdot) \in L^\infty(0,\bar{t};X^*)$; since each $u_n(\cdot)$ has norm $\le 1$ we have $\|u(\cdot)\|_{L^\infty(0,\bar{t};X^*)} \le 1$. Next, define

$$\chi_n(\sigma) = \begin{cases} B^* S^\odot(t_n-\sigma)y^* & 0 \le \sigma \le t_n \\ 0 & t_n < \sigma \le \bar{t}\,, \end{cases}$$

$$\chi(\sigma) = \begin{cases} B^* S^\odot(T-\sigma)y^* & 0 \le \sigma \le T \\ 0 & T < \sigma \le \bar{t}\,. \end{cases}$$

[2] Strong convergence $y_n \to \bar{y}$ is not necessary; $E^\odot$-weak convergence suffices. Same observation for the time optimal problem.

Continuity of $S^{\odot}(\sigma)y^*$ implies $\chi_n(\cdot) \to \chi(\cdot)$ in $L^1(0, \bar{t}; X)$. We can write (3.1.8) in the form

$$\langle y^*, y_n - S(t_n)\zeta\rangle = \left\langle y^*, \int_0^{t_n} S(t_n - \sigma)Bu_n(\sigma)d\sigma \right\rangle$$
$$= \int_0^{\bar{t}} \langle \chi_n(\sigma), u_n(\sigma)\rangle d\sigma$$

and taking limits, obtain

$$\langle y^*, \bar{y} - S(T)\zeta\rangle = \int_0^T \langle \chi(\sigma), \bar{u}(\sigma)\rangle d\sigma$$
$$= \left\langle y^*, \int_0^T S(T - \sigma)B\bar{u}(\sigma)d\sigma \right\rangle$$

which again implies

$$\bar{y} - S(T)\zeta = \int_0^T S(T - \sigma)Bu(\sigma)d\sigma$$

and shows that $\bar{u}(t)$ drives $\zeta$ to $\bar{y}$ time optimally.

The result below gives an extra bit of information about the spaces $R^\infty(T)$ based on the norm optimal part of Theorem 3.1.2. To stay in the generality of (3.1.1) we introduce the space $R^\infty(t; B)$ consisting of all

$$y = \int_0^t S(t - \sigma)Bu(\sigma)d\sigma, \quad u(\cdot) \in L^\infty(0, t; F)$$

equipped with the norm

$$\|y\|_{R^\infty(t;B)} = \inf \left\{ \|u(\cdot)\|_{L^\infty(0,t;F)}; \int_0^t S(t - \sigma)Bu(\sigma)d\sigma = y \right\}.$$

The fact that $R^\infty(t; B)$ is a Banach space is proved exactly as in Lemma 2.1.1.

**Corollary 3.1.3.** *Let $F = X^*$ satisfy the assumptions in Theorem* 3.1.2. *Then for every $y \in R^\infty(T; B)$ there exists $\bar{u}(\cdot) \in L^\infty(0, T; X^*)$ such that*

$$y = \int_0^T S(T - \sigma)B\bar{u}(\sigma)d\sigma, \quad \|y\|_{R^\infty(T;B)} = \|\bar{u}(\cdot)\|_{L^\infty(0,T;X^*)}.$$

It suffices to take a control $\bar{u}(\cdot) \in L^\infty(0, T; X^*)$ driving $\zeta = 0$ norm optimally to $\bar{y}$. We shall see in Chapters 4 and 5 examples (with $B = I$) where the conclusion of Corollary 3.1.3 does not hold.

For the treatment of uniqueness, we only consider the case $E = F$, $B = I$ in (3.1.1), so that the equation is the familiar

$$y'(t) = Ay(t) + Bu(t)\,, \quad y(0) = \zeta \tag{3.1.9}$$

with controls in $L^\infty(0, T; E)$. For this equation, we have already covered uniqueness of optimal controls in various places of Chapter 2; for easy reference, we list the results below. The first, for time optimal controls, is the simplest and most inclusive as it does not require the maximum principle, just the bang-bang property $\|\bar{u}(t)\| = 1$ a. e. which was shown in Theorem 2.1.3 with no extra assumptions on the space or the semigroup.

**Theorem 3.1.4.** *Let $E$ be strictly convex. Then time optimal controls are unique.*

This is Theorem 2.1.7. In contrast, uniqueness for the norm optimal problem needs special assumptions. The result below is Corollary 2.5.4.

**Theorem 3.1.5.** *Assume that $E$ is strictly convex, that $S(t)^*$ satisfies*

$$z \neq 0 \implies S(t)^*z \neq 0 \quad (t > 0) \quad (z \in Z_w) \tag{3.1.10}$$

*and*

$$\bar{y} - S(T)\zeta \in \overline{D(A)}\,. \tag{3.1.11}$$

*Then norm optimal controls are unique.*

**Remark 3.1.6.** Condition (3.1.10) is satisfied by groups or, more generally, by semigroups $S(t)$ such that $S(t)E = E$ for $t > 0$ (this implies each $S(t)^*$ is one-to-one). Condition (3.1.10) is also satisfied by analytic semigroups (Remark 2.5.8).

In some situations, we can ignore (3.1.10). In fact, assume that (3.1.11) holds (so that we have the maximum principle for the optimal control) and that, additionally, the target satisfies $\bar{y} \in S(T)E$. Then Theorem 2.6.3 takes over and says that $S(T - \sigma)^*z \neq 0$ in the whole control interval, which is enough for the conclusion of Corollary 2.5.4.

In general, if either (3.1.10) or (3.1.11) are given up, uniqueness for the norm optimal problem collapses. That the lack of (3.1.10) produces nonuniqueness has been shown (for a particular space and semigroup) in Theorem 2.6.1, even if (3.1.11) is reinforced to $\bar{y} - S(T)\zeta \in D(A^\infty) = \cap_{n\geq 1} D(A^n)$. (Corollary 2.6.8). The following result goes in the same direction, but it works for an arbitrary strongly continuous semigroup $S(t)$ in a Hilbert space $E$. Extensions to other Banach spaces are possible (see Remark 3.1.8).

**Theorem 3.1.7.** *Let $E$ be a Hilbert space and assume that* (3.1.10) *fails. Then, given $T > 0$ there exists a target $\bar{y}$ such that $0$ can be driven to $\bar{y}$ norm optimally in the interval $0 \leq t \leq T$ by different controls.*

*Proof.* Let $z \in Z$, $z \neq 0$ be such that

$$S(t)^* z \neq 0 \quad (0 \le t < t_0)\,, \quad S(t)^* z = 0 \quad (t \ge t_0)\,. \tag{3.1.12}$$

Replacing $z$ by $S(a)^* z$ for suitable $a$ we may always assume that $t_0 < T$ in (3.1.12). Define a control $\bar{u}(t)$ by

$$\bar{u}(t) = \frac{S(T-t)^* z}{\|S(T-t)^* z\|} \qquad (T - t_0 < t \le T)\,. \tag{3.1.13}$$

In the interval $0 \le t \le T - t_0$ the choice of $\bar{u}(t)$ is arbitrary as long as $\bar{u}(t)$ is strongly measurable and $\|\bar{u}(\cdot)\|_{L^\infty(0,T-t_0;E)} \le 1$. In Hilbert space, (3.1.13) is equivalent to

$$\langle S(T-t)^* z, \bar{u}(t)\rangle = \max_{\|u\| \le 1} \langle S(T-t)^* z, u\rangle \quad \text{a. e. in } 0 \le t \le T \tag{3.1.14}$$

and it has been proved in Theorem 2.5.5 that $\bar{u}(t)$ drives any $\zeta$ norm optimally to $y(T, \zeta, \bar{u})$ in $0 \le t \le T$. We set

$$\bar{u}(t) = u \neq 0\,, \quad \|u\| < 1 \quad (0 \le t \le T - t_0) \tag{3.1.15}$$

and take $\zeta = 0$, $\bar{y} = y(T, 0, \bar{u})$. By the choice (3.1.15) of $u$ and Theorem 2.1.3 the control $\bar{u}(t)$ does not drive $\zeta = 0$ *time* optimally to $y(T - t_0, 0, \bar{u})$ in the interval $0 \le t \le T - t_0$, so there exists a control $v(t)$, $\|v(t)\| \le 1$ a. e. that improves the driving time to $T - t_0 - \delta < T - t_0$ :

$$y(T - t_0 - \delta, 0, v) = y(T - t_0, 0, \bar{u})\,.$$

Consider the control

$$\bar{v}(t) = \begin{cases} 0 & 0 \le t < \delta \\ v(t - \delta) & \delta \le t < T - t_0 \\ \bar{u}(t) & T - t_0 \le t \le T\,. \end{cases}$$

This control does nothing in the first interval, drives to $y(T - t_0, 0, \bar{u})$ at time $T - t_0$ and then drives to $\bar{y}$, where it arrives at time $T$. We have $\bar{u}(t) = u \neq 0$, $v(t) = 0$ in $0 \le t \le \delta$, thus $\bar{u}(t)$ and $v(t)$ are different. This ends the proof.[3]

**Remark 3.1.8.** The only reason we need $E$ to be Hilbert is in the construction of $\bar{u}(t)$ from (3.1.14), considered as an equation with unknown $\bar{u}(t)$; the "solution formula" is (3.1.13). Similar formulas exist in more general spaces. Assume $E$ is

---

[3] Theorem 3.1.7 includes Theorem 2.6.1 but is not more general that Corollary 2.6.8.; in fact, although the latter deals only with a particular space and semigroup, it allows the choice of a target $\bar{y} \in D(A^\infty)$. In Theorem 3.1.7 we don't even know if $\bar{y} \in D(A)$.

a reflexive Banach space. Then, for every $y^* \in E^*$, $y^* \neq 0$ there exists $\Phi(y^*) \in E$ such that

$$\|\Phi(y^*)\|_E = 1\,, \quad \langle y^*, \Phi(y^*)\rangle = \|y^*\|_{E^*}\,. \tag{3.1.16}$$

Assume $E$ is *strictly convex,* that is, if $\|x\| = \|y\| = 1$, $x \neq y \Rightarrow \|x + y\| < 2$. Then the function

$$\Phi : (E^* \setminus \{0\}) \to \{y \in E; \|y\| = 1\} \tag{3.1.17}$$

is single valued: in fact, if two different $\Phi(y^*)_1$, $\Phi(y^*)_2$ achieve (3.1.16) then

$$\Big\|\frac{1}{2}(\Phi(y^*)_1 + \Phi(y^*)_2)\Big\|_E < 1\,, \quad \Big\langle y^*, \frac{1}{2}(\Phi(y^*)_1 + \Phi(y^*)_2)\Big\rangle = \|y^*\|_{E^*}\,,$$

which is impossible. The solution formula for (3.1.14) in a reflexive, strictly convex space is then

$$\bar{u}(t) = \Phi(S(T-t)^*z) \qquad (T - t_0 < t \leq T)\,. \tag{3.1.18}$$

If we assume in addition that the map (3.1.17) is continuous for $y^* \neq 0$, then (3.1.18) gives $\bar{u}(t)$ continuous in $T - t_0 < t < T$ since the adjoint semigroup $S(t)^*$ is continuous due to reflexiveness. We can accordingly extend Theorem 3.1.7 to this situation: the proof is exactly the same. Among the spaces that qualify are $E = L^p(\Omega)$ ($\Omega$ a $m$-dimensional domain, $1 < p < \infty$); here $E^* = L^q(\Omega)$, $1/p + 1/q = 1$ and, if $z(\cdot) \in L^q(\Omega)$ we have

$$\Phi(z)(x) = \frac{|z(x)|^{q-2}z(x)}{\|z(\cdot)\|^{q-1}_{L^q(\Omega)}} \tag{3.1.19}$$

($|z(x)|^{q-2}z(x) = 0$ *if* $z(x) = 0$). In fact,

$$\begin{aligned}
\|\Phi(z)(\cdot)\|^p_{L^p(\Omega)} &= \frac{1}{\|z(\cdot)\|^{(q-1)p}_{L^q(\Omega)}} \int_\Omega |z(x)|^{(q-2)p+p}dx \\
&= \frac{1}{\|z(\cdot)\|^{q}_{L^q(\Omega)}} \int_\Omega |z(x)|^{q}dx = 1\,, \\
\langle z, \Phi(z)\rangle &= \frac{1}{\|z(\cdot)\|^{q-1}_{L^q(\Omega)}} \int_\Omega |z(x)|^{q-2}z(x)z(x)dx \\
&= \frac{\|z(\cdot)\|^{q}_{L^q(\Omega)}}{\|z(\cdot)\|^{q-1}_{L^q(\Omega)}} = \|z(\cdot)\|_{L^q(\Omega)}\,.
\end{aligned}$$

A Banach space $E$ is *uniformly convex* if, for every $\epsilon > 0$ there exists $\delta > 0$ with

$$\|x\| = \|y\| = 1\,, \ \|x + y\| \geq 2 - \delta \implies \|x - y\| \leq \epsilon \tag{3.1.20}$$

Obviously, uniform convexity implies strict convexity. That the spaces $L^p(\Omega)$, $1 < p < \infty$ are uniformly convex is proved in Adams [1975, pp. 34-38]. The continuity requirement on the function $\Phi(y^*)$ is taken care of by

**Lemma 3.1.9.** *Let $E$ be a uniformly convex Banach space. Then the function $\Phi$ defined by* (3.1.16) *is uniformly continuous in $\|y^*\| \geq r$ for every $r > 0$.*

*Proof.* Let $\epsilon > 0$ be arbitrary and let $\delta > 0$ be given by (3.1.20) from $\epsilon$. Pick $\|y^*\|, \|y'^*\| \geq r$, $\|y'^* - y^*\| \leq \rho = r\delta/2$. We have

$$\begin{aligned}\Big\langle y^*, \Phi(y^*) + \Phi(y'^*)\Big\rangle &= \Big\langle y^*, \Phi(y^*)\Big\rangle + \Big\langle y'^*, \Phi(y'^*)\Big\rangle + \Big\langle y^* - y'^*, \Phi(y'^*)\Big\rangle \\ &\geq \|y^*\| + \|y'^*\| - \|y^* - y'^*\| \geq \|y^*\| + (\|y^*\| - \rho) - \rho \\ &= 2\|y^*\| - 2\rho = \Big(2 - \frac{2\rho}{\|y^*\|}\Big)\|y^*\| \geq \Big(2 - \frac{2\rho}{r}\Big)\|y^*\| \geq (2 - \delta)\|y^*\|\end{aligned}$$

so that

$$\|\Phi(y^*) + \Phi(y'^*)\| \geq 2 - \delta \implies \|\Phi(y^*) - \Phi(y'^*)\| \leq \epsilon\,.$$

The following result does not require any special assumption on the space $E$.

**Theorem 3.1.10.** *Let $S(t)$ be a strongly continuous semigroup in a Banach space $E$. Assume that there exists $\mu > \omega$ such that $S(t)^*$ does not satisfy $\mathcal{P}(a, b, \mu)$ for any $a > 0, b > 0$. Then, given $T > 0$ there exists a target $\bar{y}$ such that $0$ can be driven to $\bar{y}$ norm optimally in the interval $0 \leq t \leq T$ by different controls.*

*Proof.* We modify the norm optimal control $\bar{u}(\cdot)$ given by (2.8.28) as follows. Given $0 < t_0 < T$ we define

$$\bar{u}_0(t) = \begin{cases} u & 0 \leq t < t_0 \\ \bar{u}(t) & t_0 \leq t \leq T \end{cases}$$

where $u \neq 0, \|u\| < 1$. Irrespective of the choice of $t_0$, $u$, $\bar{u}_0(\cdot)$ is a norm optimal control. The proof is the same than the proof for $\bar{u}(t)$ in Theorem 2.8.6; the estimates there, beginning with (2.8.24) are asymptotic, that is, they are only used for $n$ large enough. Again we take $\zeta = 0$, $\bar{y} = y(T, 0, \bar{u}_0)$ and the proof ends just like that of Theorem 3.1.7.

**Remark 3.1.11.** Condition (3.1.11) must fail for the target $\bar{y} = y(T, 0, \bar{u}_0)$ in Theorem 3.1.10; otherwise, the control $\bar{u}_0(t)$ would satisfy the maximum principle (3.1.14) (Theorem 2.5.1). This has been ruled out for $\bar{u}(\cdot)$ (even with the multiplier $z$ in a bigger space) in Theorem 2.8.6, and the proof works as well for $\bar{u}_0(t)$ since it only uses $\bar{u}(t)$ for $t$ near $T$.

Theorems 3.1.7 and 3.1.10 produce, (under different hypotheses) very different examples of nonuniqueness; the controls in Theorem 3.1.7 qualify as regular, while those in Theorem 3.1.10 are (strongly) singular. Note also that Theorem 3.1.10, unlike Theorem 3.1.7, requires no special assumptions on the Banach space $E$.

**Miscellaneous notes.** The first infinite dimensional existence result for the equation (3.1.1) (with $B = I$) seems to be Theorem 1.2 in the author [1964]. The theorems here follow its pattern.

The assumptions in Lemma 3.1.2 do not cover all situations of interest. They are too demanding, for instance, in the modeling of

$$\frac{\partial y(t,x)}{\partial t} = \frac{\partial^2 y(t,x)}{\partial x^2} + u(t,x)\,, \quad y(t,a) = y(t,b) = 0\,, \quad u(0,x) = \zeta(x) \quad (3.1.21)$$

in $C_0[a,b]$. As a first try, we take controls in $L^\infty(0,T;C_0[a,b])$ that is, $F = C_0[a,b]$, $B = I$. This approach doesn't succeed, since $C_0[a,b]$ is not the dual of any Banach space; neither $(i)$ or $(ii)$ hold, and there are cases (see Chapters 4 and 5) where minimizing sequences exist, but there are no optimal controls.

As a second try we take controls in $L^\infty(0,T;L^\infty(a,b))$; here, $F = L^\infty(a,b)$, which is the dual of $L^1(a,b)$. This choice of $F$ causes difficulties in defining the solutions, since controls take values in a space bigger than the state space $C_0[a,b]$. These difficulties can be overcome, but $L^\infty(a,b)$ misses $(i)$ (it's not reflexive) and $(ii)$ (it's not separable), thus we fall again into nonexistence problems.

The third try is the most natural for the equation (3.1.21) and consists in using controls in $L^\infty((0,T)\times(a,b))$. To fit this into the model (3.1.1) we must write this control space as a space of vector-valued functions of $t$. For this, we use the equality

$$L^\infty((0,T)\times(a,b)) = L^\infty_w(0,T;L^\infty(a,b))$$

where $L^\infty_w(0,T;X^*)$ is defined as $L^\infty(0,T;X^*)$ but with the only requirement of $X$-weak measurability if its elements. Existence can be extended to this situation as shown in Chapters 4 and 5.

**3.2. The weak maximum principle and the time optimal problem.** Given a Banach space $E$ and a strongly continuous semigroup $S(t)$, a *multiplier space* $\mathcal{K}$ (as defined in **2.8**) is any space such that $(a)$ there exists an imbedding $E^* \hookrightarrow \mathcal{K}$, $(b)$ $S(t)^*$ is defined in $\mathcal{K}$ and $S(t)^*\mathcal{K} \subseteq E^*$ $(t>0)$. The elements of $\mathcal{K}$ are called *multipliers.* As in **2.8** we say that $\bar u(\cdot) \in L^\infty(0,T;E)$ satisfies the *weak maximum principle* if there exists a multiplier $\psi$ such that[4]

$$\langle S(T-t)^*\psi, \bar u(t)\rangle = \max_{\|u\|\le\rho} \langle S(T-t)^*\psi, u\rangle \quad \text{a. e. in } 0 \le t \le T\,, \quad (3.2.1)$$

$$S(t)^*\psi \quad \text{is not identically zero} \quad \text{a. e. in } 0 \le t \le T\,. \quad (3.2.2)$$

If we have

$$\int_0^T \|S(t)^*\psi\|_{E^*} dt < \infty \quad (3.2.3)$$

then (3.2.1) is the *strong maximum principle* or simply the *maximum principle.* As seen in Lemma 2.8.1, (3.2.3) is equivalent to the existence of $z \in Z_w(T)$ with

[4] (3.2.2) means $S(t)^*\psi \ne 0$ in a set of positive measure. The semigroup equation $S(t)^*\psi = S(t-h)^*S(h)^*\psi$ implies that the set where $S(t)^*\psi \ne 0$ is an interval $(0,a)$, and (3.2.2) says that $a > 0$.

$S(t)^*\psi = S(t)^*z$ so the maximum principle is (3.2.1)-(3.2.2)-(3.2.3) or, equivalently, (3.2.1) plus the condition that $\psi = z \in Z_w(T)$, $z \neq 0$. Calling an *extremal* a norm or time optimal control, the definitions in **2.8** say: an extremal is strongly singular if it does *not* satisfy the weak maximum principle for any multiplier $\psi$; an extremal is weakly singular if it satisfies the weak maximum principle but not the maximum principle. A *regular* extremal is one that satisfies the maximum principle. We visualize all of this in "thought balloon" form:

$$\left\{\begin{matrix}\text{strongly singular}\\ \text{extremals}\end{matrix}\right\} \Longleftrightarrow \left\{\begin{matrix}\text{do not satisfy the}\\ \text{weak maximum principle}\end{matrix}\right\}$$

$$\left\{\begin{matrix}\text{weakly singular}\\ \text{extremals}\end{matrix}\right\} \Longleftrightarrow \left\{\begin{matrix}\text{satisfy the weak maximum principle}\\ \text{but not the strong maximum principle}\end{matrix}\right\}$$

$$\left\{\begin{matrix}\text{regular}\\ \text{extremals}\end{matrix}\right\} \Longleftrightarrow \left\{\begin{matrix}\text{satisfy the strong}\\ \text{maximum principle}\end{matrix}\right\}$$

So far, we have a single example of strongly singular time optimal control (the one constructed in Theorem 2.7.4) and numerous examples of strongly singular norm optimal controls (the same in Theorem 2.7.4 and all the controls in Theorems 2.8.4 and 2.8.6). Weakly singular norm and time optimal controls are yet to be constructed later in this chapter. We show in this section that strongly singular time optimal controls don't exist at all if $E$ is a Hilbert space and $S(t)$ is an analytic semigroup (with generalizations to certain Banach spaces). The situation is totally different with norm optimal controls; the construction in Theorem 2.8.6 works, for instance, with an arbitrary analytic semigroup in a Banach space (see the comments after the proof of Theorem 2.9.2).

The results in this section are based on regularity properties of the map

$$(\mathcal{S}u)(t) = \int_0^t S(t-\sigma)u(\sigma)d\sigma\,. \tag{3.2.4}$$

**Theorem 3.2.1.** *Let $E$ be a Hilbert space, $1 < p < \infty$, $u(\cdot) \in L^p(0, T; E)$. Then*

$$(\mathcal{S}u)(t) \in D(A) \quad \textit{a. e. in } 0 \leq t \leq T \tag{3.2.5}$$

*and there exists a constant $C_p$ such that*

$$\|A(\mathcal{S}u)(\cdot)\|_{L^p(0,T;E)} \leq C_p\|u(\cdot)\|_{L^p(0,T;E)}\,. \tag{3.2.6}$$

The proof is in De Simon [1964]. The full result uses the theory of singular integrals of Hilbert space valued functions, but the case $p = 2$ is reasonably elementary; since we only need one value of $p$, we sketch the proof below. It is based on the theory of the Fourier-Plancherel transform $\mathcal{F}$ in Hilbert spaces. The transform applies to functions $u(\cdot) \in L^2(-\infty, \infty; E)$ :

$$(\mathcal{F}u)(\tau) = \lim_{a\to\infty} \frac{1}{\sqrt{2\pi}} \int_{-a}^{a} e^{i\tau t} u(t)dt \,,$$

the limit taken in the norm of $L^2(-\infty, \infty; E)$. It is shown exactly as in the scalar case that the limit exists and that $\mathcal{F} : L^2(-\infty, \infty; E) \to L^2(-\infty, \infty; E)$ is an isometry. Its inverse is given by

$$(\mathcal{F}^{-1}u)(t) = \lim_{a\to\infty} \frac{1}{\sqrt{2\pi}} \int_{-a}^{a} e^{-i\tau t} u(\tau)d\tau \,,$$

the limit in the norm of $L^2(-\infty, \infty; E)$ as well. When the integrands are in $L^1(-\infty, \infty; E)$, the limits exist pointwise in $\tau$ in the norm of $E$, that is, the integrals are regular integrals in $(-\infty, \infty)$.

A preliminary simplification. We have

$$(\mathcal{S}u)(t) = e^{\mu t} \int_0^t e^{-\mu(t-\sigma)} S(t-\sigma)(e^{-\mu\sigma} u(\sigma))d\sigma$$

for $\mu > \omega =$ the constant in (2.3.1). Multiplication by an exponential is a bounded invertible operator in $L^p(0, T; E)$, hence to establish the claims on the integral operator (3.2.4) it is enough to prove the same claims with the semigroup $e^{-\mu t}S(t)$ ($\mu$ arbitrary) instead of $S(t)$. Equivalently, we may assume arbitrarily fast negative exponential decay of $\|S(t)\|$ at infinity. Finally, the semigroup being analytic, we may also take for granted that the resolvent $R(\lambda; A) = (\lambda I - A)^{-1}$ exists in a sector $\{\lambda; |\arg(\lambda + c)| \le \phi\}$ with $c > 0$, $\phi > \pi/2$, and that

$$\|R(\lambda; A)\| \le \frac{C}{|\lambda + c|} \quad (|\arg(\lambda + c)| \le \phi)\,, \tag{3.2.7}$$

$$\|S(t)\| \le Ce^{-ct} \quad (t \ge 0)\,. \tag{3.2.8}$$

Let $u(\cdot) \in L^2(0, T; E)$. Extend $u(t)$ to $t < 0$ and $t > T$ by $u(t) = 0$ and extend $S(t)$ to $t < 0$ in a similar fashion, so that the integral in (3.2.4) can be written

$$(\mathcal{S}u)(t) = \int_{-\infty}^{\infty} S(t-\sigma)u(\sigma)d\sigma \,.$$

Let $\{u_n(\cdot)\}$ be a sequence of infinitely differentiable $E$-valued functions with compact support and such that

$$u_n(\cdot) \to u(\cdot) \quad \text{in } L^2(-\infty, \infty; E)\,. \tag{3.2.9}$$

We have

$$\begin{aligned}\mathcal{F}(\mathcal{S}u_n)(\tau) &= \frac{1}{\sqrt{2\pi}}\int_{-\infty}^{\infty} e^{i\tau t}(\mathcal{S}u_n)(t)dt \\ &= \frac{1}{\sqrt{2\pi}}\int_{-\infty}^{\infty}\left(\int_{-\infty}^{\infty} e^{i\tau(t-\sigma)}S(t-\sigma)dt\right)e^{i\tau\sigma}u_n(\sigma)d\sigma \\ &= R(-i\tau; A)(\mathcal{F}u_n)(\tau)\,. \end{aligned} \tag{3.2.10}$$

On the other hand, due to (3.2.7),

$$\|AR(-i\tau;A)\| = \|-i\tau R(i\tau;A) - I\| \le C' \quad (\infty < \tau < \infty)\,.$$

Applying $A$ to both sides of (3.2.10) and then using the equality for $u_n(\cdot) - u_m(\cdot)$ we see that the sequence $\{A\mathcal{F}(\mathcal{S}u_n)(\cdot)\}$ is Cauchy in the space $L^2(-\infty,\infty;E)$, hence

$$A\mathcal{F}(\mathcal{S}u_n)(\cdot) \to v(\cdot) \quad \text{in } L^2(-\infty,\infty;E)\,. \tag{3.2.11}$$

It follows from (3.2.8), (3.2.9) and Young's theorem on convolutions (Adams [1975, p. 90, Theorem 4.30]) that $(\mathcal{S}u_n)(\cdot) \to (\mathcal{S}u)(\cdot)$ in $L^2(-\infty,\infty;E)$, thus

$$\mathcal{F}(\mathcal{S}u_n)(\cdot) \to \mathcal{F}(\mathcal{S}u)(\cdot) \quad \text{in } L^2(-\infty,\infty;E)\,. \tag{3.2.12}$$

Going to subsequences, we may assume that convergence in (3.2.11) and (3.2.12) is pointwise for $t$ in a set of full measure in $(-\infty,\infty)$. Using closedness of $A$ we get (3.2.5) and (3.2.6) for $p=2$ at the same time. This ends the proof.

**Lemma 3.2.2.** *Let $S(t)$ be an analytic semigroup in a Hilbert space $E$, and let $\bar{u}(t)$ be a time optimal control in the interval $0 \le t \le T$. Then there exists a sequence $\{t_n\}$ such that*

$$0 < t_1 < t_2 < \ldots < t_n < T, \quad t_n \to T\,, \tag{3.2.13}$$

*and a sequence $\{y_n\} \subset E$ (each $y_n \neq 0$) such that*

$$(S(t_n - t)^* y_n, \bar{u}(t)) = \max_{\|u\|\le 1}(S(t_n - t)^* y_n, u) \quad \text{a. e. in } 0 \le t \le t_n\,. \tag{3.2.14}$$

*Proof.* Theorem 3.2.1 (we only need (3.2.5)) says it is possible to select a sequence $\{t_n\}$ satisfying (3.2.13) such that $y(t_n, 0, \bar{u}) = (\mathcal{S}\bar{u})(t_n) = \bar{y}_n \in D(A)$, hence

$$\bar{y}_n - S(t_n)\zeta \in D(A)\,. \tag{3.2.15}$$

By virtue of the optimality principle (see **1.3**), the control $\bar{u}(t)$ is time optimal in each interval $0 \le t \le t_{n+1}$, and (3.2.15) is just what is needed to apply Theorem 2.5.3 ($b$). For each $n$ we obtain[5] $z_n \in Z(T)$, $z_n \neq 0$ such that

[5] The space $E$ is Hilbert, thus reflexive; the adjoint semigroup is strongly continuous and $Z_w(T) = Z(T)$. See **2.4**, especially the comments after (2.4.19).

$$(S(t_{n+1} - t)^* z_n, \bar{u}(t)) = \max_{\|u\| \le 1} (S(t_{n+1} - t)^* z_n, u) \quad \text{a. e. in } 0 \le t \le t_{n+1}\,,$$

which implies in particular

$$\begin{aligned} &(S(t_n - t)^* S(t_{n+1} - t_n)^* z_n, \bar{u}(t)) \\ &= \max_{\|u\| \le 1} (S(t_{n+1} - t)^* S(t_{n+1} - t_n)^* z_n, u) \quad \text{a. e. in } 0 \le t \le t_n\,, \end{aligned}$$

and the result follows taking $y_n = S(t_{n+1} - t_n)^* z_n$. That $y_n \ne 0$ results from the fact that $S(t)$ is an analytic semigroup, thus

$$z \in Z(T),\ z \ne 0 \implies S(t)^* z \ne 0 \quad (t > 0)\,. \tag{3.2.16}$$

**Theorem 3.2.3.** *Assume that $S(t)$ is an analytic semigroup in a Hilbert space $E$. Then there exists a multiplier space $\mathcal{Z}$ such that, if $\bar{u}(t)$ is time optimal in the interval $0 \le t \le T$ then $\bar{u}(t)$ satisfies the weak maximum principle* (3.2.1) *with $\psi \in \mathcal{Z}$ and*

$$S(t)^* \psi \ne 0 \quad (t > 0)\,. \tag{3.2.17}$$

*Proof.* According to Lemma 3.2.2 and recalling we are in a Hilbert space and the semigroup is analytic (so that denominators can't vanish) we have

$$\bar{u}(t) = \frac{S(t_n - t)^* y_n}{\|S(t_n - t)^* y_n\|} \quad (0 \le t \le t_n)$$

and

$$\bar{u}(t) = \frac{S(t_{n+1} - t)^* y_{n+1}}{\|S(t_{n+1} - t)^* y_{n+1}\|} \quad (0 \le t \le t_{n+1})\,,$$

so that

$$\frac{S(t_n - t)^* y_n}{\|S(t_n - t)^* y_n\|} = \frac{S(t_{n+1} - t)^* y_{n+1}}{\|S(t_{n+1} - t)^* y_{n+1}\|} \quad (0 \le t \le t_n)\,.$$

Taking $t = t_n$ we deduce that

$$\frac{S(t_{n+1} - t_n)^* y_{n+1}}{\|S(t_{n+1} - t_n)^* y_{n+1}\|} = \frac{y_n}{\|y_n\|}\,,$$

hence, multiplying successively $y_2, y_3 \ldots$ by suitable constants, we may "calibrate" the sequence $\{y_n\}$ in such a way that

$$S(t_{n+1} - t_n)^* y_{n+1} = y_n\,. \tag{3.2.18}$$

The space $\mathcal{Z}$ consists of all sequences

$$\psi = \{(t_1, y_1), (t_2, y_2), \ldots\}\,, \quad (t_n \in I\!R, y_n \in E)$$

where $\{t_n\}$ satisfies (3.2.13) and $\{y_n\}$ satisfies (3.2.18). In accordance with the definition of multiplier space, we don't define operations in, or give a norm to $\mathcal{Z}$ (see **2.8**). We extend $S(t)^*$ to an operator $S(t)^* : \mathcal{Z} \to E = E^*$ as follows:

$$S(t)^*\psi = S(t - (T - t_n))^* y_n \quad (t \geq T - t_n)\,. \tag{3.2.19}$$

If $t \geq T - t_{n+1}$ we have

$$\begin{aligned} S(t)^*\psi &= S(t - (T - t_{n+1}))^* y_{n+1} \\ &= S(t - (T - t_n))^* S(t_{n+1} - t_n)^* y_{n+1} = S(t - (T - t_n))^* y_n\,, \end{aligned}$$

thus the definition is consistent for all $t > 0$. It follows from (3.2.16) (applied only to $z \in E$) that (3.2.17) holds. Moreover,

$$S(T - t)^*\psi = S(T - t - (T - t_n))^* y_n = S(t_n - t)^* y_n \quad (t \leq t_n)\,. \tag{3.2.20}$$

Finally, we must exhibit an imbedding $E = E^* \hookrightarrow \mathcal{Z}$ (the imbedding just a 1-1 map) such that the definition of $S(t)^* y$ in $\mathcal{Z}$ matches the definition in $E$ when $y \in E$. For this, we fix a sequence $\{t_n\}$ satisfying (3.2.13), and define a sequence $\psi = \{(t_n, y_n)\} \in \mathcal{Z}$ setting

$$y_n = S(T - t_n)^* y \quad (n = 1, 2, \ldots)\,.$$

This map is 1-1; if $y \neq z$ then $S(T - t_n)^* y \neq S(T - t_n)^* z$ for all $n$. We have

$$S(t_{n+1} - t_n)^* y_{n+1} = S(t_{n+1} - t_n)^* S(T - t_{n+1})^* y = S(T - t_n)^* y = y_n$$

so that (3.2.18) holds. Applying the definition (3.2.19),

$$\begin{aligned} S(t)^*\psi &= S(t - (T - t_n))^* y_n \\ &= S(t - (T - t_n))^* S(T - t_n)^* y = S(t) y \quad (t \geq t_n) \end{aligned}$$

and we obtain the matching of the definitions. That the weak maximum principle (3.2.1) holds is a consequence of (3.2.14) and (3.2.20). This ends the proof.

Theorem 3.2.3 is susceptible of generalization to Banach spaces. All we need is a space and semigroup where the regularity property (3.2.5) holds. A Banach space $E$ is $\zeta$-*convex* (see Bourgain [1983] Burkholder [1983]) if there exists a function $\zeta : E \times E \to \mathbb{R}$ such that $(a)$ to $(d)$ below hold:

$(a)$ $\zeta(x, y)$ is convex in $y$ for $x$ fixed and convex in $x$ for $y$ fixed:

$$\begin{aligned} \zeta(\alpha x_1 + (1 - \alpha) x_2, y) &\leq \alpha \zeta(x_1, y) + (1 - \alpha)\zeta(x_2, y) \\ \zeta(x, \alpha y_1 + (1 - \alpha) y_2) &\leq \alpha \zeta(x, y_1) + (1 - \alpha)\zeta(x, y_2) \end{aligned} \qquad (0 \leq \alpha \leq 1)$$

$(b)$ $\zeta(x, y)$ is symmetric: $\zeta(x, y) = \zeta(y, x)$

(*c*) $\zeta(0,0) > 0$

(*d*) $\zeta(x,y) \le \|x+y\|$ if $\|x\| \le 1 \le \|y\|$ .

A Hilbert space $E$ is $\zeta$-convex; we may set $\zeta(x,y) = 1 + (x,y)$. Conditions (*a*), (*b*) and (*c*) are obvious; as for (*d*),

$$\begin{aligned}\zeta(x,y)^2 &= (1+(x,y))^2 \le 1 + 2(x,y) + \|x\|^2\|y\|^2 \\ &= (x+y,x+y) + \|x\|^2\|y\|^2 - \|x\|^2 - \|y\|^2 + 1 \\ &= \|x+y\|^2 + (1-\|x\|^2)(1-\|y\|^2)\,.\end{aligned}$$

Spaces $L^p(\Omega)$ are also $\zeta$-convex for $1 < p < \infty$.

Let $S(t)$ be a strongly continuous semigroup of negative exponential growth at infinity,

$$\|S(t)\| \le Ce^{-ct} \quad (t \ge 0)\,.$$

We define fractional powers $(-A)^{-z}$ by the formula

$$(-A)^{-z} = \frac{1}{\Gamma(z)}\int_0^\infty t^{z-1}S(t)dt \qquad (\mathrm{Re}\, z > 0)\,.$$

Clearly, $(-A)^{-z}$ is an $\mathcal{L}(E,E)$-valued analytic function in the open half-plane $\mathrm{Re}\, z > 0$, and the definition is consistent with integer powers in the sense that

$$(-A)^{-n} = \frac{1}{\Gamma(n)}\int_0^\infty t^{n-1}S(t)dt = \frac{1}{(n-1)!}\int_0^\infty t^{n-1}S(t)dt\,.$$

**Theorem 3.2.4.** *Assume that $E$ is a $\zeta$-convex space and that $S(t)$ is an analytic semigroup such that $(-A)^{-z}$ can be extended to the closed half plane $\mathrm{Re}\, z \ge 0$ and*

$$\|(-A)^{i\tau}\| \le Ce^{\beta|\tau|} \quad (-\infty < \tau < \infty) \tag{3.2.21}$$

*for some $\beta < \pi/2$. Then, if $1 < p < \infty$,*

$$(\mathcal{S}u)(t) \in D(A) \quad a.\ e.\ in\ 0 \le t \le T \tag{3.2.22}$$

*for any $u(\cdot) \in L^p(0,T;E)$ and there exists a constant $C_p$ such that*[6]

$$\|A(\mathcal{S}u)(\cdot)\|_{L^p(0,T;E} \le C_p\|u(\cdot)\|_{L^p(0,T;E)}\,. \tag{3.2.23}$$

For the proof see Dore - Venni [1987, Theorem 3.2, p. 196]. On the basis of (3.2.22) Lemma 3.2.2 generalizes as follows:

[6] Strictly speaking, Theorem 3.2.4 is not a a generalization of Theorem 3.2.1. since the latter does not require the estimate (3.2.21) on imaginary powers.

**Lemma 3.2.5.** *Assume the space $E$ and the semigroup $S(t)$ satisfy the assumptions of Theorem* 3.2.4 *and let $\bar{u}(t)$ be time optimal in the interval* $0 \le t \le T$. *Then there exists a sequence $\{t_n\}$, $0 < t_1 < t_2 < \ldots < t_n < T$, $t_n \to T$ and a sequence $\{y_n^*\} \subset E^*$ (each $y_n^* \neq 0$) such that*

$$\langle S(t_n - t)^* y_n^*, \bar{u}(t)\rangle = \max_{\|u\|\le 1} \langle S(t_n - t)^* y_n^*, u\rangle \quad \text{a. e. in } 0 \le t \le t_n\,. \tag{3.2.24}$$

The proof needs no changes.

A $\zeta$-convex space is reflexive[7] (see Aldous [1979]) so that for every $y^* \in E^*$, $y^* \neq 0$ there exists $\Phi(y^*) \in E$ such that $\|\Phi(y^*)\|_E = 1$, $\langle y^*, \Phi(y^*)\rangle = \|y^*\|_{E^*}$. We have proved in Remark 3.1.8 that if the space $E$ is strictly convex then the map $\Phi : (E^* \setminus \{0\}) \to \{y \in E; \|y\| = 1\}$ is single valued. This map obviously satisfies

$$\Phi(\lambda y^*) = \Phi(y^*) \qquad (\lambda > 0)\,. \tag{3.2.25}$$

If the dual $E^*$ is strictly convex as well, the converse

$$\Phi(y_1^*) = \Phi(y_2^*) \Longrightarrow y_2^* = \lambda y_1^* \quad (\lambda > 0) \tag{3.2.26}$$

is also true. In view of (3.2.25), to prove this implication we may assume that $\|y_1^*\|_{E^*} = \|y_2^*\|_{E^*} = 1$. Let $\Phi(y_1^*) = \Phi(y_2^*) = y$. If $y_1^* \neq y_2^*$ then

$$\Big\|\frac{1}{2}(y_1^* + y_2^*)\Big\|_{E^*} < 1, \quad \Big\langle \frac{1}{2}(y_1^* + y_2^*), y\Big\rangle = 1 = \|y\|_E\,,$$

a contradiction.

**Theorem 3.2.6.** *Assume that the space $E$ and the semigroup $S(t)$ satisfy the assumptions of Theorem* 3.2.4 *and that $E$ and $E^*$ are strictly convex. Then there exists a multiplier space $\mathcal{Z}$ such that, if $\bar{u}(t)$ is time optimal in $0 \le t \le T$ then $\bar{u}(t)$ satisfies the weak maximum principle* (3.2.1) *with $\psi \in \mathcal{Z}$ and*

$$S(t)^*\psi \neq 0 \quad (t > 0)\,. \tag{3.2.27}$$

*Proof.* The sequence of maximum principles (3.2.24) imply

$$\bar{u}(t) = \Phi(S(t_n - t)^* y_n^*) \quad (0 \le t \le t_n)$$

and

$$\bar{u}(t) = \Phi(S(t_{n+1} - t)^* y_{n+1}^*) \quad (0 \le t \le t_{n+1})\,,$$

so that

$$\Phi(S(t_n - t)^* y_n^*) = \Phi(S(t_{n+1} - t)^* y_{n+1}^*) \quad (0 \le t \le t_n)\,.$$

---

[7] In fact, *superreflexive* according to James' definition [1972].

Taking $t = t_n$ we deduce that

$$\Phi(S(t_{n+1} - t_n)^* y_{n+1}^*) = \Phi(y_n^*),$$

thus, using (3.2.26) and multiplying successively $y_2^*, y_3^* \ldots$ by suitable constants we may achieve

$$S(t_{n+1} - t_n)^* y_{n+1}^* = y_n^* . \tag{3.2.28}$$

The space $\mathcal{Z}$ consists of all sequences $\psi = \{(t_1, y_1^*), (t_2, y_2^*), \ldots\}$ where $\{t_n\}$ satisfies (3.2.13) and $\{y_n^*\} \subset E^*$ satisfies (3.2.28). From there on, the proof of Theorem 3.2.6 is exactly the same as that of Theorem 3.2.3 and we omit the details.

**Miscellaneous notes.** Theorem 3.2.3 is in the author [2000]. The rather obvious generalization to $\zeta$-convex spaces (Theorem 3.2.6) seems to be new.

The results proved in this section combined with those in **2.2**, **2.8** and **2.9** can be jointly pictured as

$$\left\{\begin{array}{c}\text{Analytic semigroups in Hilbert space}\\ \text{and analytic semigroups satisfying}\\ \text{(3.2.21) in } \zeta\text{-convex Banach spaces}\end{array}\right\} \Longrightarrow \left\{\begin{array}{c}\text{All time optimal controls}\\ \text{(but not all norm optimal controls)}\\ \text{satisfy the weak maximum principle}\end{array}\right\}$$

$$\left\{\begin{array}{c}\text{Strongly continuous semigroups in}\\ \text{Banach spaces satisfying } S(t)E = E\end{array}\right\} \Longrightarrow \left\{\begin{array}{c}\text{All time and norm optimal controls}\\ \text{satisfy the strong maximum principle}\\ \text{with multiplier space } \mathcal{K} = E^*\end{array}\right\}$$

In particular, these results say that there is a very narrow window where to look for strongly singular time optimal controls; we must avoid analytic semigroups in a large class of Banach spaces, as well as (on the other end of the spectrum) reversible semigroups. This explains *a posteriori* the choice of the right translation semigroup

$$S(t)y(x) = \begin{cases} y(x-t) & (x \geq t), \\ 0 & (x < t), \end{cases} \tag{3.2.29}$$

as the carrier of the strongly singular time optimal controls constructed in 2.7.4. Apparently, no such examples are known for any other semigroup.

The properties of the map $\mathcal{S}$ in (3.2.4) and of its Banach space counterpart (called *maximal regularity* properties) were not fully used in the results proved in this section. What we actually needed was the following *minimal regularity property:* for every $u(\cdot) \in L^\infty(0, T; E)$ there exists a sequence $\{t_n\} \subset [0, T]$, $t_n \to T$ (depending on $u(\cdot)$) such that

$$(\mathcal{S}u)(t_n) \in D(A).$$

**Problem 3.2.7.** *What semigroups have the minimal regularity property?*

We don't know of any result beyond the fact that the semigroup (3.2.29) in $L^2(0,\infty)$ does not have the minimal regularity property. To check this, recall (formula (2.6.5) and Figure 2.6.1) that (3.2.4) for this semigroup is

$$(\mathcal{S}u)(t)(x) = \int_0^t u(\sigma, x-(t-\sigma))d\sigma \quad (0 \le x < \infty)\,, \tag{3.2.30}$$

and that $Ay(x) = -y'(x)$ with maximal domain, boundary condition $y(0) = 0$. To produce $u(\cdot\,,\cdot)$ with $(\mathcal{S}u)(t) \notin D(A)$ $(t > 0)$, let $\chi(x)$ be the characteristic function of $[0,1]$. Define $u(t,x) = \chi(x-t)$. We have $u(\sigma, x-(t-\sigma)) = \chi(x-(t-\sigma)-\sigma) = \chi(x-t)$, so that (3.2.30) gives $(\mathcal{S}u)(t)(x) = t\chi(x-t)$ for $x \ge t$.

For the related question of what does maximal regularity imply for $A$, $S(t)$ see Dore [2000]; among the results there it is shown that maximal regularity implies the semigroup $S(t)$ is analytic.

**3.3. Modeling of parabolic equations.** Let $\Omega \in \mathbb{R}^m$ be a bounded $C^\infty$ domain with boundary $\Gamma$, $A$ the differential operator

$$Ay(x) = \sum_{j=1}^m \sum_{k=1}^m \frac{\partial}{\partial x_j}\Big(a_{jk}(x)\frac{\partial}{\partial x_k}y(x)\Big) + \sum_{j=1}^m b_j(x)\frac{\partial}{\partial x_j}y(x) + c(x)y(x)\,. \tag{3.3.1}$$

We assume that $a_{jk}(x) = a_{kj}(x)$ and that $A$ is uniformly elliptic:

$$\sum_{j=1}^m \sum_{k=1}^m a_{jk}(x)\xi_j\xi_k \ge c\sum_{j=1}^m \xi_j^2 \quad (x \in \overline{\Omega},\ c > 0)\,.$$

We denote by $\beta$ a boundary condition on the boundary $\Gamma$, either Dirichlet ($\beta y(x) = y(x) = 0$) or variational,

$$\begin{aligned} \beta y(x) &= \partial^\nu y(x) - \gamma(x)y(x) \\ &= \sum_{j=1}^m \sum_{k=1}^m a_{jk}(x)\nu_j(x)\frac{\partial}{\partial x_k}y(x) - \gamma(x)y(x) = 0 \quad (x \in \Gamma) \end{aligned} \tag{3.3.2}$$

where $\nu(x) = (\nu_1(x), \dots, \nu_m(x))$ is the outer normal vector at the boundary; $\partial^\nu y$ is called the *conormal derivative* of $y(x)$ at the boundary $\Gamma$. For simplicity, we assume $C^\infty$ coefficients for the operator and the boundary condition. The objective is to model the parabolic distributed parameter system

$$\begin{aligned} \frac{\partial y(t,x)}{\partial t} &= Ay(t,x) + u(t,x) && (0 \le t \le T,\ x \in \Omega) \\ y(0,x) &= \zeta(x) && (x \in \Omega) \\ \beta y(t,x) &= 0 && (0 \le t \le T,\ x \in \Gamma) \end{aligned} \tag{3.3.3}$$

with initial condition $\zeta(\cdot) \in L^p(\Omega)$ $(1 < p < \infty)$ and $L^p$ constraint

$$\int_\Omega |u(t,x)|^p dx \le C \quad (0 \le t \le T)$$

on the controls; this means controls are in the space $L^\infty(0,T;L^p(\Omega))$. The state space is $E = L^p(\Omega)$ and the abstract model for (3.3.3) is

$$y'(t) = A_p y(t) + u(t)\,, \quad y(0) = \zeta \tag{3.3.4}$$

where $A_p$ is the semigroup generator in $L^p(\Omega)$ determined by $A$ and $\beta$. We can define the operator $A_p$ in two equivalent ways: $(s)$ (strong) and $(w)$ (weak).

$(s)$ *Strong definition of* $A_p$ : $D(A_p) = W^{p,2}_\beta(\Omega)$ and $A_p y = Ay$.

Here, $W^{p,k}(\Omega)$ is the Sobolev space of all functions $y(\cdot)$ having distributional derivatives of order $\le k$ in $L^p(\Omega)$, the norm defined by

$$\|y\|_{W^{p,k}(\Omega)} = \Big(\sum_{|\alpha|\le k} \|D^\alpha y\|^p_{L^p(\Omega)}\Big)^{1/p} = \Big(\sum_{|\alpha|\le k} \int_\Omega |D^\alpha y(x)|^p dx\Big)^{1/p}$$

where $\alpha = (\alpha_1, \dots, \alpha_m), \alpha_j$ integer $\ge 0\,, |\alpha| = \alpha_1 + \dots + \alpha_m$ and

$$D^\alpha y(x) = D^{(\alpha_1,\dots,\alpha_m)} y(x) = \frac{\partial^{\alpha_1+\dots+\alpha_m} y(x)}{\partial x_1^{\alpha_1} \dots \partial x_m^{\alpha_m}}\,.$$

The space $W^{p,2}_\beta(\Omega)$ is the subspace of $W^{p,2}(\Omega)$ of all $y$ that satisfy the boundary condition $\beta$ on the boundary $\Gamma$. Application of the differential operator $\partial^\nu$ involves first order derivatives on the boundary and is justified via trace theorems that require Sobolev spaces of fractional orders: see Adams [1975, Chapter VII]. For the Dirichlet boundary condition, the usual notation is $W^{p,2}_\beta(\Omega) = W^{p,2}_0(\Omega)$.

The weak definition needs adjoints. The *formal adjoint* $A'$ of $A$ is defined by

$$A'y(x) = \sum_{j=1}^m \sum_{k=1}^m \frac{\partial}{\partial x_j}\Big(a_{jk}(x)\frac{\partial}{\partial x_k} y(x)\Big) - \sum_{j=1}^m \frac{\partial}{\partial x_j}(b_j(x)y(x)) + c(x)y(x)\,. \tag{3.3.5}$$

If $\beta$ is Dirichlet, the *adjoint boundary condition* is $\beta' = \beta$; otherwise, $\beta'$ is

$$\begin{aligned}\beta' y(x) &= \partial^\nu y(x) - (\gamma(x) + b(x))y(x) \\ &= \sum_{j=1}^m \sum_{k=1}^m a_{jk}(x)\nu_j(x)\frac{\partial}{\partial x_k} y(x) - (\gamma(x) + b(x))y(x) = 0 \quad (x \in \Gamma)\end{aligned} \tag{3.3.6}$$

with $b(x) = \sum_{j=1}^m b_j(x)\nu_j(x)$. Of course, the definitions are such that, if $y(\cdot), z(\cdot)$ are smooth in $\overline{\Omega}$ and $y(\cdot)$ (resp. $z(\cdot)$) satisfies $\beta$ (resp. $\beta'$) then

$$\int_\Omega Ay(x)v(x)dx = \int_\Omega y(x)A'v(x)dx\,. \tag{3.3.7}$$

(*w*) *Weak definition of* $A_p$ : $y \in D(A_p)$ if and only if there exists $z$ $(= A_p y)$ in $L^p(\Omega)$ such that

$$\int_\Omega y(x)(A'v)(x)dx = \int_\Omega z(x)v(x)dx$$

for every $v \in W^{2,q}_{\beta'}(\Omega)$, where $1/p + 1/q = 1$.

If $p < r$ we have $L^r(\Omega) \subseteq L^p(\Omega)$ and it follows from either definition that

$$A_r \subseteq A_p\,, \quad A_r = A_p|_{L^r(\Omega)}\,. \tag{3.3.8}$$

The operator $A_p$ is the infinitesimal generator of a semigroup $S_p(t)$ in $L^p(\Omega)$ analytic in $\mathrm{Re}\,t > 0$ (Amann - Hieber - Simonett [1994]) so that if $0 < \phi < \pi$ the resolvent $R(\lambda; A_p) = (\lambda I - A_p)^{-1}$ exists in a sector $|\arg(\lambda - c(\phi))| \le \phi$ and

$$\|(R(\lambda; A_p)\| \le \frac{C}{|\lambda - c(\phi)\,|}$$

in the sector. The semigroup can be retrieved from the resolvent via inverse Laplace transform

$$S_p(t) = \frac{1}{2\pi i}\int_\Gamma R(\lambda; A_p)e^{\lambda t}d\lambda = \frac{1}{2\pi i}\int_\Gamma (\lambda I - A_p)^{-1}e^{\lambda t}d\lambda \quad (t > 0)\,, \tag{3.3.9}$$

where $\Gamma$ is any contour $|\arg(\lambda - c')| = \phi$ $(\pi/2 < \phi < \pi,\ c' > c(\phi))$ oriented from $-i\infty$ to $+i\infty$. Both relations in (3.3.8) hold for $\lambda I - A_r$, $\lambda I - A_p$. Taking inverses,

$$(\lambda I - A_r)^{-1} \subseteq (\lambda I - A_p)^{-1}, \quad (\lambda I - A_r)^{-1} = (\lambda I - A_p)^{-1}|_{L^r(\Omega)}\,,$$

and using (3.3.9),

$$\begin{aligned} S_r(t) &= \frac{1}{2\pi i}\int_\Gamma (\lambda I - A_r)^{-1}e^{\lambda t}d\lambda \subseteq S_p(t) = \frac{1}{2\pi i}\int_\Gamma (\lambda I - A_p)^{-1}e^{\lambda t}d\lambda\,, \\ S_r(t) &= \frac{1}{2\pi i}\int_\Gamma (\lambda I - A_r)^{-1}e^{\lambda t}dt \\ &= \frac{1}{2\pi i}\int_\Gamma (\lambda I - A_p)^{-1}|_{L^r(\Omega)}e^{\lambda t}dt = S_p(t)|_{L^r(\Omega)}\,. \end{aligned} \tag{3.3.10}$$

The operator $A'_q$ is defined in a similar way, using the formal adjoint $A'$ in (3.3.5) and the adjoint boundary condition $\beta'$ in (3.3.6). The weak definition of $A'_q$ and (3.3.7) say that if $1/p + 1/q = 1$ we have

$$A'_q = A^*_p\,. \tag{3.3.11}$$

The operator $A'_q$ generates an analytic semigroup $S'_q(t)$ in $L^q(\Omega)$. We have

$$((\lambda I - A_p)^{-1})^* = (\bar\lambda I^* - A^*_p)^{-1} = (\bar\lambda I^* - A'_q)^{-1}\,,$$

hence, using (3.3.9),

$$
\begin{aligned}
S_p(t)^* &= -\frac{1}{2\pi i}\int_\Gamma ((\bar\lambda I - A_p)^{-1})^* e^{\bar\lambda t} d\bar\lambda = -\frac{1}{2\pi i}\int_\Gamma (\bar\lambda I^* - A'_q)^{-1} e^{\bar\lambda t} d\bar\lambda \\
&= \frac{1}{2\pi i}\int_\Gamma (\lambda I^* - A'_q)^{-1} e^{\lambda t} d\lambda = S_q(t)\,. \qquad (3.3.12)
\end{aligned}
$$

The $-$ sign in the first and second integrals comes from taking the conjugate of $1/2\pi i$. To go from the second to the third integral we do the change of variable $\lambda \to \bar\lambda$, which reverses the orientation of the contour and neutralizes the $-$ sign. The inclusion-restriction relations (3.3.8) and (3.3.10) are of course valid for the operators $A'_q(t)$ and the semigroups $S'_1(t)$.

The space $E = L^p(\Omega)$ satisfies either of the two conditions $(i)$ or $(ii)$ required in Theorem 3.1.2, hence existence of minimizing sequences implies existence of (time, norm) optimal controls $\bar u(t, x)$. $L^p(\Omega)$ spaces are also $\zeta$-convex for $1 < p < \infty$ (Aldous [1979], Bourgain [1983], Burkholder [1983]) and the operator $A_p$ satisfies the assumptions in Theorem 3.2.4 (Amann - Hieber - Simonett [1994])[8] thus the results in **3.2** apply. In particular, Theorem 3.2.6 says that every time optimal control $\bar u(t)$ satisfies the weak maximum principle (3.2.1),

$$
\langle S'_q(T-t)\psi, \bar u(t)\rangle = \max_{\|u\|_{L^p(\Omega)}\le 1} \langle S'_q(T-t)\psi, u\rangle \quad \text{a. e. in } 0 \le t \le T\,, \qquad (3.3.13)
$$

where $\psi$ is a multiplier in $\mathcal{Z}$ (see **3.2** for a description of this space) such that $S(t)^*\psi \neq 0$ for all $t$. We can get an explicit formula from (3.3.13) using the function $\Phi : (L^q(\Omega) \setminus \{0\}) \to \{y \in L^p(\Omega); \|y\|_{L^p(\Omega)} = 1\}$ in Remark 3.1.8:

$$
\Phi(z)(x) = \frac{|z(x)|^{q-2} z(x)}{\|z(\cdot)\|^{q-1}_{L^q(\Omega)}} \qquad (z(\cdot) \in L^q(\Omega),\ \|z(\cdot)\|_{L^q(\Omega)} \neq 0)
$$

($|z(x)|^{q-2}z(x) = 0$ *if* $z(x) = 0$). It follows from (3.3.13) that every time optimal control satisfies

$$
\bar u(t) = \Phi(S'_q(T-t)\psi) \quad \text{a. e. in } 0 \le t \le T \qquad (3.3.14)
$$

or, setting $z(t, x) = (S'_q(T-t)\psi)(x)$,

$$
\bar u(t, x) = \frac{|z(t,x)|^{q-2} z(t,x)}{\|z(t,\cdot)\|^{q-1}_{L^q(\Omega)}} \quad \text{a. e. in } 0 \le t \le T\,, \qquad (3.3.15)
$$

where the denominator is guaranteed nonzero (see Theorem 3.2.6). More information can be coaxed out of this formula via

---

[8] Under the present hypotheses ($C^\infty$ domain, $C^{(\infty)}$ coefficients) it is enough to use Seeley [1971]. See Dore - Venni [1987].

**Theorem 3.3.1.** *After eventual modification in a set of null measure in* $[0,T)\times\overline{\Omega}$ *the function* $z(t,x)$ *is infinitely differentiable in* $[0,T)\times\overline{\Omega}$ *and satisfies the reverse adjoint equation*

$$\begin{aligned} \frac{\partial z(t,x)}{\partial t} &= -A'z(t,x) \quad (0 \le t < T,\ x \in \Omega) \\ \beta' y(t,x) &= 0 \qquad\qquad\quad (0 \le t < T,\ x \in \Gamma) \end{aligned} \tag{3.3.16}$$

*in the classical sense.*

The proof of Theorem 3.3.1 is based on the following regularity result for the infinitesimal generator $A_2$ in $L^2(\Omega)$ (Theorem 3.3.2) and on Sobolev's Imbedding Theorem 3.3.3 below.

**Theorem 3.3.2.** *Let* $\mu$ *be large enough. Then, for* $y \in D(A_2)$, $k$ *integer* $\ge 0$,

$$\begin{aligned} (\mu I - A_2)y \in W^{2,k}(\Omega) &\implies y \in W^{2,k+2}(\Omega)\,, \\ \|y\|_{W^{2,k+2}(\Omega)} &\le C\|(\mu I - A_2)y\|_{W^{2,k}(\Omega)}\,. \end{aligned} \tag{3.3.17}$$

This result can be found in Gilbarg - Trudinger [1977, Theorem 8.4, p. 176], or the author [1983, Theorem 4.7.14 p. 234].[9] Iterating (3.3.17) (applied to $A_2'$) we obtain, with $C$ a generic constant,

$$\begin{aligned} y \in D({A_2'}^k) &\implies y \in W^{2,2k}(\Omega)\,, \\ \|y\|_{W^{2,2k}(\Omega)} &\le C\|(\mu I - A_2')y\|_{W^{2,2(k-1)}(\Omega)} \\ &\le C\|(\mu I - A_2')^2 y\|_{W^{2,2(k-2)}(\Omega)} \le \cdots \\ &\le C\|(\mu I - A_2')^k y\|_{L^2(\Omega)} \quad (y \in D({A'}_2^k))\,. \end{aligned} \tag{3.3.18}$$

**Theorem 3.3.3.** *Let* $q \ge 1$, $k$ *an integer* $\ge 1$. $(a)$ *If* $kq < m$ *then*

$$W^{q,k}(\Omega) \hookrightarrow L^r(\Omega)\,, \qquad 1 \le r \le \frac{mq}{m-kq}\,. \tag{3.3.19}$$

$(b)$ *If* $kq = m$ *then* (3.3.19) *holds for* $1 \le r < \infty$. $(c)$ *If* $(k-j)q > m$ *for a nonnegative integer* $j$*, then after eventual modification in a null set,*

$$W^{q,k}(\Omega) \hookrightarrow C^{(j)}(\overline{\Omega})\,. \tag{3.3.20}$$

For a proof see Adams [1975, p. 97].[10] In (3.3.20) the space $C^{(j)}(\overline{\Omega})$ consists of all $y(\cdot)$ with partial derivatives of order $\le j$ in $\Omega$ having continuous extensions to $\overline{\Omega}$.

[9] The standing assumptions are: the domain $\Omega$ is bounded and of class $C^\infty$ and the coefficients of $A$ belong to $C^{(\infty)}(\overline{\Omega})$. This is in excess of what is needed for Theorem 3.3.2, but it is about right if we want the result to hold for $k$ arbitrary.

[10] The proof of $(a)$ and $(b)$ there is for arbitrary domains satisfying the *cone property* (Adams [1975, pp.76 and 66]). The proof of $(c)$ requires the *strong local Lipschitz property* (Adams [1975, p. 66]). Both properties are satisfied by a domain of class $C^1$.

*Proof of Theorem* 3.3.1. We begin by showing that, for any $\epsilon > 0$ we have[11]

$$S'_q(\epsilon)\psi \in L^2(\Omega)\,. \tag{3.3.21}$$

This is obvious if $q \geq 2$, since $S'_q(\epsilon)\psi \in L^q(\Omega) \subseteq L^2(\Omega)$. If $q < 2$ we divide the proof in two cases.

$$(i) \quad q \geq \frac{m}{2}\,.$$

Take $\delta = \epsilon/2$. Since $S'_q(\delta)\psi \in L^q(\Omega)$, $S'_q(\epsilon) = S'_q(\delta)S'_q(\delta)\psi \in D(A'_q) = W^{q,2}(\Omega)$. If $q > m/2$ we have $2q > m$ and we are in case $(c)$ of Theorem 3.3.3 with $k = 2$, $j = 0$; $S'_q(\epsilon)\psi \in C(\overline{\Omega})$, in excess of (3.3.21). If $q = m/2$ we are in case $(b)$ of Theorem 3.3.3 so that $S'_q(\epsilon)\psi \in L^p(\Omega)$ for all $p \geq 1$.

$$(ii) \quad q < \frac{m}{2}\,.$$

Here $m > 2q$ and we select an integer $n \geq 2$ such that

$$2(n-1)q < m \leq 2nq\,. \tag{3.3.22}$$

We take $\delta > 0$, write

$$S'_q((n+1)\delta)\psi = \Big(\prod_{j=1}^{n} S'_q(\delta)\Big)S'_q(\delta)\psi$$

and deal with the product by induction. Since $S'_q(\delta)\psi \in L^q(\Omega)$ we have

$$S'_q(\delta)S'_q(\delta)\psi \in D(A'_q) \subseteq W^{q,2}(\Omega) \hookrightarrow L^{q_1}(\Omega)\,, \quad q_1 = \frac{mq}{m-2q} > q$$

hence, due to (3.3.10) applied to the semigroup $S'_q(t)$,

$$\begin{aligned} &S'_q(\delta)S'_q(\delta)S'_q(\delta)\psi = S'_{q_1}(\delta)S'_q(\delta)S'_q(\delta)\psi \in D(A'_{q_1}) \\ &\subseteq W^{q_1,2}(\Omega) \subseteq L^{q_2}(\Omega)\,, \quad q_2 = \frac{mq_1}{m-2q_1} = \frac{mq}{m-4q} > q_1\,. \end{aligned}$$

Arguing like this another $n-2$ times,

$$\begin{aligned} S'_q(n\delta)\psi = \Big(\prod_{j=1}^{n-1} S'_q(\delta)\Big)S'_q(\delta)\psi \in L^r(\Omega)\,, \\ r = q_{n-1} = \frac{mq}{m-2(n-1)q} > q_{n-2} > \ldots > q_1 > q\,. \end{aligned} \tag{3.3.23}$$

[11] The reason for proving (3.3.21) is to be able to use the $L^2$-regularity results in Theorem 3.3.2. $L^p$-regularity is less elementary.

Now, the second inequality (3.3.22) is $0 \geq m - 2nq$, hence $2q \geq m + 2(1-n)q$, or

$$2r = 2q_{n-1} = \frac{2mq}{m - 2(n-1)q} \geq m\,. \tag{3.3.24}$$

Take $\delta = \epsilon/(n+1)$. Since $S'_q(n\delta)\psi \in L^{q_{n-1}}(\Omega) = L^r(\Omega)$ and $r > q$ (from (3.3.23)) we have $S'_q(\epsilon)\psi = S'_q(\delta)S'_q(n\delta)\psi = S_r(\delta)S'_q(n\delta)\psi$ with $r \geq m/2$ (from 3.3.24)), thus the proof of (3.3.21) ends as that of case $(i)$.

With (3.3.21) granted, we write

$$z(t,\cdot) = S'_q(T-t)\psi = S'_q(T-\epsilon-t)S_q(\epsilon)\psi = S'_2(T-\epsilon-t)S_q(\epsilon)\psi \tag{3.3.25}$$

after (3.3.10). Since $S'_2(t)$ is an analytic semigroup, $S'_2(t)L^2(\Omega) \subseteq D({A'}_2^k)$ for any integer $k \geq 1$ and the function

$$t \to (\mu I - A'_2)^k S'_2(t) \in \mathcal{L}(L^2(\Omega), L^2(\Omega)) \tag{3.3.26}$$

is analytic for $t > 0$ in the norm of $\mathcal{L}(L^2(\Omega), L^2(\Omega))$. The integer $k$ is arbitrary, thus, given an integer $j \geq 0$ we may select $k$ with $2(2k - j) > m$ and deduce from (3.3.18) and Theorem 3.3.3 $(c)$ that the function

$$t \to S'_2(t) \in \mathcal{L}(L^2(\Omega), W^{2,2k}(\Omega)) \hookrightarrow \mathcal{L}(L^2(\Omega), C^{(j)}(\overline{\Omega})) \tag{3.3.27}$$

is continuous in the norm of the space $\mathcal{L}(L^2(\Omega), W^{2,2k}(\Omega))$, thus also in the norm of $\mathcal{L}(L^2(\Omega), C^{(j)}(\overline{\Omega}))$ for $t > 0$. Finally, making use of (3.3.25) we deduce that $z(t,\cdot) = S'_q(T-t)\psi$ (after eventual modification in a $x$-null set for each $t$, or in a null $(t,x)$-set) is continuous in the norm of $C^{(j)}(\overline{\Omega})$ in $t \leq T - \epsilon$ for all $\epsilon > 0$, thus it is infinitely differentiable in $[0,T) \times \overline{\Omega}$. To show that $z(t,x)$ is a solution of (3.3.16) in the classical sense, note first that, since $z(t,\cdot) \in C^{(\infty)}(\overline{\Omega})$ then $A'_q z(t,\cdot) = Az(t,\cdot)$ in the classical sense; moreover,

$$\begin{aligned}&\frac{z(t+h,\cdot) - z(t,\cdot)}{h} - A'_q z(t,\cdot)\\ &= \frac{S'_q(T-t-h)\psi - S'_q(T-t)\psi}{h} - A'_q S(T-t)\psi\,.\end{aligned}$$

Because of the smoothness of $z(t,x)$ the limit on the left can be taken in the space $C(\overline{\Omega})$; on the other hand, the limit on the right exists in $L^q(\Omega)$, hence both are the same. Also because of smoothness, the boundary condition is assumed in the classical sense. This ends the proof of Theorem 3.3.1.

**Remark 3.3.4.** The smoothness conclusions in Theorem 3.3.1 can be improved to analyticity in $t$. If $\phi < \pi$, $S_q(t)$ can be extended to a $\mathcal{L}(L^q(\Omega), L^q(\Omega))$-valued function $S_q(\zeta)$ analytic in the sector $|\arg\zeta| < \phi$. Then the function (3.3.26) can be similarly extended to the same sector, and so does (3.3.27). We deduce that $z(t,x)$

can be extended to an analytic function $z(\zeta, x)$ in the sector $|\arg(T-\zeta)| < \phi$ for each $x$ fixed.

Let $\bar{u}(t)$ satisfy the maximum principle (3.3.13) or its equivalent formulations (3.3.14) $\Leftrightarrow$ (3.3.15). What characterizes a multiplier $\psi \in Z(T)$ is

$$\int_0^T \|S_p(t)^*\psi\|_{L^q(\Omega)}dt = \int_0^T \|S_q'(t)\psi\|_{L^q(\Omega)}dt < \infty\,,$$

which, in terms of $z(t,x)$ reduces to

$$\int_0^T \left(\int_\Omega |z(t,x)|^q dx\right)^{1/q} dt < \infty\,. \tag{3.3.28}$$

We can formulate the necessary conditions for optimality (Theorem 2.5.3) and the sufficient conditions (Theorems 2.5.5 and 2.5.7) as follows.

**Theorem 3.3.5.** *(a) Let $\bar{u}(t,x)$ be a control that drives $\zeta(\cdot) \in L^p(\Omega)$ to*

$$\bar{y}(\cdot) = y(T,\zeta,\cdot) \in \overline{D(A_p)} \tag{3.3.29}$$

*norm optimally in $0 \le t \le T$. Then*

$$\bar{u}(t,x) = \rho\frac{|z(t,x)|^{q-2}z(t,x)}{\|z(t,\cdot)\|^{q-1}_{L^q(\Omega)}}\,, \tag{3.3.30}$$

*where $z(t,x)$ is a nontrivial solution of the reverse adjoint equation (3.3.16) such that (3.3.28) holds, and $\rho$ is the minimum norm,*

$$\rho = \operatorname*{ess.\,sup}_{0\le t\le T}\left(\int_\Omega |\bar{u}(t,x)|^p dx\right)^{1/p}.$$

*(b) If $\bar{u}(t,x)$ is time optimal and (3.3.29) holds then $\bar{u}(t,x)$ is given by (3.3.30) with $\rho = 1$ and $z(t,x)$ satisfying (3.3.16) and (3.3.28).*

Theorem 3.3.5 is a straight restatement of Theorem 2.5.3. The condition that enables Theorem 2.5.3 is $\bar{y} - S(T)\zeta \in \overline{D(A)}$ (closure in $R^\infty(T)$), and $S_p(t)$ is analytic so that $S_p(T)\zeta \in D(A_p)$ is automatic. Theorems 2.5.5 and 2.5.7 yield

**Theorem 3.3.6.** *Assume $\bar{u}(t,x)$ satisfies (3.3.30), where $z(t,x)$ is a solution of the reverse adjoint equation (3.3.16) satisfying (3.3.28). Then $\bar{u}(t,x)$ drives any $\zeta \in L^p(\Omega)$ to $y(T,\zeta,\cdot)$ norm optimally. If $\rho = 1$ and*

$$\zeta(\cdot) \in D(A_p)\,,\ \ \|A_p\zeta\| < 1 \quad or \quad y(T,\zeta,\cdot) \in D(A_p)\,,\ \ \|A_p y(T)\| < 1\,, \tag{3.3.31}$$

*then the drive is time optimal.*

Note that, in Theorems 3.3.5 and 3.3.6 the element $\psi = z \in Z(T)$ in the maximum principle evaporates; it is the "final condition" $z(T, x)$ of the costate, which doesn't show up in the statements.

**Remark 3.3.7.** One of the time-honored uses of the maximum principle is to show smoothness of optimal controls. Are optimal controls, defined by (3.3.30), "as smooth as" $z(t, x)$? This is certainly the case when $p = 2$ as (3.3.30) is

$$\bar{u}(t,x) = \rho \frac{z(t,x)}{\|z(t,\cdot)\|_{L^2(\Omega)}} = \frac{S_2(T-t)z(x)}{\|S_2(T-t)z(\cdot)\|_{L^2(\Omega)}} \tag{3.3.32}$$

where the numerator has the smoothness of $z(t, x)$ (analytic in $t$ and infinitely differentiable in $x$) and the denominator is real analytic in $t$. The latter follows from the fact that if a Hilbert space valued function $f(t)$ is real analytic and $\|f(t)\| > 0$, then $\|f(t)\|$ is real analytic. To see this, we write $\|f(t)\| = \sqrt{(f(t), f(t))}$ and prove that $(f(t), f(t))$ and its square root are real analytic, just as for a product of ordinary functions. Note that this result and formula (3.3.32) produce smoothness only in $0 \le t < T$; for $t = T$ the control may be singular.

If $p \neq 2$, smoothness of $z(t, x)$ may not "communicate" wholly to the optimal control $\bar{u}(t, x)$. For an example, consider the one-dimensional heat equation

$$\frac{\partial y(t,x)}{\partial t} = \frac{\partial^2 y(t,x)}{\partial x^2} + u(t,x)$$

in the domain $\Omega = (0, \pi)$ with Dirichlet boundary condition $u(t, 0) = u(t, \pi) = 0$. The state space is $E = L^3(0, \pi)$. The reverse adjoint equation is

$$\frac{\partial z(t,x)}{\partial t} = -\frac{\partial^2 z(t,x)}{\partial x^2} \tag{3.3.33}$$

with the same boundary condition, and since for $p = 3$ we have $q = 3/2$, the costate $z(t, \cdot)$ lives in the space $L^{3/2}(0, \pi)$. The function

$$z(t,x) = e^{4t} \sin 2x$$

satisfies (3.3.33). and (3.3.28), the latter with largesse since the costate $z(t, x)$ is smooth in $[0, T] \times \overline{\Omega}$. According to Theorem 3.3.6 the control

$$\bar{u}(t,x) = \frac{1}{\sqrt{\|z(t,\cdot)\|_{L^{3/2}(0,\pi)}}} \cdot \frac{z(t,x)}{\sqrt{|z(t,x)|}}$$

is norm optimal and time optimal (the latter under (3.3.31)). Now,

$$\sqrt{\|z(t,\cdot)\|_{L^{3/2}(0,\pi)}} = e^{2t}\left( \int_0^{\pi} |\sin 2x|^{3/2} dx \right)^{1/3} = 1.2046\, e^{2t} ,$$

and

$$\frac{z(t,x)}{\sqrt{|z(t,x)|}} = \begin{cases} e^{2t}\sqrt{\sin 2x} & 0 \le x \le \pi/2\,, \\ -e^{2t}\sqrt{-\sin 2x} & \pi/2 < x \le \pi\,. \end{cases}$$

Accordingly, the optimal control is

$$\bar{u}(t,x) = 0.83014\,\mathrm{sign}\Big(\frac{\pi}{2} - x\Big)\sqrt{|\sin 2x|}$$

which has infinite derivative with respect to $x$ at the lines $x = 0, \pi/2, \pi$.

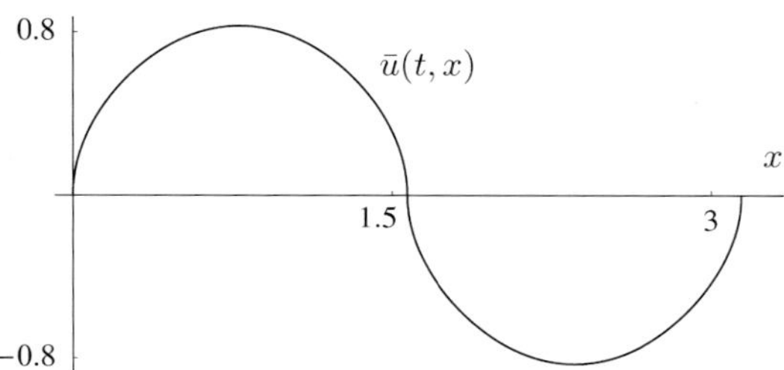


Figure 3.3.1

**Miscellaneous notes.** The control theory component of this section is new. The material on the parabolic equations (3.3.3) and (3.3.16) is known. In particular, results of the type of Theorem 3.3.1 (semigroup solution $\Rightarrow$ classical solution) have been around since the beginning of semigroup theory in one form or another.

From the $L^p$ theory with $1 < p < \infty$, the case $p = 2$ is the most interesting in applications. In the full range of $p$, the case $p = \infty$ is even more compelling but not included here; it needs a separate treatment (in Chapter 4). On the other hand, $p = 1$ is crucial for applications to controlled diffusions, but requires controls whose values are measures. This setup is that of Chapter 5.

**3.4. Weakly singular extremals.** We produce in this section a weakly singular (time, norm) optimal control $\bar{u}(t)$; it satisfies (3.2.1) with a multiplier $\psi$ such that $S(t)^*\psi \neq 0$ for $t > 0$ but it does *not* satisfy (3.2.1) with $\psi = z \in Z(T), z \neq 0$. The state space is $E = L^2(0,\infty)$ and the operator

$$A = A_c y(\lambda) = -(\lambda + c)y(\lambda) \tag{3.4.1}$$

with maximal domain, with $c > 0$ fixed. The self adjoint operator $A_c$ generates the self adjoint analytic semigroup

$$S_c(t)y(\lambda) = e^{-(\lambda+c)t}y(\lambda)\,. \tag{3.4.2}$$

We have $\|S_c(t)\| = e^{-ct}$. The equation is

$$y'(t) = A_c y(t) + u(t)\,, \quad y(0) = \zeta\,. \tag{3.4.3}$$

If $\bar{u}(\cdot)$ satisfies the weak maximum principle (3.2.1) with $S(\sigma)^*\psi \neq 0$ then

$$\bar{u}(\sigma) = \frac{S_c(T-\sigma)^*\psi}{\|S_c(T-\sigma)^*\psi\|} = \frac{S_c(T-\sigma)\psi}{\|S_c(T-\sigma)\psi\|}\,, \tag{3.4.4}$$

and the corresponding trajectory with initial condition $\zeta = 0$ is

$$y(t,0,\bar{u}) = \int_0^t S_c(t-\sigma)\bar{u}(\sigma)d\sigma = \int_0^t S_c(t-\sigma)\frac{S_c(T-\sigma)\psi}{\|S_c(T-\sigma)\psi\|}d\sigma$$
$$= \int_0^t \frac{S_c(T+t-2\sigma)\psi}{\|S_c(T-\sigma)\psi\|}d\sigma = S(T-t)\int_0^t \frac{S_c(2(t-\sigma))\psi}{\|S_c(T-\sigma)\psi\|}d\sigma\,.$$

In particular, the final point $y(T,0,\bar{u})$ of a trajectory is given by

$$y(T,0,\bar{u}) = \int_0^T \frac{S_c(2(T-\sigma))\psi}{\|S_c(T-\sigma)\psi\|}d\sigma = \int_0^T \frac{S_c(2\sigma)\psi}{\|S_c(\sigma)\psi\|}d\sigma\,. \tag{3.4.5}$$

We denote by $\mathcal{H} \supset L^2(0,\infty)$ the space of all measurable functions $\psi(\lambda)$ (not necessarily in $L^2(0,\infty)$) such that

$$S_c(t)^*\psi(\cdot) = S_c(t)\psi(\cdot) = e^{-(\cdot\,+c)t}\psi(\cdot) \in L^2(0,\infty) \quad (t>0)\,.$$

According to the definition in **2.8**, $\mathcal{H}$ is a multiplier space.[12]

The results in this and next section depend on computing

$$y_{0,\infty}(\psi) = \int_0^\infty \frac{S_c(2\sigma)\psi}{\|S_c(\sigma)\psi\|}d\sigma \tag{3.4.6}$$

as a Laplace transform for certain specific choices of $\psi \in \mathcal{H}$. After this, we can do asymptotics for $y(T,0,\bar{u})$ in (3.4.5) "disposing of" the integral

$$y_{T,\infty}(\psi) = \int_T^\infty \frac{S_c(2\sigma)\psi}{\|S_c(\sigma)\psi\|}d\sigma\,. \tag{3.4.7}$$

**Lemma 3.4.1.** *Let $\psi$ be a multiplier such that*

$$\frac{1}{\|S_c(t)\psi\|} \leq C(1+t)^\gamma e^{ct} \quad (t \geq 0)\,,$$

[12] Working along the lines of Remark 2.8.3 we can show that $\mathcal{H} = \mathcal{M} =$ space of all multipliers. This has no particular relevance.

*where* $\gamma \geq 0$. *Then, for every* $\epsilon > 0$ *we have* $y_{T,\infty}(\psi) = S_c(2(T-\epsilon))\eta_{T,\infty,\epsilon}(\psi)$ *with*

$$\|\eta_{T,\infty,\epsilon}(\psi)\| \leq C_{\gamma,\epsilon}\|S_c(2\epsilon)\psi\|(1+T)^{\gamma}e^{cT} \quad (T \geq \epsilon)\,.$$

*Proof.* We have

$$y_{T,\infty}(\psi) = S_c(2(T-\epsilon))\int_T^{\infty}\frac{S_c(2(\sigma-T))S_c(2\epsilon)\psi}{\|S_c(\sigma)\psi\|}d\sigma$$

and

$$\begin{aligned}\left\|\int_T^{\infty}\frac{S_c(2(\sigma-T))S_c(2\epsilon)\psi}{\|S_c(\sigma)\psi\|}d\sigma\right\| &\leq \|S_c(2\epsilon)\psi\|\int_T^{\infty}\frac{\|S_c(2(\sigma-T))\|}{\|S_c(\sigma)\psi\|}d\sigma \\ &\leq C\|S_c(2\epsilon)\psi\|\int_T^{\infty}(1+\sigma)^{\gamma}e^{c\sigma}e^{-2c(\sigma-T)}d\sigma \\ &= C\|S_c(2\epsilon)\psi\|e^{cT}\int_0^{\infty}(1+\tau+T)^{\gamma}e^{-c\tau}d\tau \\ &= C\|S_c(2\epsilon)\psi\|(1+T)^{\gamma}e^{cT}\int_0^{\infty}\left(\frac{1+\tau+T}{1+T}\right)^{\gamma}e^{-c\tau}d\tau \\ &\leq C\|S_c(2\epsilon)\psi\|(1+T)^{\gamma}e^{cT}\int_0^{\infty}(1+\tau)^{\gamma}e^{-c\tau}d\tau\end{aligned}$$

since $1+\tau+T \leq (1+\tau)(1+T)$ for $\tau, T \geq 0$. This ends the proof of Lemma 3.4.1.

The computations below are all based on the Laplace transform

$$\int_0^{\infty}e^{-\lambda t}\lambda^{\alpha}d\lambda = \frac{\Gamma(1+\alpha)}{t^{1+\alpha}}\,, \quad \alpha > -1\,. \tag{3.4.8}$$

The function $\psi_\alpha(\lambda) = \lambda^\alpha$ belongs to $\mathcal{H}$ (thus is a multiplier) if $\alpha > -1/2$; for $\alpha < -1/2$ $\psi_\alpha(\lambda)$ is not a multiplier since $S_c(t)\psi(\lambda) \notin L^2(0,\infty)$ ($e^{-2(\lambda+c)t}\lambda^{2\alpha}$ has a nonintegrable singularity at $\lambda = 0$). We define a control $\bar{u}_\alpha(t) = \bar{u}_\alpha(t,\cdot) \in L^2(0,\infty)$ of the form (3.4.4) in the interval $0 \leq t \leq T$ for $\alpha > -1/2$ by

$$\bar{u}_\alpha(t,\lambda) = \frac{S_c(T-t)\psi_\alpha(\lambda)}{\|S_c(T-t)\psi_\alpha(\cdot)\|}\,. \tag{3.4.9}$$

Using (3.4.8) we have

$$\|S_c(t)\psi_\alpha\|^2 = e^{-2ct}\int_0^{\infty}e^{-2\lambda t}\lambda^{2\alpha}d\lambda = \frac{\Gamma(1+2\alpha)}{e^{2ct}(2t)^{1+2\alpha}}\,, \tag{3.4.10}$$

so that

$$\frac{1}{\|S_c(t)\psi_\alpha\|} = \frac{2^{1/2+\alpha}}{\sqrt{\Gamma(1+2\alpha)}}e^{ct}t^{1/2+\alpha}\,. \tag{3.4.11}$$

By (3.4.5) and (3.4.11) we have

$$\begin{aligned} y(T,0,\bar{u}_\alpha,\lambda) &= \int_0^T \frac{S_c(2\sigma)\psi_\alpha}{\|S_c(\sigma)\psi_\alpha\|}d\sigma \\ &= \int_0^T e^{-(\lambda+c)(2\sigma)}\lambda^\alpha \frac{2^{1/2+\alpha}}{\sqrt{\Gamma(1+2\alpha)}}e^{c\sigma}\sigma^{1/2+\alpha}d\sigma \\ &= y_{0,\infty}(\psi_\alpha,\lambda) - y_{T,\infty}(\psi_\alpha,\lambda) \end{aligned} \tag{3.4.12}$$

(for the last line, see (3.4.6) and (3.4.7)). We have

$$\begin{aligned} y_{0,\infty}(\psi_\alpha,\lambda) &= \frac{2^{1/2+\alpha}}{\sqrt{\Gamma(1+2\alpha)}}\lambda^\alpha \int_0^\infty e^{-(2\lambda+c)\sigma}\sigma^{1/2+\alpha}d\sigma \\ &= \frac{2^{1/2+\alpha}}{\sqrt{\Gamma(1+2\alpha)}}\frac{\Gamma(3/2+\alpha)}{(2\lambda+c)^{3/2+\alpha}}\lambda^\alpha \\ &= \frac{\Gamma(3/2+\alpha)}{2\sqrt{\Gamma(1+2\alpha)}}\frac{\lambda^\alpha}{(\lambda+c/2)^{3/2+\alpha}} \approx \frac{C}{\lambda^{3/2}} \quad \text{as } \lambda\to\infty\,, \end{aligned}$$

so that $y_{0,\infty}(\psi_\alpha) \notin D(A)$. On the other hand, we obtain from Lemma 3.4.1 that $y_{T,\infty}(\psi_\alpha) \in S_c(2(T-\epsilon))E$ $(\epsilon > 0)$, hence $y_{T,\infty}(\psi_\alpha) \in D(A)$ and it follows from (3.4.12) that $y(T,0,\bar{u}_\alpha) \notin D(A)$. In view of (3.4.10),

$$\int_0^T \|S_c(\sigma)\psi_\alpha\|d\sigma = \frac{\sqrt{\Gamma(1+2\alpha)}}{2^{1/2+\alpha}}\int_0^T \frac{e^{-c\sigma}}{\sigma^{1/2+\alpha}}d\sigma \begin{cases} <\infty & (\alpha < \frac{1}{2}) \\ =\infty & (\alpha \ge \frac{1}{2}) \end{cases}$$

thus

$$\psi_\alpha \begin{cases} \in Z(T) & \alpha < \frac{1}{2}\,, \\ \notin Z(T) & \alpha \ge \frac{1}{2}\,. \end{cases} \tag{3.4.13}$$

The first line of (3.4.13), Theorems 2.5.5 (for norm optimality) and 2.5.7 (for time optimality) imply the "$\alpha < 1/2$" part of

**Theorem 3.4.2.** *If $\alpha \le 1/2$, the control $\bar{u}_\alpha(t)$ drives $0$ to $y(T,0;\bar{u}_\alpha)$ both time and norm optimally in the interval $0 \le t \le T$ for any $T > 0$.*

*Proof.* Only the case $\alpha = 1/2$ is pending, but the proof works for $\alpha \le 1/2$, so we obtain for free a new proof for the full range. Also, time optimality $\Rightarrow$ norm optimality, thus we need only show time optimality.

Let $t > 0$, $u(\cdot) \in L^\infty(0,T;L^2(0,\infty))$ and let $u(\sigma,\lambda) = u(\sigma)(\lambda)$. The function $\sigma \to u(\sigma,\cdot) \in L^2(0,\infty)$ is strongly measurable, thus it can be approximated in the norm of $L^\infty(0,T;L^2(0,\infty))$ by functions of the form $\sum \chi_j(\sigma)f_j(\lambda)$ (the $\chi_j(\sigma)$ and

$f_j(\lambda)$ measurable) it follows that $u(\sigma, \lambda)$ is jointly measurable in both variables. Moreover, $u(\sigma, \cdot) \in L^2(0, \infty)$ a. e. in $\sigma$ with

$$\int_0^\infty |u(\sigma, \lambda)|^2 d\lambda = \|u(\sigma, \cdot)\|_{L^2(0,\infty)} \leq \|u(\cdot)\|_{L^\infty(0,T;L^2(0,\infty))},$$

so that $u(\cdot, \cdot) \in L^2((0, T) \times (0, \infty))$. We have

$$y(t, 0, u)(\cdot) = \int_0^t S_c(t - \sigma) u(\sigma, \cdot) d\sigma = \int_0^t e^{-(\lambda+c)(t-\sigma)} u(\sigma, \cdot) d\sigma,$$

the integral understood as a $L^2(0, \infty)$-valued Lebesgue - Bochner integral, and we show below that this integral can be computed "pointwise" in the sense that $y(t, 0, u)(\lambda)$ equals

$$y(t, 0, u, \lambda) = \int_0^t e^{-(\lambda+c)(t-\sigma)} u(\sigma, \lambda) d\sigma \tag{3.4.14}$$

a. e. in $0 \leq \lambda < \infty$. To check this, note that, by Schwarz's inequality

$$\begin{aligned} \int_0^\infty |y(t, 0, u, \lambda)|^2 d\lambda &\leq \int_0^\infty \left( \int_0^t e^{-2(\lambda+c)(t-\sigma)} d\sigma \right) \left( \int_0^\infty |u(\sigma, \lambda)|^2 d\sigma \right) d\lambda \\ &\leq t \int_0^\infty \int_0^\infty |u(\sigma, \lambda)|^2 d\sigma d\lambda \end{aligned}$$

so that $y(t, 0, u, \cdot) \in L^2(0, \infty)$, and to prove (3.4.14) it is enough to show that

$$\int_0^\infty y(t, 0, u)(\lambda) \phi(\lambda) d\lambda = \int_0^\infty y(t, 0, u, \lambda) \phi(\lambda) d\lambda$$

for every $\phi(\cdot) \in L^2(0, \infty)$. The right side of this equality is

$$\int_0^\infty \int_0^t e^{-(\lambda+c)(t-\sigma)} u(\sigma, \lambda) \phi(\lambda) d\sigma d\lambda, \tag{3.4.15}$$

and the left side

$$\begin{aligned} \left\langle \int_0^t e^{-(\cdot\, + c)(t-\sigma)} u(\sigma, \cdot) d\sigma, \phi(\cdot) \right\rangle &= \int_0^t \langle e^{-(\cdot\, + c)(t-\sigma)} u(\sigma, \cdot), \phi(\cdot) \rangle d\sigma \\ &= \int_0^t \int_0^\infty e^{-(\lambda+c)(t-\sigma)} u(\sigma, \lambda) \phi(\lambda) d\lambda d\sigma, \end{aligned}$$

which coincides with (3.4.15) by Fubini's theorem.

With (3.4.14) granted, let $T' < T$ and let $u(\cdot) \in L^\infty(0, T'; L^2(0, \infty))$ be a control that satisfies $\|u(\cdot)\|_{L^\infty(0,T';L^2(0,\infty))} \leq 1$. To lighten up the writing, we set $v(\sigma, \lambda) = u(T' - \sigma, \lambda)$, so that

$$y(T', 0, u, \lambda) = \int_0^{T'} e^{-(\lambda+c)(T'-\sigma)} u(\sigma, \lambda) = \int_0^{T'} e^{-(\lambda+c)\sigma} v(\sigma, \lambda).$$

If $-1/2 < \beta < 1/2$ we have, using Schwarz's inequality,

$$\begin{aligned}\int_0^\infty \lambda^\beta y(T', 0, u, \lambda) d\lambda &\leq \int_0^\infty \lambda^\beta \Big( \int_0^{T'} e^{-(\lambda+c)\sigma} |v(\sigma, \lambda)| d\sigma \Big) d\lambda \\ &= \int_0^{T'} \Big( \int_0^\infty \lambda^\beta e^{-(\lambda+c)\sigma} |v(\sigma, \lambda)| d\lambda \Big) d\sigma \\ &\leq \int_0^{T'} e^{-c\sigma} \Bigg( \sqrt{\int_0^\infty \lambda^{2\beta} e^{-2\lambda\sigma} d\lambda} \sqrt{\int_0^\infty |v(\sigma, \lambda)|^2 d\lambda} \Bigg) d\sigma \\ &= \frac{\sqrt{\Gamma(1+2\beta)}}{2^{1/2+\beta}} \int_0^{T'} \frac{e^{-c\sigma} d\sigma}{\sigma^{1/2+\beta}} \end{aligned} \tag{3.4.16}$$

in view of (3.4.8) (this computation, as well as the one below, is justified by convergence of the $d\lambda\, d\sigma$ integral and Tonelli's theorem). On the other hand, combining (3.4.6) and (3.4.11) we obtain

$$y(T, 0, \bar{u}_\alpha, \lambda) = \frac{\lambda^\alpha\, 2^{1/2+\alpha}}{\sqrt{\Gamma(1+2\alpha)}} \int_0^T e^{-2(\lambda+c)\sigma} e^{c\sigma} \sigma^{1/2+\alpha} d\sigma\,,$$

thus, again by (3.4.8),

$$\begin{aligned}\int_0^\infty \lambda^\beta y(T, 0, \bar{u}_\alpha, \lambda) d\lambda &= \frac{2^{1/2+\alpha}}{\sqrt{\Gamma(1+2\alpha)}} \int_0^\infty \lambda^\beta \lambda^\alpha \Big( \int_0^T e^{-2(\lambda+c)\sigma} e^{c\sigma} \sigma^{1/2+\alpha} d\sigma \Big) d\lambda \\ &= \frac{2^{1/2+\alpha}}{\sqrt{\Gamma(1+2\alpha)}} \int_0^T e^{-c\sigma} \sigma^{1/2+\alpha} \Big( \int_0^\infty e^{-2\lambda\sigma} \lambda^{\alpha+\beta} d\lambda \Big) d\sigma \\ &= \frac{2^{1/2+\alpha}}{\sqrt{\Gamma(1+2\alpha)}} \Gamma(1+\alpha+\beta) \int_0^T \frac{\sigma^{1/2+\alpha} e^{-c\sigma}}{(2\sigma)^{1+\alpha+\beta}} d\sigma \\ &= \frac{1}{2^{1/2+\beta}} \frac{\Gamma(1+\alpha+\beta)}{\sqrt{\Gamma(1+2\alpha)}} \int_0^T \frac{e^{-c\sigma} d\sigma}{\sigma^{1/2+\beta}}\,. \end{aligned} \tag{3.4.17}$$

Assume now we can drive the origin to $y(T, 0, \bar{u}_\alpha)$ in time $T' < T$ with a control $u(\cdot) \in L^\infty(0, T'; L^2(0, \infty))$, $\|u(\cdot)\|_{L^\infty(0,T';L^2(0,\infty))} \leq 1$. This means

$$y(T, 0, \bar{u}_\alpha, \lambda) = y(T', 0, u, \lambda)\,.$$

We multiply both sides by $\lambda^\beta$, integrate in $0 \leq \lambda < \infty$ and make use of (3.4.17) and (3.4.16). The result is

$$\frac{\Gamma(1+\alpha+\beta)}{\sqrt{\Gamma(1+2\alpha)}\sqrt{\Gamma(1+2\beta)}} \int_0^T \frac{e^{-c\sigma} d\sigma}{\sigma^{1/2+\beta}} \leq \int_0^{T'} \frac{e^{-c\sigma} d\sigma}{\sigma^{1/2+\beta}}\,. \tag{3.4.18}$$

If $\alpha < 1/2$ we set $\beta = \alpha$, thus the coefficient of the integral on the left of (3.4.18) becomes 1 and we have an instant contradiction. For $\alpha = 1/2$ we can't take $\beta = \alpha$, since this would blow up the integrals; here, (3.4.18) becomes

$$\frac{\Gamma(3/2+\beta)}{\sqrt{\Gamma(1+2\beta)}} = \frac{\Gamma(3/2+\beta)}{\sqrt{\Gamma(2)}\,\sqrt{\Gamma(1+2\beta)}} \le \frac{\Phi(c,\beta,T')}{\Phi(c,\beta,T)}\,, \quad -\frac{1}{2} < \beta < \frac{1}{2}\,, \tag{3.4.19}$$

where

$$\begin{aligned}\Phi(c,\beta,T) &= \int_0^T \frac{e^{-c\sigma}d\sigma}{\sigma^{1/2+\beta}} = \sum_{n=0}^{\infty}\frac{(-1)^n c^n}{n!}\int_0^T \sigma^{n-1/2-\beta}d\sigma \\ &= \sum_{n=0}^{\infty}\frac{(-1)^n c^n}{n!}\cdot\frac{T^{n+1/2-\beta}}{n+1/2-\beta} \\ &= \frac{T^{1/2-\beta}}{1/2-\beta}\left(1+\left(\frac{1}{2}-\beta\right)\sum_{n=1}^{\infty}\frac{(-1)^n c^n T^n}{n!(n+1/2-\beta)}\right) \\ &= \frac{T^{1/2-\beta}}{1/2-\beta}\left(1+\left(\frac{1}{2}-\beta\right)\Psi(c,\beta,T)\right).\end{aligned}$$

Note that, although $\Phi(c,\beta,T)$ is only defined for $\beta < 1/2$, the function

$$\Psi(c,\beta,T) = \sum_{n=1}^{\infty}\frac{(-1)^n c^n T^n}{n!(n+1/2-\beta)}$$

is defined for $\beta < 3/2$; hence, simplifying the denominators $1/2-\beta$, the quotient $\Phi(c,\beta,T')/\Phi(c,\beta,T)$ is defined for $\beta < 3/2$. We have

$$\Psi(c,1/2,T) = \sum_{n=1}^{\infty}\frac{(-1)^n c^n T^n}{n!\,n}\,,$$

thus

$$\frac{\partial}{\partial T}\Psi(c,1/2,T) = \frac{1}{T}\sum_{n=1}^{\infty}\frac{(-1)^n c^n T^n}{n!} = -\frac{1-e^{-cT}}{T}\,. \tag{3.4.20}$$

Now,

$$\begin{aligned}\frac{\Phi(c,\beta,T')}{\Phi(c,\beta,T)} &= \left(\frac{T'}{T}\right)^{1/2-\beta}\left(\frac{1+(1/2-\beta)\Psi(c,\beta,T')}{1+(1/2-\beta)\Psi(c,\beta,T)}\right) \\ &= \left(\frac{T'}{T}\right)^{1/2-\beta}\Omega(c,\beta,T',T)\,,\end{aligned} \tag{3.4.21}$$

and we have

$$\frac{\partial}{\partial\beta}\left(1+\left(\frac{1}{2}-\beta\right)\Psi(c,\beta,T)\right) = -\Psi(c,\beta,T)+\left(\frac{1}{2}-\beta\right)\frac{\partial\Psi}{\partial\beta}(c,\beta,T)\,,$$

hence
$$\frac{\partial}{\partial\beta}\bigg|_{\beta=1/2}\Big(1+\Big(\frac{1}{2}-\beta\Big)\Psi(c,\beta,T)\Big) = -\Psi(c,1/2,T)\,.$$
Taking into account that
$$\Big(1+\Big(\frac{1}{2}-\beta\Big)\Psi(c,\beta,T)\Big)\bigg|_{\beta=1/2} = 1$$
we obtain from the quotient rule that
$$\frac{\partial}{\partial\beta}\bigg|_{\beta=1/2}\Omega(c,\beta,T',T) = \Psi(c,1/2,T)-\Psi(c,1/2,T')\,. \tag{3.4.22}$$
On the other hand
$$\frac{\partial}{\partial\beta}\Big(\frac{T'}{T}\Big)^{1/2-\beta} = \frac{\partial}{\partial\beta}\Big(\frac{T}{T'}\Big)^{\beta-1/2} = \Big(\log\frac{T}{T'}\Big)\Big(\frac{T}{T'}\Big)^{\beta-1/2},$$
thus
$$\frac{\partial}{\partial\beta}\bigg|_{\beta=1/2}\Big(\frac{T'}{T}\Big)^{1/2-\beta} = \log\frac{T}{T'}\,. \tag{3.4.23}$$
Noticing that
$$\Big(\frac{T}{T'}\Big)^{1/2-\beta}\bigg|_{\beta=1/2} = 1, \quad \Omega(c,1/2,T',T) = 1\,,$$
and using (3.4.20), (3.4.22), (3.4.23) and the product rule in (3.4.21) we obtain
$$\begin{aligned}\frac{\partial}{\partial\beta}\bigg|_{\beta=1/2}\frac{\Phi(c,\beta,T')}{\Phi(c,\beta,T)} &= \log T-\log T'+\Psi(c,1/2,T)-\Psi(c,1/2,T')\\ &= \int_{T'}^{T}\Big(\frac{1}{s}-\frac{1-e^{-cs}}{s}\Big)ds = \int_{T'}^{T}\frac{e^{-cs}}{s}ds = a>0\,.\end{aligned}$$
Accordingly,
$$\frac{\Phi(c,\beta,T')}{\Phi(c,\beta,T)} = 1-a\Big(\frac{1}{2}-\beta\Big)+O\Big(\Big(\frac{1}{2}-\beta\Big)^2\Big)\,. \tag{3.4.24}$$
On the other hand, we have
$$\frac{\partial}{\partial\beta}\frac{\Gamma(3/2+\beta)}{\sqrt{\Gamma(1+2\beta)}} = \frac{1}{\Gamma(1+2\beta)}\bigg(\Gamma'(3/2+\beta)\sqrt{\Gamma(1+2\beta)}-\frac{\Gamma(3/2+\beta)\Gamma'(1+2\beta)}{\sqrt{\Gamma(1+2\beta)}}\bigg)$$
so that
$$\frac{\partial}{\partial\beta}\bigg|_{\beta=1/2}\frac{\Gamma(3/2+\beta)}{\sqrt{\Gamma(1+2\beta)}} = 0\,,$$

which implies

$$\frac{\Gamma(3/2+\beta)}{\sqrt{\Gamma(1+2\beta)}} = 1 + O\Big(\Big(\frac{1}{2}-\beta\Big)^2\Big). \tag{3.4.25}$$

In view of (3.4.19), estimates (3.4.24) and (3.4.25) can be combined into

$$O\Big(\Big(\frac{1}{2}-\beta\Big)^2\Big) \le -a\Big(\frac{1}{2}-\beta\Big) + O\Big(\Big(\frac{1}{2}-\beta\Big)^2\Big)$$

with $a > 0$, absurd near $1/2$. The graph below shows the functions on both sides of (3.4.19) with $c = 1$, $T = 1$, $T' = 0.9$ for $0 \le \beta \le 1/2$; although initially

$$\frac{\Gamma(3/2+\beta)}{\sqrt{\Gamma(1+2\beta)}} < \frac{\Phi(1,\beta,0.9)}{\Phi(1,\beta,1)},$$

the function on the left overtakes the one on the right at $\beta = 0.3729$ and

$$\frac{\Gamma(3/2+\beta)}{\sqrt{\Gamma(1+2\beta)}} > \frac{\Phi(1,\beta,0.9)}{\Phi(1,\beta,1)}$$

for $0.3729 < \beta \le 1/2$.

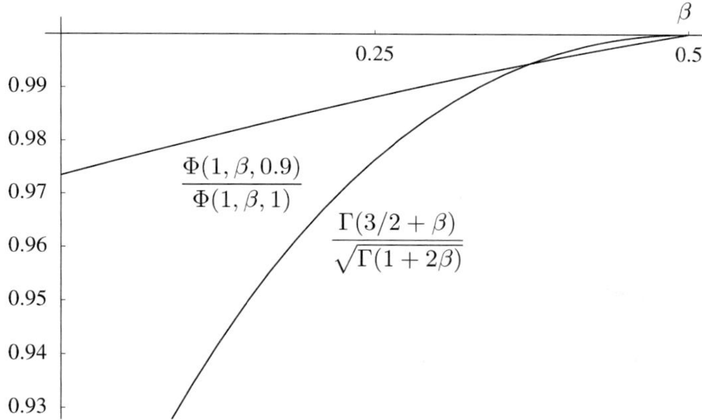


Figure 3.4.1

**Corollary 3.4.3.** *The control $\bar{u}_{1/2}(\cdot)$ is a weakly singular extremal.*

*Proof.* We must show that $\bar{u}_{1/2}(t)$ cannot be a regular extremal. If it were, there would be a $z \in Z(T)$ such that

$$\bar{u}(t) = \frac{S_c(T-t)\psi_{1/2}}{\|S_c(T-t)\psi_{1/2}\|} = \frac{S_c(T-t)z}{\|S_c(T-t)z\|}. \tag{3.4.26}$$

We have

$$(-A_c)^{-\gamma}\psi_{1/2}(\lambda) = \frac{\sqrt{\lambda}}{(\lambda + c)^\gamma}$$

so that $(-A_c)^{-\gamma}\psi_{1/2} \in L^2(0,\infty)$ for $\gamma > 1$. On the other hand, since $Z(T) \subseteq E_{-1}$ we have $A_c^{-1}z \in L^2(0,\infty)$, thus we may apply $(-A_c)^{-\gamma}$ to both sides of (3.4.26):

$$\frac{S_c(T-t)(-A)^{-\gamma}\psi_{1/2}}{\|S_c(T-t)\psi_{1/2}\|} = \frac{S_c(T-t)(-A)^{-\gamma}z}{\|S_c(T-t)z\|}.$$

Using this equality for any fixed $t < T$ we obtain

$$(-A_c)^{-\gamma}\psi_{1/2} = C(-A_c)^{-\gamma}z = (-A_c)^{-\gamma}Cz\,,$$

($C \neq 0$ a constant) thus $\psi_{1/2} = Cz$, a contradiction since $\psi_{1/2} \notin Z(T)$.

**Miscellaneous notes.** We have just shown that the control system (3.4.3) possesses weakly singular time and norm optimal controls. It does also possess strongly singular norm optimal controls by Theorem 2.8.6; in fact, since $S(t)$ is an analytic semigroup with unbounded infinitesimal generator, $S(t)^* = S(t)$ does not satisfy $\mathcal{P}(a,b,\mu)$ for any $a > 0, b > 0, \mu > -c$ (see the remarks after Theorem 2.9.2). On the other hand, Theorem 3.2.3 says that (3.4.3) does not have any strongly singular time optimal controls. We can picture all of this as

$$\text{The system}\quad y'(t) = A_c y(t) + u(t)$$

$$\left\{\begin{array}{c}\text{has weakly singular}\\ \text{norm optimal controls}\end{array}\right\} \qquad \left\{\begin{array}{c}\text{has weakly singular}\\ \text{time optimal controls}\end{array}\right\}$$

$$\left\{\begin{array}{c}\text{has strongly singular}\\ \text{norm optimal controls}\end{array}\right\} \qquad \left\{\begin{array}{c}\text{has \textit{no} strongly singular}\\ \text{time optimal controls}\end{array}\right\}$$

The construction of the weakly singular extremal is from the author [2001:2]. Apparently, no other example is known.

**3.5. More on the weak maximum principle.** What the controls

$$\bar{u}_\alpha(t,\lambda) = \frac{S_c(T-t)\psi_\alpha(\lambda)}{\|S_c(T-t)\psi_\alpha(\cdot)\|} \tag{3.5.1}$$

in (3.4.9) do for $\alpha > 1/2$ is revealed in

**Theorem 3.5.1.** *Let $\alpha > 1/2$. Then the control $\bar{u}_\alpha(t)$ is not norm optimal (thus not time optimal) in any interval $0 \leq t \leq T$.*

*Proof.* According to the first two lines of (3.4.12),

$$y(T,0,\bar{u}_\alpha,\lambda) = \frac{2^{1/2+\alpha}}{\sqrt{\Gamma(1+2\alpha)}}\,\mathcal{F}(T,\alpha,\lambda)\,, \tag{3.5.2}$$

while a function $u(t)$ with functional values produces a function of two variables reading (4.1.15) backwards, that is

$$u(t,x) = u(t)(x) \tag{4.1.16}$$

and it is in this sense that the spaces in $(a)$ and $(b)$ are equal: we can go from the space on the right to the space on the left with (4.1.15) and back with (4.1.16). The identification is algebraic and metric.

We begin with $(a)$. Let $v(\cdot,\cdot) \in L^p((0,T) \times \Omega)$. By Fubini's theorem the function $v(t)(x)$ defined by (4.1.15) belongs to $L^p(\Omega)$ for almost all $t$. The function $v(t,x)$ can be approximated in the $L^p((0,T) \times \Omega)$ norm by functions of the form

$$v_n(t,x) = \sum_{\text{finite}} \psi_{nj}(t)\phi_{nj}(x) \tag{4.1.17}$$

with the $\psi_{nj}(\cdot) \in L^p(0,T)$, $\phi_{nj}(\cdot) \in L^p(\Omega)$. Each function $v_n(t)(x) = v_n(t,x)$ is in $L^p(0,T;L^p(\Omega))$ and the $v_n(\cdot)$ converge to $v(\cdot)$ in the norm of $L^p(0,T;L^p(\Omega))$. Since

$$\|v_n(\cdot)\|_{L^p(0,T;L^p(\Omega))} = \|v_n(\cdot,\cdot)\|_{L^p((0,T)\times\Omega)}$$

the inclusion $\subseteq$ in $(a)$ follows. This argument works backwards; in fact, a function $v(\cdot) \in L^p(0,T;L^p(\Omega))$ can be approximated in the $L^p(0,T;L^p(\Omega))$ norm by functions of the form (4.1.17), which gives the opposite inclusion.

To show $(b)$, assume that $u(\cdot,\cdot) \in L^\infty((0,T)\times\Omega)$. The function $u(t)(x)$ defined by (4.1.15) takes values in $L^\infty(\Omega)$ a. e. If $v(\cdot) \in L^1(\Omega)$ then

$$t \to \langle u(t,\cdot), v(\cdot)\rangle = \int_\Omega u(t,x)v(x)dx \tag{4.1.18}$$

is measurable and $\leq \|u(\cdot,\cdot)\|_{L^\infty((0,T)\times\Omega)}\|v(\cdot)\|_{L^1(\Omega)}$ a. e. in $t$; it then follows that $u(\cdot) \in L^\infty_w(0,T;L^\infty(\Omega))$ and

$$\|u(\cdot)\|_{L^\infty_w(0,T;L^\infty(\Omega))} \leq \|u(\cdot,\cdot)\|_{L^\infty((0,T)\times\Omega)}\,. \tag{4.1.19}$$

Conversely, let $u(\cdot) \in L^\infty_w(0,T;L^\infty(\Omega))$. Then $u(\cdot)$ produces a function $u(t,x)$ of two variables through (4.1.16). This function makes (4.1.18) measurable for every $v(\cdot) \in L^1(\Omega)$ and

$$\int_\Omega u(t,x)v(x)dx \leq \|u(\cdot)\|_{L^\infty_w(0,T;L^\infty(\Omega))}\|v(\cdot)\|_{L^1(\Omega)} \quad \text{a. e.}$$

Take $p$ with $1 < p < \infty$, $1/q + 1/p = 1$. We may think of $u(\cdot)$ as a $L^p(\Omega)$-valued function and (4.1.18) is measurable for $v(\cdot) \in L^q(\Omega) \subset L^1(\Omega)$, so that $u(\cdot)$ is $L^q(\Omega)$-weakly measurable, hence strongly measurable as an $L^p(\Omega)$-valued function.[1] It then follows from the $L^p$ result $(a)$ that the function $u(t,x)$ in (4.1.16) is measurable. To show that (4.1.19) is actually an equality assume

$$\|u(\cdot,\cdot)\|_{L^\infty((0,T)\times\Omega)} > \|u(\cdot)\|_{L^\infty_w(0,T;L^\infty(\Omega))}\,.$$

Then there exists $\epsilon > 0$ and a set $e \subseteq (0,T)\times\Omega$ of positive measure such that

$$|u(t,x)| \ge \|u\|_{L^\infty_w(0,T;L^\infty(\Omega))} - \epsilon \quad ((t,x)\in e)\,.$$

Let $v(t,x) = \chi_e(t,x)\operatorname{sign} u(t,x)$, $\chi_e(t,x)$ the characteristic function of $e$. We have

$$\begin{aligned}\int_{(0,T)\times\Omega} u(t,x)v(t,x)dtdx &= \int_e |u(t,x)|dtdx \\ &\ge |e|\big(\|u\|_{L^\infty(0,T;L^\infty(\Omega))} - \epsilon\big) \end{aligned} \tag{4.1.20}$$

($|\cdot|$ indicates measure) while

$$\begin{aligned}\int_\Omega u(t,x)v(t,x)dx &\le \|u\|_{L^\infty_w(0,T;L^\infty(\Omega))}\int_\Omega |v(t,x)|dx \\ &= \|u\|_{L^\infty_w(0,T;L^\infty(\Omega))}|e\cap(\{t\}\times\Omega)|\end{aligned}$$

which, upon integration in $0\le t\le T$ contradicts (4.1.20). This ends the proof of $(b)$ and of Lemma 4.1.1.

We define an extension of $S_c(t)$ by[2]

$$S_\infty(t) = S_1'(t)^*\,. \tag{4.1.21}$$

Since $S_1'(t)$ is an analytic semigroup, $S_\infty(t)\in\mathcal{L}(L^\infty(\Omega),L^\infty(\Omega))$ and is analytic for $\operatorname{Re} t > 0$ (but not strongly continuous at $t=0$). We obtain from (4.1.9) for $S_1'(t)$ that

$$\|S_\infty(t)\|_{\mathcal{L}(L^\infty(\Omega),L^\infty(\Omega))} \le Ce^{\omega t} \quad (t\ge 0)\,. \tag{4.1.22}$$

The "correct" version of the variation-of-constants formula turns out to be

$$y(t,\zeta,u) = S_\infty(t)\zeta + \int_0^t S_\infty(t-\sigma)u(\sigma)d\sigma\,. \tag{4.1.23}$$

However, much work remains to be done; we have to show that $y(t,\zeta,u)$ takes values in $C(\overline{\Omega})$ rather than in $L^\infty(\Omega)$. Also, in view that $u(\cdot)$ is merely weakly measurable, we have to justify the integral in (4.1.23).

---

[1] If $u(\cdot)$ takes values in a separable Banach space $E$ and $\langle u(\cdot),u^*\rangle$ is measurable for any $u^*\in E^*$ then $u(\cdot)$ is strongly measurable. See Hille - Phillips [1957, Theorem 3.5.3, p. 72].

[2] At this point, it is not obvious that $S_\infty(t)$ is an extension of $S_c(t)$. This result will be proved in **4.3**.

**Miscellaneous notes.** The proof of the relation (4.1.14) for general spaces is tricky, as is the very definition of the space $L^\infty_w(0, T; X^*)$. The equivalence relation in this space is

$$u_1(\cdot) \approx u_2(\cdot) \iff \langle u_1(t), v\rangle = \langle u_2(t), v\rangle \quad \text{a. e.}$$

(a. e. depending on $v$). To see how this relation works, consider the space $X$ of all real valued families $u = \{u_s;\ 0 \le s \le 1\}$ such that

$$\|u\|_X = \sum_{0\le s\le 1} |u_s|^2 < \infty$$

(this condition implies of course that $u_s = 0$ except for $s$ in a finite or countable set depending on $u$). The space $X$ equipped with $\|\cdot\|_X$ is a (nonseparable) Hilbert space, with scalar product

$$(u, v) = \sum_{0\le s\le 1} u_s v_s\,,$$

so that $X^* = X$. Let $\phi(t)$, $0 \le t \le 1$ be an arbitrary scalar function. Define a $X$-valued function $u(t)$ by

$$u(t)_s = \begin{cases} \phi(t) & (s = t)\,, \\ 0 & (s \ne t)\,. \end{cases}$$

If $v \in X$ then $\langle u(t), v\rangle = \sum u(t)_s v_s$ can be $\ne 0$ only for those values of $t$ where $v_t \ne 0$, Since the number of these values are countable or finite, $u(\cdot)$ is $X$-weakly measurable, and in the same equivalence class as the zero function. On the other hand, $\|u(t)\|_X = \phi(t)$ so the norm of a $X^*$-valued $X$-weakly measurable function is undefined in the equivalence class, and since $\phi(t)$ is arbitrary, it may not even be measurable, so that the $L^\infty_w(0, T; X^*)$ and the $L^\infty(0, T; X^*)$ norm are entirely different things.

Weirdness dissipates under separability assumptions. If $X$ is separable, then the norm $\|u(t)\|_{X^*}$ is well defined and measurable, as

$$\|u(t)\|_{X^*} = \sup_n \langle u(t), v_n\rangle \quad \text{a. e.}$$

where $\{v_n\}$ is a countable dense set in the unit ball of $X$. The norm of $L^\infty_w(0, T; X^*)$ and the norm of $L^\infty(0, T; X^*)$ are the same, but the two spaces may be different, that is, the inclusion $L^\infty(0, T; X^*) \subset L^\infty_w(0, T; X^*)$ may be strict. In fact there may be functions $u(\cdot) \in L^\infty_w(0, T; X^*)$ not strongly measurable. An example is the translation function $u(t) = u(t + \cdot)$ in $X^* = L^1(-\infty, \infty)^* = L^\infty(-\infty, \infty)$; this function $u(\cdot)$ is always $X = L^1(-\infty, \infty)$-weakly measurable, but it is not strongly measurable, for example, if $u(x)$ is the characteristic function of $t \ge 0$.

Finally, if $X^*$ itself is separable, every $X^*$-valued $X$-weakly measurable function is strongly measurable, thus we have

$$L^\infty(0, T; X^*) = L^\infty_w(0, T; X^*) .$$

The variation-of-constants formula (4.1.23) is in the author [2001:4] and [2002:2].

**4.2. Adjoints.** We need some linear adjoint theory in multivalued generality. Let $A, B$ be arbitrary unbounded operators in a Banach space $X = \{y, z, \ldots\}$ with dual $X^* = \{y^*, z^*, \ldots\}$. We do *not* assume that the domains $D(A)$, $D(B)$ are dense in $X$. The adjoint of $A^*$ is (as usual) defined by: $y^* \in D(A^*)$ if the functional

$$y \to \langle y^*, Ay\rangle$$

is bounded in the norm of $E$ in $D(A)$; if it is, it can be extended by the Hahn-Banach theorem to a bounded linear functional $z^* \in E^*$ and we define $A^*y^* = z^*$. This can be summarized as

$$\langle z^*, y\rangle = \langle y^*, Ay\rangle \quad (y \in D(A)) . \tag{4.2.1}$$

If $D(A)$ is not dense, the extension $z^*$ is not uniquely defined and (4.2.1) does not determine $z^*$ uniquely: this means that the adjoint is multivalued. Any two values $z^*$, $\zeta^*$ of $A^*y^*$ satisfy

$$\langle h^*, y\rangle = 0 \quad (y \in D(A)) , \tag{4.2.2}$$

where $h^* = \zeta^* - z^*$, and, conversely, if $h^*$ satisfies (4.2.2) and $z^*$ is a value of $A^*y^*$ then $\zeta^* = z^* + h^*$ is a value as well.

We denote by $[A^*y^*]$ the set of all values of $A^*y^*$. It follows from the preceding arguments that, if $z^* \in [A^*y^*]$ then

$$[A^*y^*] = z^* + \mathcal{N}_A , \tag{4.2.3}$$

where

$$\mathcal{N}_A = \{y^* \in E^*; \langle y^*, y\rangle = 0 , \ y \in D(A)\} . \tag{4.2.4}$$

Given the two (possibly) multivalued operators $A^*$ and $B^*$, the composition $A^*B^*$ is defined as usual: the domain is

$$D(A^*B^*) = \{y^* \in E^*; B^*y^* \in D(A^*)) ,$$

and

$$A^*B^*y^* = A^*(B^*y^*) .$$

Taking multivaluedness into account, these definitions can be rewritten

$$D(A^*B^*) = \{y^* \in E^*; [B^*y^*] \cap D(A^*) \neq \emptyset\},$$
$$[A^*B^*y^*] = \bigcup_{z^* \in [B^*y^*] \cap D(A^*)} [A^*z^*]. \tag{4.2.5}$$

Let $y^* \in D(A^*B^*)$, and let $B^*y^*$ be an element of $[B^*y^*] \cap D(A^*)$. Then, for any element $A^*B^*y^*$ of $[A^*B^*y^*]$ we have

$$\langle A^*B^*y^*, y\rangle = \langle B^*y^*, Ay\rangle \quad (y \in D(A)).$$

If, in addition, $Ay \in D(B)$, then

$$\langle A^*B^*y^*, y\rangle = \langle B^*y^*, Ay\rangle = \langle y^*, BAy\rangle \quad (y \in D(A)),$$

which means $y^* \in D((BA)^*)$ and $A^*B^*y^* \in [(BA)^*y^*]$. On account of possible multivaluedness, all of this can be reformulated as follows:

$$[A^*B^*y^*] = \bigcup_{z^* \in [B^*y^*] \cap D(A^*)} [A^*z^*] \subseteq [(BA)^*y^*] \quad (y^* \in D(A^*B^*)).$$

In shorthand, this is the familiar inclusion

$$A^*B^* \subseteq (BA)^*. \tag{4.2.6}$$

An important particular case is that where $B$ is everywhere defined and bounded. Here, $D(BA) = D(A)$. If $y^* \in D((BA)^*)$ then

$$y \to \langle y^*, BAy\rangle = \langle B^*y^*, Ay\rangle$$

is a bounded functional of $y$, hence $B^*y^* \in D(A^*)$, or $y^* \in D(A^*B^*)$ with

$$\langle y^*, BAy\rangle = \langle A^*B^*y^*, y\rangle,$$

so that

$$[(BA)^*y^*] \subseteq [A^*B^*y^*] = \bigcup_{z^* \in [B^*y^*] \cap D(A^*)} [A^*z^*] \quad (y^* \in D((BA)^*))$$

or, in brief,

$$(BA)^* \subseteq A^*B^*.$$

Combining with (4.2.6), if $B$ is everywhere defined and bounded we have

$$[(BA)^*y^*] = [A^*B^*y^*] \quad (y^* \in D((BA)^*),$$

or

$$(BA)^* = A^* B^* . \tag{4.2.7}$$

Another useful relation is

$$B \subseteq A \implies A^* \subseteq B^* . \tag{4.2.8}$$

To see this, let $y^* \in D(A^*)$. Then, since $B \subseteq A$,

$$\langle y^*, By \rangle = \langle y^*, Ay \rangle = \langle A^* y^*, y \rangle \quad (y \in D(B))$$

for any $A^* y^* \in [A^* y^*]$. This means, $A^* y^* \in [B^* y^*]$, thus

$$[A^* y^*] \subseteq [B^* y^*] \quad (y^* \in D(B^*)) ,$$

which is (4.2.8).

Let $A$ be an operator (not necessarily densely defined) having a bounded, everywhere defined inverse $A^{-1}$. Applying (4.2.6),

$$(A^{-1})^* A^* \subseteq (AA^{-1})^* = I^* , \tag{4.2.9}$$

which inclusion, taking into account possible multivaluedness of $A^*$ means

$$(A^{-1})^* A^* y^* = y^* \quad (A^* y^* \in [A^* y^*]) .$$

Now, we write $A^{-1} A \subseteq I$ in the form $A^{-1} A = I|_{D(A)}$ and use (4.2.7), obtaining

$$[A^* (A^{-1})^* y^*] = \left[ \left( I|_{D(A)} \right)^* y^* \right] = y^* + \mathcal{N}_A , \tag{4.2.10}$$

with $\mathcal{N}_A$ given by (4.2.4). We have $D(A) = A^{-1} E$, thus $y^* \in \mathcal{N}_A$ is equivalent to $\langle y^*, y \rangle = \langle y^*, A^{-1} z \rangle = \langle (A^{-1})^* y^*, z \rangle = 0$ for all $z \in E$; this means

$$\mathcal{N}_A = \{ y^* ; (A^{-1})^* y^* = 0 \} , \tag{4.2.11}$$

that is, $\mathcal{N}_A$ is the nullspace of $(A^{-1})^*$. We can condense (4.2.9) and (4.2.10) as

$$(A^*)^{-1} = (A^{-1})^* , \tag{4.2.12}$$

although (4.2.10) must be kept in mind. If $D(A)$ is dense, (4.2.10) is interpreted in the standard way $A^* (A^{-1})^* = I$, thus (4.2.12) is also interpreted classically.

Let $y^* \in D(A^* + B^*) = D(A^*) \cap D(B^*)$. If $y \in D(A + B) = D(A) \cap D(B)$,

$$\begin{aligned} \langle (A^* + B^*) y^*, y \rangle &= \langle A^* y^*, y \rangle + \langle B^* y^*, y \rangle \\ &= \langle y^*, Ay \rangle + \langle y^*, By \rangle = \langle y^*, (A + B) y \rangle , \end{aligned}$$

so that $y^* \in D(A+B)^*$. We then have

$$[A^*y^*] + [B^*y^*] \subseteq [(A+B)y^*] \quad (y^* \in D(A^* + B^*)),$$

which, in shorthand, is

$$A^* + B^* \subseteq (A+B)^* . \tag{4.2.13}$$

If $B$ is everywhere defined and bounded, we write $A = (A+B) - B$ and use (4.2.13): since $D(B^*) = E^*$, if $y^* \in D((A+B)^* - B^*) = D((A+B)^*)$ we have[3]

$$[(A+B)^*y^*] - [B^*y^*] \subseteq [A^*y^*] \quad (y^* \in D(A+B)^*),$$

which can be rearranged as

$$[A^*y^*] + [B^*y^*] \supseteq [(A+B)^*y^*] \quad (y^* \in D(A+B)^*),$$

or $(A+B)^* \supseteq (A+B)^*$. Combining with (4.2.13),

$$[A^*y^*] + [B^*y^*] = [(A+B)^*y^*] \quad (y^* \in D(A^* + B^*) = D((A+B)^*))$$

or, in different notation,

$$A^* + B^* = (A+B)^* . \tag{4.2.14}$$

Finally, an observation about inverses. Let $E \subseteq F$ be two linear spaces, $A$ (resp. $B$) an operator with domain $D(A) \subseteq E$ and range in $E$ (resp. $D(B) \subseteq F$ and range in $F$). Assume $A$ (resp. $B$) has an inverse $A^{-1} : E \to E$ (resp. an inverse $B^{-1} : F \to F$). Then

$$A \subseteq B \implies A^{-1} \subseteq B^{-1}, \quad A^{-1} = B^{-1}|_E . \tag{4.2.15}$$

In fact, let $y \in D(A)$. Then $y = A^{-1}Ay = A^{-1}By = B^{-1}By$ so that $A^{-1}$ and $B^{-1}$ coincide in $AD(A) = E$. The equality in (4.2.15) follows from the inclusion.

The space $\Sigma(\overline{\Omega})$ consists of all finite regular Borel measures in $\overline{\Omega}$ such that

$$\|\mu\|_{\Sigma(\overline{\Omega})} = |\mu|(\overline{\Omega}) < \infty$$

where $|\mu|$, the *total variation* of $\mu$ is the Borel measure defined by

$$|\mu|(e) = \sup \sum |\mu(e_j)| ,$$

the supremum taken over all finite partitions of $e$ in measurable sets $\{e_j\}$. Under the norm $\|\cdot\|_{\Sigma(\overline{\Omega})}$, $\Sigma(\overline{\Omega})$ is a Banach space (for the proof see Dunford - Schwartz [1958, p. 97]). If $y(\cdot) \in C(\overline{\Omega})$ we have

$$\left| \int_{\overline{\Omega}} y(x)\mu(dx) \right| \le \int_{\overline{\Omega}} |y(x)||\mu|(dx) \le \|y\|_{C(\overline{\Omega})} \|\mu\|_{\Sigma(\overline{\Omega})} .$$

[3] Of course, $[B^*y^*]$ contains only one element since $B^*$ is single valued.

Interpreting functions $f(\cdot) \in L^1(\Omega)$ as measures $\mu(e) = \int_e f(x)dx$ we have

$$L^1(\Omega) \overset{i}{\hookrightarrow} \Sigma(\overline{\Omega}) \tag{4.2.16}$$

( $\overset{i}{\hookrightarrow}$ means isometric imbedding). The space $\Sigma(\overline{\Omega})$ is algebraically and metrically isomorphic to the dual $C(\overline{\Omega})^*$, the duality between both spaces given by

$$\langle \mu, y \rangle = \int_{\overline{\Omega}} y(x)\mu(dx) \tag{4.2.17}$$

(see Dunford - Schwartz [1958, Theorem 3, p. 265]. The dual of $C_0(\overline{\Omega})$ is the subspace $\Sigma_0(\overline{\Omega}) \subseteq \Sigma(\overline{\Omega})$ defined by $|\mu|(\Gamma) = 0$.

We apply duality to the operators $A_c, A'_c, A_p, A'_p$ in **4.1**. Define an operator in the space $\Sigma(\overline{\Omega})$ by

$$A'_\Sigma = A^*_c\,. \tag{4.2.18}$$

It is clear from the definition of $A^*_c$ and the weak definition of $A'_1$ that

$$A'_1 \subseteq A'_\Sigma\,, \quad A'_1 = A'_\Sigma\big|_{L^1(\Omega)}\,. \tag{4.2.19}$$

Let $\lambda > \omega$, $\omega$ the constant in (4.1.7), so that $R(\lambda; A_c) = (\lambda I - A_c)^{-1}$ exists. Using (4.2.12), (4.2.14) and the fact that $D(A_c)$ is densely defined we see that $(\lambda I^* - A_\Sigma)^{-1} = (\lambda I^* - A^*_c)^{-1} = ((\lambda I - A_c)^*)^{-1}$ exists with

$$(\lambda I^* - A_\Sigma)^{-1} = ((\lambda I - A_c)^{-1})^*\,. \tag{4.2.20}$$

The inclusion and the equality in (4.2.19) for $A'_1$, $A'_\Sigma$ imply the corresponding statements for $\lambda I - A'_1$, $\lambda I - A'_\Sigma$. Using (4.2.15) we get

$$(\lambda I^* - A'_1)^{-1} \subseteq (\lambda I^* - A'_\Sigma)^{-1}\,, \quad (\lambda I^* - A'_1)^{-1} = (\lambda I^* - A'_\Sigma)^{-1}\big|_{L^1(\Omega)}\,. \tag{4.2.21}$$

The operator $A'_\Sigma$ is not densely defined in $\Sigma(\overline{\Omega})$ or in $\Sigma_0(\overline{\Omega})$ since

$$D(A'_\Sigma) \subseteq L^1(\Omega) \tag{4.2.22}$$

and $L^1(\Omega)$ is not dense[4] in $\Sigma(\overline{\Omega})$. To prove (4.2.22), we use the following result:

**Lemma 4.2.1.** *Let $\mu \in \Sigma(\overline{\Omega})$. Then there exists a sequence $f_n(\cdot) \in L^1(\Omega)$ with*

$$\|f_n(\cdot)\|_{L^1(\Omega)} \le C\|\mu\|_{\Sigma(\overline{\Omega})}\,, \quad \langle y, f_n\rangle \to \langle y, \mu\rangle \quad (y(\cdot) \in C(\overline{\Omega}))\,,$$

*where the constant $C$ is independent of $\mu$.*

---

[4] If it were, it would follow from (4.2.16) that $L^1(\Omega) = \Sigma(\overline{\Omega})$.

where the control in the last line of (4.4.15) is

$$w(\sigma) = \frac{S_c(\sigma)(S_c(t-s)-I)\zeta}{s} + \frac{S_c(\sigma)}{s}\int_0^{t-s} S_\infty(t-s-\tau)u(\tau)d\tau + u(\sigma+t-s)$$

and we have the estimate

$$\|w(\sigma)\| \le \frac{M(s)}{s}\|S_c(t-s)\zeta - \zeta\| + \Big(1 + \frac{t-s}{s}M(s)\,M(t-s)\Big)\rho \quad (0 \le \sigma \le t) \tag{4.4.16}$$

(see (4.4.8)). According to (4.4.15) the control $w(\cdot)$ drives $\zeta$ to $\bar{y}$ in time $s < t$ and, since $\rho < 1$, it follows from (4.4.16) that, if $s$ is sufficiently close to $t$ then $\|w(\cdot)\|_{L^\infty(0,s;E)} \le 1$, which shows that $t$ is not the optimal time.

Assume finally that $\zeta \in L^\infty(\Omega)$, and let $0 < h < t$. Using the control $u(\cdot)$ in (4.4.14) and the fact that $y(h,\zeta,u) \in C(\overline{\Omega})$ we can construct a control $w(\cdot)$ that improves the driving time in $h \le \sigma \le t$, which, by the optimality principle, implies that $v(\cdot)$ is not time optimal.

**Theorem 4.4.3.** *Let $\bar{u}(t)$ be a time optimal control in $0 \le t \le T$. Then*

$$\|\bar{u}(t)\|_{L^\infty(\Omega)} = 1 \quad \text{a. e. in } 0 \le t \le T\,. \tag{4.4.17}$$

The proof is exactly the same as that of Theorem 2.1.3, as strong continuity of the semigroup at $t = 0$ plays no role. We don't bother to check the details since Theorem 4.4.3 does not say much; for instance, there is no uniqueness implication for optimal controls as the space $L^\infty(\Omega)$ is not strictly convex.

The manipulations in the proof of Lemma 2.1.6 can be extended and produce an unexpected result. Given $t > 0$, let $R^\infty(t)$ be (as in Chapters 2 and 3) the space of all elements $y \in C(\overline{\Omega})$ of the form

$$y = y(t,0,u) = \int_0^t S_c(t-\sigma)u(\sigma)d\sigma\,, \quad u(\cdot) \in L^\infty(0,t;C(\overline{\Omega}))\,, \tag{4.4.18}$$

equipped with the norm

$$\|y\|_{R_s^\infty(t)} = \inf\left\{\|u\|_{L^\infty(0,t;C(\overline{\Omega}))};\ \int_0^t S_c(t-\sigma)u(\sigma)d\sigma = y\right\}. \tag{4.4.19}$$

Clearly,

$$R^\infty(t) \subseteq R_w^\infty(t)\,, \quad \|y\|_{R_w^\infty(t)} \le \|y\|_{R^\infty(t)}\,, \tag{4.4.20}$$

but we have a lot more.

**Theorem 4.4.4.** $R^\infty(t) = R^\infty_w(t)$ *with equivalent norms. If* $\|S(t)\| \le 1$ *for* $t \ge 0$,

$$\|y\|_{R^\infty_w(t)} = \|y\|_{R^\infty(t)} \quad (t > 0)\,.$$

*Proof.* Given $\epsilon > 0$ select $a > 0$ such that $e^a \le 1 + \epsilon$ and define $t_n = t(1 - e^{-an})$ $(n = 0, 1, \ldots)$. We have $0 = t_0 < t_1 < \ldots < t$, $t_n \to t$ and

$$\frac{t_n - t_{n-1}}{t_{n+1} - t_n} = \frac{e^{-a(n-1)} - e^{-an}}{e^{-an} - e^{-a(n+1)}} = e^a \le 1 + \epsilon \quad (n = 1, 2, \ldots)\,. \tag{4.4.21}$$

Let

$$y = \int_0^t S_\infty(t - \sigma)u(\sigma)d\sigma\,, \quad u(\cdot) \in L^\infty_w(0, t; L^\infty(\Omega))$$

be an element of $R^\infty(t)$. We have

$$\begin{aligned} y &= \int_0^t S_\infty(t-\sigma)u(\sigma)d\sigma \\ &= \sum_{n=0}^\infty S_\infty(t - t_{n+1}) \int_{t_n}^{t_{n+1}} S_\infty(t_{n+1} - \tau)u(\tau)d\tau \\ &= \sum_{n=0}^\infty \int_{t_{n+1}}^{t_{n+2}} S_\infty(t - \sigma) \\ &\quad \times \left( \frac{S_\infty(\sigma - t_{n+1})}{t_{n+2} - t_{n+1}} \int_{t_n}^{t_{n+1}} S_\infty(t_{n+1} - \tau)u(\tau)d\tau \right) d\sigma \\ &= \int_0^t S_\infty(t - \sigma)v(\sigma)d\sigma \end{aligned}$$

where $v(\cdot) \in L^\infty(0, T; C(\Omega))$ (in fact, $v(\sigma)$ is a continuous $C(\overline{\Omega})$-valued function in each interval $t_n < t \le t_{n+1}$), and

$$\begin{aligned} \|v(\sigma)\| &\le \frac{t_{n+1} - t_n}{t_{n+1} - t_{n+2}} M(t)^2 \|u(\cdot)\|_{L^\infty_w(0,T;L^\infty(\Omega))} \\ &\le (1 + \epsilon)M(t)^2 \|u(\cdot)\|_{L^\infty_w(0,T;L^\infty(\Omega))} \quad (t_n \le \sigma \le t_{n+1})\,, \end{aligned}$$

where $1 + \epsilon$ comes from (4.4.21) and $M(t)$ from (4.4.4). This inequality implies

$$R^\infty_w(T) \subseteq R^\infty(T)\,, \quad \|y\|_{R^\infty(t)} \le M(t)^2 \|y\|_{R^\infty_w(t)}\,,$$

which, combined with (4.4.20) shows equivalence of the norms. If $\|S(t)\| \le 1$ then $M(t) = 1$ and equality of the norms follows, ending the proof of Theorem 4.4.4.

At first sight, Theorem 4.4.4 seems to say we need not even consider weakly measurable controls at all and we may operate with $R^\infty(T)$ just the same as in Chapters 2 and 3. This is illusory, however. The main difference between $R^\infty(T)$ and $R_w^\infty(T)$ is that (as we shall see in **4.8**), given an element $y \in R^\infty(T) = R_w^\infty(T)$ there *always* exist $u(\cdot) \in L_w^\infty(0, t; L^\infty(\Omega))$ such that

$$y = \int_0^t S_\infty(t - \sigma)u(\sigma)d\sigma \tag{4.4.22}$$

and $\|u(\cdot)\|_{L_w^\infty(0,t;L^\infty(\Omega))} = \|y\|_{R_w^\infty(T)}$ (this is nothing but an existence theorem for norm optimal controls). On the other hand, there may not be any control $u(\cdot) \in L^\infty(0, T; C(\overline{\Omega}))$ satisfying (4.4.22) and $\|u(\cdot)\|_{L^\infty(0,T;C(\overline{\Omega}))} = \|y\|_{R^\infty(T)}$, just $u_\epsilon(\cdot) \in L^\infty(0, T; C(\overline{\Omega}))$ with $\|u_\epsilon(\cdot)\|_{L^\infty(0,T;C(\overline{\Omega}))} \le \|y\|_{R^\infty(T)} + \epsilon$ for each $\epsilon > 0$. In short, *the control space* $L^\infty(0, T; C(\overline{\Omega}))$ *is a bad space for existence.*

**Miscellaneous notes.** Some of the results in this section may be new but most are simple generalizations of those in the author [1964]. Theorem 4.4.4 raises the question of whether we can obtain the space $R^\infty(t) = R_w^\infty(t)$ restricting even further the class of controls, for instance taking controls in the space $C(0, T; C(\overline{\Omega}))$ of $C(\overline{\Omega})$-valued continuous functions defined in $0 \le t \le T$. The answer is an emphatic "no". In fact, if $R_c^\infty(t)$ is the reachable space based on $C(0, T; C(\overline{\Omega}))$, separability of $C(\overline{\Omega})$ implies separability of $C(0, T; C(\overline{\Omega}))$ implies separability of $R_c^\infty(t)$. On the other hand, $S_c(t)$ is an analytic semigroup with unbounded infinitesimal generator, thus $S_c(t)^*$ does not satisfy Property $\mathcal{P}(a, b, \mu)$ for any $a, b > 0$ (see the remarks preceding (2.9.7)). Accordingly, Theorem 2.8.8 says that $R^\infty(t)$ is not separable.

**4.5. The reachable space and its dual, I.** As we have seen in Chapter 2, Pontryagin's maximum principle for (time, norm) optimal controls for the system $y'(t) = Ay(t) + u(t)$, $y(0) = \zeta$ (driven by strongly measurable controls) can be guaranteed under the condition $\bar{y} - S(T)\zeta \in \overline{D(A)}$, where the bar indicates closure in $R^\infty(T)$. In fact, this condition insures separation of $\bar{y} - S(T)\zeta$ from the ball $B_\rho^\infty(T)$ by means of a regular functional. We may then attempt the theory of

$$y'(t) = A_c y(t) + u(t), \quad y(0) = \zeta \tag{4.5.1}$$

in $C(\overline{\Omega})$ with controls in $L_w^\infty(0, T; L^\infty(\Omega)) = L^\infty((0, T) \times \Omega)$ under the condition[5]

$$\bar{y} \in \overline{D(A_c)}\,.$$

However, the norm of $R_w^\infty(T)$ is elusive and closures in this norm are difficult to figure out, thus in practice we would use

$$\bar{y} \in D(A_c)\,. \tag{4.5.2}$$

[5] Equivalent to $\bar{y} - S_\infty(T)\zeta \in \overline{D(A_c)}$ since $S_\infty(T)L^\infty(\Omega) \subseteq D(A_c)$.

The handling of the maximum principle under (4.5.2) is essentially the same as in the setting of strongly measurable controls, with minor technical complications. But we must ask, how strong is (4.5.2) in the $L^\infty(\Omega)$ setting? We test this in

**Example 4.5.1.** The system is

$$\frac{\partial y(t,x)}{\partial t} = \frac{\partial^2 y(t,x)}{\partial x^2} + u(t,x)\,, \quad y(t,0) = y(t,\pi) = 0\,,$$

with initial condition $u(0,x) = \zeta(x) = 0$ and time independent control

$$u_a(t,x) = h(x-a) \quad (0 \le x \le \pi)\,,$$

where $0 < a < \pi$ and $h(x) = 1$ $(x \ge 0)$, $h(x) = 0$ $(x < 0)$. We have

$$u_a(t,x) = \sum_{n=1}^{\infty} \left( \frac{2}{\pi} \int_a^\pi \sin n\xi d\xi \right) \sin nx = \sum_{n=1}^{\infty} \frac{2(\cos na - (-1)^n)}{n\pi} \sin nx$$

(all sine Fourier series convergent in the norm of $L^2(0,\pi)$), thus

$$y(t,x) = \sum_{n=1}^{\infty} \frac{1 - e^{-n^2 t}}{n^2} \cdot \frac{2(\cos na - (-1)^n)}{n\pi} \sin nx\,,$$

and

$$\frac{\partial^2 y(t,x)}{\partial x^2} = -\sum_{n=1}^{\infty} \frac{2(\cos na - (-1)^n)}{n\pi} \sin nx$$
$$+ \sum_{n=1}^{\infty} e^{-n^2 t} \frac{2(\cos na - (-1)^n)}{n\pi} \sin nx = -h(x-a) + \phi(t,x)\,,$$

where $\phi(t,x)$ is infinitely differentiable in $t$ and $x$ for $t > 0$; accordingly, the second derivative $\partial^2 y(t,x)/\partial x^2$ is discontinuous across the line $x = a$.

In Example 4.5.1 the operator $A_c$ is defined by $A_c y(x) = y''(x)$ with domain $D(A_c) = \{$all $C^{(2)}$ functions with $y(0) = 0$, $y(\pi) = 0$, $y''(0) = 0$, $y''(\pi) = 0\}$. Condition (4.5.2) is not satisfied; in fact, $y(t,\cdot) \notin D(A_c)$ for $t > 0$. On the other hand, $A_\infty$ is defined by $A_\infty y(x) = y''(x)$ with domain $D(A_\infty) = \{$all $y(x)$ with $y'(x)$ absolutely continuous, $y''(\cdot) \in L^\infty(\Omega)$, $y(0) = 0$, $y(\pi) = 0\}$, so that $y(t,x)$ satisfies $y(t,\cdot) \in D(A_\infty)$ for $t > 0$. It then becomes clear that a theory based on

$$\bar{y} \in D(A_\infty) \tag{4.5.3}$$

will be a lot more inclusive than one based on (4.5.2).

We begin with the characterization of the dual of $D(A_\infty)$, equipped with the graph norm or with the equivalent norm

$$\|y\|_{D(A_\infty),\lambda} = \|R(\lambda; A_\infty)y\| = \|(\lambda I - A_\infty)^{-1} y\|_{L^\infty(\Omega)}\,, \tag{4.5.4}$$

where $\lambda > \omega$ ($\omega$ the constant in (4.1.9)); $(\lambda I - A_1')^{-1}$ is everywhere defined and bounded, thus the same is true of $(\lambda I - A_\infty)^{-1} = ((\lambda I - A_1')^{-1})^*$ (see (4.2.12)). Since $D(A_\infty)$ is "based on $L^\infty(\Omega)$", we first identify the dual of $L^\infty(\Omega)$.

We call $\Sigma_w(\Omega)$ the space of all finitely additive measures $\eta$ defined in the field of all Lebesgue measurable sets of $\Omega$, of bounded variation (which means $\|\eta\| = \sup \Sigma|\eta(e_j)| < \infty$, supremum over all finite partitions of $\overline{\Omega}$ in Lebesgue measurable sets) and such that

$$\eta(e) = 0 \quad \text{if } |e| = 0\,, \tag{4.5.5}$$

where $|e|$ indicates the Lebesgue measure of $e$. The space $\Sigma_w(\Omega)$ is equipped with the total variation norm $\|\eta\|_{\Sigma_w(\Omega)} = \|\eta\|$.

**Theorem 4.5.2.** *The space $\Sigma_w(\Omega)$ is algebraically and metrically isomorphic to the dual $L^\infty(\Omega)^*$. The duality between $L^\infty(\Omega)$ and $\Sigma_w(\Omega)$ is given by*

$$\langle \eta, u\rangle = \int_\Omega u(x)\eta(dx)\,. \tag{4.5.6}$$

*Proof:* From Dunford - Schwartz [1958, p. 296, Theorem 16]; there is additional information on integration with respect to finitely additive measures in various parts of Dunford - Schwartz [1958, Chapter 3]. We begin by defining the integral in (4.5.6). Let $\eta \in \Sigma_w(\Omega)$, $u(\cdot) \in L^\infty(\Omega)$. Select a sequence of *simple* functions

$$u_n(x) = \sum_{\text{finite}} c_{nk}\chi_{nk}(x) \tag{4.5.7}$$

(the $\chi_{nk}(\cdot)$ characteristic functions of pairwise disjoint measurable sets $e_{nk} \subseteq \Omega$) such that

$$\|u(\cdot) - u_n(\cdot)\|_{L^\infty(\Omega)} \to 0 \qquad (n \to \infty)\,. \tag{4.5.8}$$

For instance, we may use $u_n(x) = -c + 2kc/n$ in each set $e_{nk}$ defined by

$$-c + \frac{2kc}{n} < u(x) \le -c + \frac{2(k+1)c}{n}\,, \quad (k = 0, 1, \dots n-1)\,,$$

with $c = \|u\|_{L^\infty(\Omega)}$; the first inequality is $\le$ for $k = 0$. Define

$$\int_\Omega u(x)\eta(dx) = \lim_{n\to\infty} \sum c_{nk}\eta(e_{nk})\,. \tag{4.5.9}$$

Checking that the integral exists and does not depend on the particular sequence (4.5.7) chosen is standard. If $c_{nk}$ is one of the coefficients in (4.5.7) then it follows from (4.5.8) that $|c_{nk}| \le \|u\|_{L^\infty(\Omega)} + o(1)$, hence

$$\begin{aligned}\Big|\sum c_{nk}\eta(e_{nk})\Big| &\le \sum |c_{nk}||\eta(e_{nk})| \\ &\le (\|u\|_{L^\infty(\Omega)} + o(1))\sum |\eta(e_{nk})| \le (\|u\|_{L^\infty(\Omega)} + o(1))\|\eta\|\,,\end{aligned}$$

and (4.5.9) gives

$$\left| \int_\Omega u(x)\eta(dx) \right| \le \|u\|_{L^\infty(\Omega)} \|\eta\|_{\Sigma_w(\Omega)} \,.$$

This makes clear that $\eta$, through (4.5.6), defines a bounded linear functional in $L^\infty(\Omega)$. Moreover, if $\{e_j\}$ is a finite partition of $\Omega$ in measurable sets and we define $u(x) = \sum \operatorname{sign} \eta(e_k)\chi_k(x)$ then

$$\langle \eta, u \rangle = \int_\Omega u(x)\eta(dx) = \sum |\eta(e_k)| \,,$$

where the sum on the right can be made arbitrarily close to $\|\eta\|_{\Sigma_w(\Omega)}$, hence $\|\eta\|_{L^\infty(\Omega)^*} = \|\eta\|_{\Sigma_w(\Omega)}$ . Conversely, let $\Phi$ be a bounded linear functional in $L^\infty(\Omega)$. Define $\eta(e) = \Phi(\chi_e(\cdot))$, $\chi_e(\cdot)$ the characteristic function of $e$. Then $\eta$ is a finitely additive measure defined on Lebesgue measurable sets and satisfying (4.5.5). If $u_n(\cdot)$ is the sequence in (4.5.7), we have

$$\Phi(u_n(\cdot)) = \sum c_{nk}\Phi(\chi_{nk}(\cdot)) = \sum c_{nk}\eta(e_{nk}) \,,$$

thus, by (4.5.8), boundedness of $\Phi$ and the definition (4.5.9) of the integral,

$$\Phi(u(\cdot)) = \lim_{n\to\infty} \sum c_{nk}\eta(e_{nk}) = \int_\Omega u(x)\eta(dx) \,,$$

which ends the proof of Theorem 4.5.2.

**Remark 4.5.3.** The meaning of (4.5.5) may not be immediately apparent. For instance, let $m = 1$, $\Omega = (0, 1)$ and consider the measure

$$\eta(e) = \operatorname*{LIM}_{n\to\infty} \frac{n}{2} \left| e \cap \left( a - \frac{1}{n}, a + \frac{1}{n} \right) \right|$$

where $a \in (0, 1)$ is fixed and $\mathrm{LIM}_{n\to\infty}$ is a Banach limit (see **2.2**, Miscellaneous Notes). Clearly, $\eta$ is a finitely additive measure, and it satisfies (4.5.5). It is not a countably additive measure; in fact, for $e = (a, 1)$ we have

$$\eta((a,1)) = \lim_{n\to\infty} \frac{n}{2} \left| (a,1) \cap \left( a - \frac{1}{n}, a + \frac{1}{n} \right) \right| = \lim_{n\to\infty} \frac{n}{2} \cdot \frac{1}{n} = \frac{1}{2} \,.$$

On the other hand,

$$(a,1) = \bigcup_{m=1}^\infty \left[ a + \frac{1-a}{m+1} \,,\, a + \frac{1-a}{m} \right) = \bigcup_{m=1}^\infty e_m \,,$$

but, for each $m \ge 1$ and $n$ large enough,

$$e_m \cap \left( a - \frac{1}{n}, a + \frac{1}{n} \right) = \emptyset \,,$$

so that $\eta(e_m) = 0$. If $u(\cdot)$ is continuous in $\overline{\Omega}$ we can do (4.5.8) by means of a function (4.5.7) where the $e_{nk}$ are *intervals* covering (0, 1) and there is no loss of generality in requiring one of these intervals $e_{nj}$ to be open and to contain $a$, and setting $c_{nj} = u(a)$. We have $\eta(e_{nk}) = 1$ $(k = j)$, $\eta(e_{nk}) = 0$ $(k \neq j)$, so that

$$\int_{\Omega} u(x)\eta(dx) = \lim_{n\to\infty} \sum c_{nk}\eta(e_{nk}) = \lim_{n\to\infty} u(a) = u(a)\,,$$

and it follows that $\eta$ is an extension of the Dirac delta $\delta_a$ with mass in $a$, a measure that misses (4.5.5) totally. This "anomaly" is typical: we have

**Lemma 4.5.4.** *Let $\mu \in \Sigma(\overline{\Omega})$. Then there exists $\eta \in \Sigma_w(\Omega)$ such that*

$$\int_{\overline{\Omega}} y(x)\mu(dx) = \int_{\Omega} y(x)\eta(dx) \quad (y(\cdot) \in C(\overline{\Omega}))\,, \tag{4.5.10}$$

*and $\|\mu\|_{\Sigma(\overline{\Omega})} = \|\eta\|_{\Sigma_w(\overline{\Omega})}$. Conversely, if $\eta \in \Sigma_w(\Omega)$ there exists a unique $\mu \in \Sigma(\overline{\Omega})$ such that* (4.5.10) *holds.*

*Proof.* As a functional in $C(\overline{\Omega})$, $\mu$ can be extended by the Hahn-Banach theorem to $L^\infty(\Omega)$ (that is, to an element of $\eta \in L^\infty(\Omega)^* = \Sigma_w(\Omega)$) with the same norm. Conversely, any $\eta \in \Sigma_w(\Omega)$ defines a bounded linear functional in $C(\overline{\Omega})$, thus (4.5.10) follows. This ends the proof.

**Remark 4.5.5.** In the direct part of Lemma 4.5.4, "$\mu$ can be extended to $\eta$" does not mean that $\eta(e) = \mu(e)$ on Borel sets $e$, as $\mu = \delta_a$ and $\eta =$ the measure in Remark 4.5.3 shows; we have $\mu(\{a\}) = 1$, $\eta(\{a\}) = 0$. Note also that $\mu$ need not satisfy (4.5.5), even if $\eta$ does. As typical in applications of the Hahn - Banach theorem, we can't insure that the extension $\eta$ is unique. The converse part of Lemma 4.5.4 lacks the "equality-of-norms" statement; in fact, the measure $\mu$ may vanish even if $\eta \neq 0$. For instance, since $C(\overline{\Omega})$ is a proper closed subspace of $L^\infty(\Omega)$ we may use the Hahn-Banach theorem to construct a nonzero bounded functional in $L^\infty(\Omega)$ that vanishes on $C(\overline{\Omega})$. If $\eta$ is the measure representing this functional, (4.5.10) can only be true with $\mu = 0$.

Since $A_\infty$ is not densely defined in $L^\infty(\Omega)$, $A_\infty^*$ is multivalued in $L^\infty(\Omega)^*$. In view of (4.2.3), if $y^* \in D(A_\infty^*)$ the set $[A_\infty^* y^*]$ of values of $A_\infty^* y^*$ satisfies $[A_\infty^* y^*] = z^* + \mathcal{N}_{A_\infty}$, where $z^*$ is an arbitrary element of $[A^* y^*]$ and

$$\mathcal{N}_{A_\infty} = \{y^* \in L^\infty(\Omega)^*; \langle y^*, y\rangle = 0,\ y \in D(A_\infty)\}\,.$$

We have shown in (4.3.7) that $D(A_\infty)$ is dense in $C(\overline{\Omega})$, hence

$$\mathcal{N}_{A_\infty} = \mathcal{N}_c(\overline{\Omega}) = \{y^* \in L^\infty(\Omega)^*;\ \langle y^*, y\rangle = 0,\, y \in C(\overline{\Omega})\}\,. \tag{4.5.11}$$

Due to (4.2.12) we have $(\lambda I^* - A_\infty^*)^{-1} = ((\lambda I - A_\infty)^{-1})^*$, but, since $A_\infty$ is not densely defined, this equality must be interpreted as (4.2.9) and (4.2.10),

$$((\lambda I - A_\infty)^{-1})^*(\lambda I^* - A_\infty^*) \subseteq I\,,$$
$$[(\lambda I^* - A_\infty^*)((\lambda I - A_\infty)^{-1})^* y^*] = y^* + \mathcal{N}_c(\overline{\Omega}) \quad (y^* \in L^\infty(\Omega)^*)\,.$$

According to (4.2.11), $\mathcal{N}_c(\overline{\Omega})$ is the nullspace of $((\lambda I - A_\infty)^{-1})^*$ :

$$\mathcal{N}_c(\overline{\Omega}) = \{y^* \in L^\infty(\Omega)^*; ((\lambda I - A_\infty)^{-1})^* y^* = 0\}. \tag{4.5.12}$$

**Lemma 4.5.6.** *We have*

$$A'_1 \subseteq A^*_\infty, \quad D(A^*_\infty) \subseteq L^1(\Omega). \tag{4.5.13}$$

*Proof.* The first inclusion follows from general duality theory. In fact, let $E$ be a Banach space and $A$ a linear operator in $E$ (not necessarily densely defined). We have $E \overset{i}{\hookrightarrow} E^{**}$ ( $\overset{i}{\hookrightarrow}$ is the canonical isometric imbedding) and, if $y \in D(A)$, $\langle y, A^* y^* \rangle = \langle Ay, y^* \rangle$ for $y^* \in D(A^*)$ so that $y \in D(A^{**})$ and $A^{**}y = Ay$. This is $A \subseteq A^{**}$ where the left side is single-valued and the right side may be multivalued.

We now show the second inclusion in (4.5.13). Assume that $\eta \in D(A^*_\infty) = D(\lambda I^* - A^*_\infty) = D((\lambda I - A_\infty)^*)$ with $(\lambda I^* - A^*_\infty)\eta = (\lambda I - A_\infty)^*\eta = \kappa \in L^\infty(\Omega)^*$. This means

$$\int_\Omega (\lambda I - A_\infty) y(x) \eta(dx) = \int_\Omega y(x) \kappa(dx) \quad (y(\cdot) \in D(A_\infty)). \tag{4.5.14}$$

By Lemma 4.5.4 there exist $\mu, \nu \in \Sigma(\overline{\Omega})$ such that

$$\int_{\overline{\Omega}} y(x)\mu(dx) = \int_\Omega y(x)\eta(dx), \quad \int_{\overline{\Omega}} y(x)\nu(dx) = \int_\Omega y(x)\kappa(dx)$$

for $y(\cdot) \in C(\overline{\Omega})$, so that restricting (4.5.14) to $y(\cdot) \in D(A_c)$,

$$\int_{\overline{\Omega}} (\lambda I - A_c) y(x) \mu(dx) = \int_{\overline{\Omega}} y(x) \nu(dx) \quad (y(\cdot) \in D(A_c)). \tag{4.5.15}$$

This equation implies $\mu \in D((\lambda I^* - A^*_c)) = D(A^*_c) = D(A'_\Sigma) \subset L^1(\Omega)$ (for this inclusion, see (4.2.22)). It follows that $\mu(dx) = z(x)dx$ with $z(\cdot) \in L^1(\Omega)$, thus (4.5.15) can be rewritten

$$\int_\Omega (\lambda I - A_\infty) y(x) z(x) dx = \int_\Omega y(x) \nu(dx) \quad (y(\cdot) \in D(A_c)). \tag{4.5.16}$$

It only remains to pass from (4.5.16) to

$$\begin{aligned} \int_\Omega (\lambda I - A_\infty) y(x) z(x) dx &= \int_{\overline{\Omega}} y(x) \nu(dx) \\ &= \int_\Omega y(x) \kappa(dx) \quad (y(\cdot) \in D(A_\infty)) \end{aligned} \tag{4.5.17}$$

(where the last two integrals are equal because $y(\cdot) \in D(A_\infty) \subset C(\overline{\Omega})$). To do this, note that $(\lambda I - A_c)D(A_c) = C(\overline{\Omega})$ $((\lambda I - A_c)D(A_c) = C_0(\overline{\Omega})$ for the Dirichlet boundary condition) and $(\lambda I - A_\infty)D(A_\infty) = L^\infty(\Omega)$, so that, given $y(\cdot) \in D(A_\infty)$ we can select a sequence $\{y_n(\cdot)\} \subseteq D(A_c)$ such that

$$(\lambda I - A_c)y_n(\cdot) \to (\lambda I - A_\infty)y(\cdot) \text{ a. e.,} \quad \|(\lambda I - A_c)y_n\|_{C(\overline{\Omega})} \le C\,. \tag{4.5.18}$$

Now, the operator $(\lambda I - A_c)^{-1}$ is compact (see the comments after (4.1.9)), hence it follows from the uniform bound in (4.5.18) that a subsequence of the sequence $\{y_n(\cdot)\} = \{(\lambda I - A_c)^{-1}(\lambda I - A_c)y_n(\cdot)\}$ (equally named) is convergent in $C(\overline{\Omega})$. To show (4.5.17) we write (4.5.16) for $y(\cdot) = y_n(\cdot)$ and take limits (on the left, we use the dominated convergence theorem).

With (4.5.17) on hand we combine it with (4.5.14), obtaining

$$\int_\Omega (\lambda I - A_\infty)y(x)z(x)dx = \int_\Omega (\lambda I - A_\infty)y(x)\eta(dx) \quad (y(\cdot) \in D(A_\infty))\,,$$

thus, since $(\lambda I - A_\infty)D(A_\infty) = L^\infty(\Omega)$ we have $\eta(dx) = z(x)dx$ as claimed. This ends the proof of Lemma 4.5.6.

Since $S_1'(t)$ is analytic in $\mathcal{L}(L^1(\Omega), L^1(\Omega))$ for $\operatorname{Re} t > 0$, the semigroup

$$S_\infty(t)^* = S_1'(t)^{**} \tag{4.5.19}$$

in the space $L^\infty(\Omega)^* = \Sigma_w(\Omega)$ is an analytic $\mathcal{L}(L^\infty(\Omega)^*, L^\infty(\Omega)^*)$-valued function for $\operatorname{Re} t > 0$ but it is not strongly continuous at $t = 0$. Using the fact that $S_\infty(t)A_\infty$ (domain $D(A_\infty)$) is bounded, we use (4.2.6) and obtain $(S_\infty(t)A_\infty)^* \supseteq A_\infty^* S_\infty(t)^*$ (the operators on both sides multivalued). This shows that

$$S_\infty(t)^* L^\infty(\Omega)^* \subseteq D(A_\infty^*) \subseteq L^1(\Omega) \tag{4.5.20}$$

(the second inclusion after Lemma 4.5.6).

**Lemma 4.5.7.** *We have*

$$S_1'(t) \subseteq S_\infty(t)^*, \quad S_\infty(t)^*|_{L^1(\Omega)} = S_1'(t) \quad (t > 0)\,. \tag{4.5.21}$$

The proof of Lemma 4.5.7 comes from duality theory; when $B$ is bounded and everywhere defined in $E$, $B^*$ and $B^{**}$ are everywhere defined, single-valued and bounded, and $B \subseteq B^{**}$, $B^{**}|_E = B$.

Construction of the space $L^1(\Omega)_{-1}(\lambda)$ $(\lambda > \omega)$ parallels that of $E^*_{-1}(\mu)$ in **2.3**. $L^1(\Omega)_{-1}(\lambda)$ is the completion of $L^1(\Omega)$ under the norm

$$\|y\|_{L^1(\Omega)_{-1},\lambda} = \|R(\lambda; A_1')y\|_{L^1(\Omega)} = \|(\lambda I - A_1')^{-1}y\|_{L^1(\Omega)}\,.$$

The resolvent equation $R(\lambda; A_1') - R(\mu; A_1') = (\mu - \lambda)R(\lambda; A_1')R(\mu; A_1')$ shows that all norms $\|\cdot\|_{L^1(\Omega)_{-1},\lambda}$, $\|\cdot\|_{L^1(\Omega)_{-1},\mu}$ are equivalent for $\lambda, \mu > \omega$ and define the same space $L^1(\Omega)_{-1}(\lambda) = L^1(\Omega)_{-1}(\mu) = L^1(\Omega)_{-1}$. We have

$$L^1(\Omega) \hookrightarrow L^1(\Omega)_{-1} \tag{4.5.22}$$

and $R(\lambda; A_1')$ is extended to $L^1(\Omega)_{-1}$ as follows. If $\psi \in L^1(\Omega)_{-1}(\lambda)$ there exists a sequence $\{y_n\} \subset L^1(\Omega)$ such that $\|y_n - \psi\|_{L^1(\Omega)_{-1},\lambda} \to 0$ so that $R(\lambda; A_1')y_n$ is Cauchy in $L^1(\Omega)$. The extension is

$$R(\lambda; A_1')^\bullet \psi = \lim_{n\to\infty} R(\lambda; A_1')y_n \,,$$

and the definition implies that the operator

$$R(\lambda; A_1')^\bullet : (L^1(\Omega)_{-1}(\lambda), \|\cdot\|_{L^1(\Omega)_{-1},\lambda}) \to (L^1(\Omega), \|\cdot\|_{L^1(\Omega)}) \qquad (4.5.23)$$

is an isometry. It is also *onto,* since[6] $R(\lambda; A_1')^\bullet L^1(\Omega)_{-1} = \overline{D(A_1')} = L^1(\Omega)$. We extend $A_1'$ to $L^1(\Omega)$ setting

$$A_1'^\bullet = \lambda I - (R(\lambda; A_1')^\bullet)^{-1} \,. \qquad (4.5.24)$$

If $y \in D(A_1')$ and $z = \lambda y - A_1' y$ we have

$$\begin{aligned} A_1'^\bullet y &= \lambda y - (R(\lambda; A_1')^\bullet)^{-1} y \\ &= \lambda y - (R(\lambda; A_1')^\bullet)^{-1} R(\lambda; A_1') z = \lambda y - z = A_1' y \,, \end{aligned}$$

justifying the first equality

$$A_1'^\bullet|_{D(A_1')} = A_1' \,, \quad R(\lambda; A_1'^\bullet) = R(\lambda; A_1')^\bullet \,. \qquad (4.5.25)$$

The second follows writing (4.5.24) in the form $\lambda I - A_1'^\bullet = (R(\lambda; A_1')^\bullet)^{-1}$.

**Lemma 4.5.8.** $A_1'^\bullet$ *does not depend on* $\lambda > \omega$.

*Proof.* $D(A_1')$ is dense in $L^1(\Omega)$ in the norm of $L^1(\Omega)$ and $L^1(\Omega)$ is dense in $L^1(\Omega)_{-1}$ in the norm of $L^1(\Omega)_{-1}$, hence $D(A_1')$ is dense in $L^1(\Omega)_{-1}$ in the norm of $L^1(\Omega)_{-1}$. On the other hand, (4.5.24) implies that $\lambda I - A_1'^\bullet : L^1(\Omega)_{-1} \to L^1(\Omega)$ is bounded. The result then follows from the first equality (4.5.25).

We extend $S_1'(t)$ to $L^1(\Omega)_{-1}$ by

$$S_1'(t)^\bullet \psi = (\lambda I - A_1'^\bullet) S_1'(t) R(\lambda; A_1')^\bullet \psi \,. \qquad (4.5.26)$$

We have

$$S_1'(t)^\bullet L^1(\Omega)_{-1} \subseteq L^1(\Omega) \quad (t > 0) \,, \quad S_1'(t) = S_1'(t)^\bullet|_{L^1(\Omega)} \,. \qquad (4.5.27)$$

In fact, the inclusion follows from (4.5.26) and analyticity of the semigroup $S_1'(t)$, while the restriction is obvious taking $\psi \in L^1(\Omega)$ in (4.5.26). The inclusion justifies simplifying $S_1'(t)^\bullet \psi$ to $S_1'(t)\psi$, which we do in most (but not all) cases below.

---

[6] If the norm $\|\cdot\|_{L^1(\Omega)_{-1},\lambda}$ is replaced by the equivalent norm $\|\cdot\|_{L^1(\Omega)_{-1},\mu}$, $\mu \neq \lambda$, "isometry onto" is downgraded to "bounded operator with bounded inverse".

The space $Z^1(T)$ is the subspace of $L^1(\Omega)_{-1}$ defined by

$$\|z\|_{Z^1(T)} = \int_0^T \|S_1'(t)z\|_{L^1(\Omega)} < \infty \,. \tag{4.5.28}$$

Membership in $Z^1(T)$ doesn't depend on $T$; the spaces $Z^1 = Z^1(T)$ coincide for all $T > 0$ and all the norms $\| \cdot \|_{Z^1(T)}$ are equivalent. The proof is essentially the same as the one for $Z_w(T)$ in **2.3**; see the comments following (2.3.23).

A *multiplier space* for $S_1'(t)$ is any space such that $(a)$ there is an imbedding $L^1(\Omega) \hookrightarrow \mathcal{M}$, $(b)$ $S_1'(t)$ is defined in $\mathcal{M}$ and $S_1'(t)\mathcal{M} \subseteq L^1(\Omega)$. Elements of $\mathcal{M}$ are called *multipliers*.[7] The result below characterizes $Z^1$ among all multiplier spaces.

**Lemma 4.5.9.** *Let $\psi$ be a multiplier such that*

$$\int_0^T \|S_1'(t)\psi\|_{L^1(\Omega)} < \infty \,. \tag{4.5.29}$$

*Then there exists a unique $z \in Z^1(T)$ such that*

$$S_1'(t)\psi = S_1'(t)z \quad (t > 0).$$

*Proof.* Essentially the same as that of Lemma 2.8.1. Let $\lambda > \omega$, and let $0 < s < t$. Since $A_1' - \lambda I$ is the infinitesimal generator of $e^{-\lambda t}S_1'(t)$ we have

$$\begin{aligned} &- (R(\lambda; A_1')e^{-\lambda t}S_1'(t)\psi - R(\lambda; A_1')e^{-\lambda s}S_1'(s)\psi) \\ &= e^{-\lambda s}S_1'(s)(A_1' - \lambda I)^{-1}(e^{-\lambda(t-s)}S_1'(t-s) - I)\psi \\ &= \int_0^{t-s} e^{-\lambda(s+\sigma)}S_1'(s+\sigma)\psi d\sigma = \int_s^t e^{-\lambda\sigma}S_1'(\sigma)\psi d\sigma \,. \end{aligned} \tag{4.5.30}$$

Let $\{t_n\}$ be a decreasing sequence of positive numbers with $t_n \to 0$. Using (4.5.30) for $s = t_n$, $t = t_m$ $(m < n)$ we obtain

$$\begin{aligned} &\|R(\lambda; A_1')e^{-\lambda t_m}S_1'(t_m)\psi - R(\lambda; A_1')e^{-\lambda t_n}S_1'(t_n)\psi\| \\ &\qquad \le \int_{t_n}^{t_m} e^{-\lambda\sigma}\|S_1'(\sigma)\psi\|d\sigma \,, \end{aligned}$$

which says that $\{R(\lambda; A_1')e^{-\lambda t_n}S_1'(t_n)\psi\}$ is Cauchy, hence convergent to $y \in L^1(\Omega)$, whereas $\{e^{-\lambda t_n}S_1'(t_n)\psi\}$ is convergent to $z \in L^1(\Omega)_{-1}$ with

$$R(\lambda; A_1')^{\bullet}z = y \iff z = (\lambda I - A_1'^{\bullet})y \,.$$

[7] As in **2.8** we don't even require $\mathcal{M}$ to be a linear space. The imbedding $L^1(\Omega) \hookrightarrow \mathcal{M}$ is just a 1-1 map.

We then have

$$\lim_{n\to\infty} S_1'(t-t_n)R(\lambda; A_1')e^{-\lambda t_n}S_1'(t_n)\psi = S_1'(t)y\,,$$
$$\lim_{n\to\infty} (\lambda I - A_1')S_1'(t-t_n)R(\lambda; A_1')e^{-\lambda t_n}S_1'(t_n)\psi$$
$$= \lim_{n\to\infty} S_1'(t-t_n)e^{-\lambda t_n}S_1'(t_n)\psi = S_1'(t)\psi\,,$$

so that $S_1'(t)y \in D(A_1')$ and

$$S_1'(t)\psi = (\lambda I - A_1')S_1'(t)y = S_1'(t)(\lambda I - A_1'^{\bullet})y = S_1'(t)z$$

as claimed (that $z \in Z^1(T)$ follows from (4.5.29)).

**Corollary 4.5.10.** *Let $\eta(t)$ $(t > 0)$ be a $L^1(\Omega)$-valued function such that*

$$\eta(s+t) = S_1'(t)\eta(s) \quad (s, t > 0) \tag{4.5.31}$$

*and*

$$\int_0^T \|\eta(t)\|_{L^1(\Omega)}dt < \infty\,.$$

*Then there exists $z \in Z^1(T)$ such that*

$$\eta(t) = S_1'(t)z \quad (t > 0)\,.$$

The proof is the same as that of Corollary 2.8.2; it suffices to provide a multiplier $\psi$ such that $\eta(t) = S_1'(t)\psi$. The multiplier $\psi$ is the infinite sequence

$$\psi = \{\eta(t_1), \eta(t_2), \dots\}$$

($\{t_n\}$ a decreasing sequence of positive numbers tending to zero) and

$$S_1'(t)\psi = S_1'(t-t_n)\eta(t_n) \quad (t > t_n)\,,$$

the definition independent of $n$ due to (4.5.31).

In the following result, $[\eta]$ denotes the equivalence class of $\eta \in L^\infty(\Omega)^*$ in the quotient space $L^\infty(\Omega)^*/\mathcal{N}_c(\overline{\Omega})$ equipped with its standard norm: the subspace $\mathcal{N}_c(\overline{\Omega})$ is defined by (4.5.11) $\Leftrightarrow$ (4.5.12).

**Theorem 4.5.11.** *Every bounded linear functional $\Phi$ in $D(A_\infty)$ is given by*

$$\Phi(y) = \langle [\eta], (\lambda I - A_\infty)y\rangle = \langle \eta, (\lambda I - A_\infty)y\rangle \tag{4.5.32}$$

*for some $[\eta] \in L^\infty(\Omega)^*/\mathcal{N}_c(\overline{\Omega})$. Moreover, if $D(A_\infty)$ is given the norm (4.5.4),*

$$\|\Phi\|_{D(A_\infty)^*} = \|[\eta]\|_{L^\infty(\Omega)^*/\mathcal{N}_c(\overline{\Omega})}\,. \tag{4.5.33}$$

*Proof.* Let $\eta \in L^\infty(\Omega)^*$. Then (4.5.32) defines a bounded linear functional in $D(A_\infty)$ and it's obvious from the fact that $\mathcal{N}_{A_\infty} = \mathcal{N}_c(\overline{\Omega})$ that (4.5.32) only depends on the equivalence class of $[\eta]$. That (4.5.33) holds follows from

$$(\lambda I - A_\infty)D(A_\infty) = L^\infty(\Omega)\,, \qquad \|(\lambda I - A_\infty)y\|_{L^\infty(\Omega)} = \|y\|_{D(A_\infty)}\,.$$

Conversely, let $\Phi$ be a bounded linear functional in $D(A_\infty)$. Then $\Phi(A_\infty^{-1}y)$ is a bounded linear functional in $L^\infty(\Omega)$, so that $\Phi((\lambda I - A_\infty)^{-1}y) = \langle \eta, y\rangle$, which is equivalent to (4.5.32).

**Miscellaneous notes.** Most of the material in this section is taken (sometimes with improvements) from the author [2001:4], [2002:2], although some results are new. We note a notational inconsistency in the use of the symbol $^\bullet$ for extensions here and in Chapter 2. In fact, in **2.3** $R(\lambda; A)^\bullet$ is an extension of $R(\lambda; A)^*$, while in this section $R(\lambda; A_1')^\bullet$ is an extension of $R(\lambda; A_1')$. The same observation can be made about the extensions of $A_1'$ and the semigroup $S_1'(t)$.

**4.6. The reachable space and its dual, II.** An element $z \in Z^1(T)$ defines a functional $\xi_z \in R_w^\infty(T)^\star$ according to the formula

$$\left\langle\!\!\left\langle \xi_z\,, \int_0^T S_\infty(T-\sigma)u(\sigma)d\sigma \right\rangle\!\!\right\rangle = \int_0^T \langle S_1'(T-\sigma)z\,, u(\sigma)\rangle d\sigma \tag{4.6.1}$$

where $\langle\!\langle\cdot,\cdot\rangle\!\rangle$ indicates the duality of $R_w^\infty(T)$ and $R_w^\infty(T)^\star$ and $S_1'(T-\sigma)z$ is shorthand for $S_1'(T-\sigma)^\bullet z$ (see the comments after (4.5.27)). The integral is a particular case of (4.1.11) and is interpreted along the same lines. First of all we must check that (4.6.1) heeds the equivalence relation in $R_w^\infty(T)$, that is,

$$\int_0^T S_\infty(T-\sigma)u(\sigma) = 0 \implies \left\langle\!\!\left\langle \xi_z\,, \int_0^T S_\infty(T-\sigma)u(\sigma)d\sigma \right\rangle\!\!\right\rangle = 0\,. \tag{4.6.2}$$

This follows from an application of the dominated convergence theorem:

$$\begin{aligned}
&\int_0^T \langle S_1'(T-\sigma)z\,, u(\sigma)\rangle d\sigma \\
&\quad = \lim_{h\to 0}\int_0^T \langle S_1'(T-\sigma+h)z\,, u(\sigma)\rangle d\sigma \\
&\quad = \lim_{h\to 0}\int_0^T \langle S_1'(T-\sigma)S_1'(h)z\,, u(\sigma)\rangle d\sigma \\
&\quad = \lim_{h\to 0}\int_0^T \langle S_1'(h)z\,, S_\infty(T-\sigma)u(\sigma)\rangle d\sigma \\
&\quad = \lim_{h\to 0}\left\langle S_1'(h)z\,, \int_0^T S_\infty(T-\sigma)u(\sigma)\right\rangle d\sigma\,.
\end{aligned} \tag{4.6.3}$$

Equation (4.6.3), valid for any element of $R_w^\infty(T)$, has other uses. Assuming that

$$\int_0^T S_\infty(T-\sigma)u(\sigma)d\sigma \in D(A_\infty)\,, \tag{4.6.4}$$

we obtain from (4.6.3) and (4.5.26) that

$$\begin{aligned}
&\left\langle\!\!\left\langle \xi_z\,, \int_0^T S_\infty(T-\sigma)u(\sigma)d\sigma \right\rangle\!\!\right\rangle \\
&\quad = \lim_{h\to 0}\left\langle S_1'(h)z\,, \int_0^T S_\infty(T-\sigma)u(\sigma)d\sigma \right\rangle \\
&\quad = \lim_{h\to 0}\left\langle (\lambda I - A_1')S_1'(h)R(\lambda;A_1')^\bullet z\,, \int_0^T S_\infty(T-\sigma)u(\sigma)d\sigma \right\rangle \\
&\quad = \lim_{h\to 0}\left\langle S_1'(h)R(\lambda;A_1')^\bullet z\,,\ (\lambda I - A_\infty)\int_0^T S_\infty(T-\sigma)u(\sigma)d\sigma \right\rangle \\
&\qquad = \left\langle R(\lambda;A_1')^\bullet z\,,\ (\lambda I - A_\infty)\int_0^T S_\infty(T-\sigma)u(\sigma)d\sigma \right\rangle .
\end{aligned} \tag{4.6.5}$$

It follows from (4.6.1) that

$$\begin{aligned}
\left\langle\!\!\left\langle \xi_z\,, \int_0^T S_\infty(T-\sigma)u(\sigma)d\sigma \right\rangle\!\!\right\rangle &\le \|u(\cdot)\|_{L_w^\infty(0,T;L^\infty(\Omega))}\int_0^T \|S_1'(T-\sigma)z\|_{L^1(\Omega)}d\sigma \\
&= \|z\|_{Z^1(T)}\|u(\cdot)\|_{L_w^\infty(0,T;L^\infty(\Omega))}\,,
\end{aligned}$$

so that $\|\xi_z\|_{R_w^\infty(T)^\star} \le \|z\|_{Z^1(T)}$. In fact, we have

$$\|\xi_z\|_{R_w^\infty(T)^\star} = \|z\|_{Z^1(T)}\,, \tag{4.6.6}$$

as we see setting[8] $\bar u(\sigma)(x) = \operatorname{sign} S_1'(T-\sigma)z(x)$ a. e. in (4.6.1). A functional of the form (4.6.1) is called *regular* and the lot of them is $\mathcal{R}_w(T)$; by (4.6.6), we have the isometric imbedding

$$\mathcal{R}_w(T) \overset{i}{\hookrightarrow} R_w^\infty(T)^\star. \tag{4.6.7}$$

According to (4.4.13),

$$D(A_\infty) \hookrightarrow R_w^\infty(\Omega)\,, \tag{4.6.8}$$

and a functional $\xi_s \in R_w^\infty(T)^\star$ is *singular* if it vanishes in $D(A_\infty)$, that is,

$$\xi_s|_{D(A_\infty)} = 0\,. \tag{4.6.9}$$

---

[8] The statement that $\bar u(\sigma)(x) = \operatorname{sign} S_1'(T-\sigma)z(x)$ should be accompanied by "where $S_1'(T-\sigma)z(x) \ne 0$". If $S_1'(T-\sigma)z(x) = 0$ in a set $e \subseteq (0,T)\times\Omega$ of positive measure then $\bar u(\sigma)(x)$ is undetermined in $e$ except for the condition that $\|\bar u(\cdot)\|_{L_w^\infty(0,T;L^\infty(\Omega))} \le 1$, which indetermination does not affect (4.6.6) or any other statements obtained by means of $\bar u(\sigma)(x) = \operatorname{sign} S_1'(T-\sigma)z(x)$. This is a moot point, however, since we shall see in **4.7** that if $z \ne 0$ then $S_1'(T-\sigma)z(x) \ne 0$ a. e.

The space of all singular functionals is $\mathcal{S}_w(T)$.

**Lemma 4.6.1.** *Singular functionals $\xi_s$ are localized*[9] *at $T$ in the sense that*

$$\left\langle\!\!\left\langle \xi_s\,, \int_0^T S_\infty(T-\sigma)u(\sigma)d\sigma \right\rangle\!\!\right\rangle = \left\langle\!\!\left\langle \xi_s\,, \int_0^T S_\infty(T-\sigma)v(\sigma)d\sigma \right\rangle\!\!\right\rangle$$

*if $u(\sigma) = v(\sigma)$ in $T-\delta \le \sigma \le T$ for some $\delta > 0$.*

*Proof.* We have

$$\begin{aligned}&\int_0^T S_\infty(T-\sigma)u(\sigma)d\sigma \\ &\quad = S_c(\delta)\int_0^{T-\delta} S_\infty(T-\delta-\sigma)u(\sigma)d\sigma + \int_{T-\delta}^T S_\infty(T-\sigma)u(\sigma)d\sigma\,,\end{aligned}$$

where the first term on the right belongs to $D(A_c) \subset D(A_\infty)$, hence

$$\left\langle\!\!\left\langle \xi_s\,, \int_0^T S_\infty(T-\sigma)u(\sigma)d\sigma \right\rangle\!\!\right\rangle = \left\langle\!\!\left\langle \xi_s\,, \int_{T-\delta}^T S_\infty(T-\sigma)u(\sigma)d\sigma \right\rangle\!\!\right\rangle,$$

and a similar equality holds for $v(\cdot)$. This ends the proof.

**Theorem 4.6.2.** *The subspaces $\mathcal{R}_w(T)$, $\mathcal{S}_w(T)$ are closed. We have*

$$R_w^\infty(T)^\star = \mathcal{R}_w(T) \oplus \mathcal{S}_w(T) \tag{4.6.10}$$

*with $L^1$ direct sum: if $\xi = \xi_z + \xi_s$ then*

$$\|\xi\|_{R_w^\infty(T)^\star} = \|\xi_z\|_{R_w^\infty(T)^\star} + \|\xi_s\|_{R_w^\infty(T)^\star}\,. \tag{4.6.11}$$

*Proof.* In view of (4.6.8), a functional $\xi \in R_w^\infty(T)^\star$ is also a functional in $D(A^\infty)$. By Theorem 4.5.11 there exists $\eta \in L^\infty(\Omega)^*$ such that

$$\left\langle\!\!\left\langle \xi\,, \int_0^T S_\infty(T-\sigma)u(\sigma)d\sigma \right\rangle\!\!\right\rangle = \left\langle \eta\,, (\lambda I - A_\infty)\int_0^T S_\infty(T-\sigma)u(\sigma)d\sigma \right\rangle \tag{4.6.12}$$

whenever (4.6.4) holds. Among controls $u(\cdot)$ satisfying (4.6.4) are those with $u(\sigma) = 0$ $(T-\delta \le t \le T)$ and, when using these controls we may introduce $\lambda I - A_\infty$ inside the integral on the right of (4.6.12). Using Lemma 4.5.7 we obtain

$$\begin{aligned}\langle \eta\,, (\lambda I - A_\infty)S_\infty(t)y\rangle &= \langle \eta\,, S_\infty(t/2)(\lambda I - A_\infty)S_\infty(t/2)y\rangle \\ &= \langle S_\infty(t/2)^*\eta\,, (\lambda I - A_\infty)S_\infty(t/2)y\rangle \\ &= \langle(\lambda I - A_1')S_\infty(t/2)^*\eta\,, S_\infty(t/2)y\rangle \\ &= \langle S_\infty(t/2)^*(\lambda I - A_1')S_\infty(t/2)^*\eta\,, y\rangle \\ &= \langle(\lambda I - A_1')S_\infty(t)^*\eta\,, y\rangle \quad (t>0)\,,\end{aligned}$$

[9] The converse is also true: localized functionals are singular. See Theorem 4.6.5.

so that (4.6.12) and the fact that $\xi$ is bounded in $R_w^\infty(T)$ give

$$\int_0^{T-\delta} \langle (\lambda I - A_1')S_\infty(T-\sigma)^*\eta\,,\, u(\sigma)\rangle d\sigma$$
$$= \left\langle\!\!\left\langle \xi\,, \int_0^{T-\delta} S_\infty(T-\sigma)u(\sigma)d\sigma \right\rangle\!\!\right\rangle \le \|\xi\|_{R_w^\infty(T)^\star}\|u(\cdot)\|_{L_w^\infty(0,T;L^\infty(\Omega))}\,,$$

for any $u(\cdot) \in L_w^\infty(0,T;L^\infty(\Omega))$. Since $u(\cdot)$ is arbitrary,

$$\int_0^{T-\delta} \|(\lambda I - A_1')S_\infty(T-\sigma)^*\eta\|_{L^1(\Omega)}d\sigma \le \|\xi\|_{R_w^\infty(T)^\star}\,,$$

and, since $\delta > 0$ is arbitrary,

$$\int_0^{T} \|(\lambda I - A_1')S_\infty(\sigma)^*\eta\|_{L^1(\Omega)}d\sigma$$
$$= \int_0^{T} \|(\lambda I - A_1')S_\infty(T-\sigma)^*\eta\|_{L^1(\Omega)}d\sigma \le \|\xi\|_{R_w^\infty(T)^\star}\,. \tag{4.6.13}$$

On the other hand, using (4.5.20) and (4.5.21) we deduce that

$$S_\infty(s)^*L^\infty(\Omega)^* = S_1'(s/2)S_\infty(s/2)^*L^\infty(\Omega)^* \subseteq D(A_1') \quad (s>0)\,,$$

so that

$$(\lambda I - A_1')S_\infty(t+s)^*\eta = (\lambda I - A_1')S_1'(t)^*S_\infty(s)^*\eta$$
$$= S_1'(t)(\lambda I - A_1')S_\infty(s)^*\eta\,, \tag{4.6.14}$$

and (4.6.13) and (4.6.14) are the hypotheses necessary to apply Corollary 4.5.10 to the function $\eta(t) = (\lambda I - A_1')S_\infty(t)^*\eta$. We deduce in this way that there exists $z \in Z^1(T)$ such that

$$\eta(t) = (\lambda I - A_1')S_\infty(t)^*\eta = S_1'(t)z\,.$$

Using this $z$ we define a functional $\xi_z$ in all of $R_w^\infty(T)$ by

$$\left\langle\!\!\left\langle \xi_z\,, \int_0^{T} S_\infty(T-\sigma)u(\sigma)d\sigma \right\rangle\!\!\right\rangle = \int_0^T \langle S_1'(T-\sigma)z\,,\, u(\sigma)\rangle d\sigma\,.$$

Under (4.6.4) we can use both (4.6.5) and (4.6.12). We obtain

$$\left\langle\!\!\left\langle \xi_z\,, \int_0^{T} S_\infty(T-\sigma)u(\sigma)d\sigma \right\rangle\!\!\right\rangle$$
$$= \left\langle R(\lambda; A_1')^\bullet z\,,\, (\lambda I - A_\infty)\int_0^T S_\infty(T-\sigma)u(\sigma)\right\rangle d\sigma$$
$$= \left\langle \eta\,, (\lambda I - A_\infty)\int_0^T S_\infty(T-\sigma)u(\sigma)d\sigma\right\rangle \tag{4.6.15}$$

for all $u(\cdot)$ such that (4.6.4) is satisfied, or, in view of the first imbedding (4.4.13),

$$\langle\!\langle \xi_z, y\rangle\!\rangle = \langle R(\lambda; A_1')^\bullet z\,,\,(\lambda I - A_\infty)y\rangle = \langle \eta\,,\,(\lambda I - A_\infty)y\rangle \\ (y \in R_w^\infty(T)\cap D(A_\infty) = D(A_\infty))\,. \tag{4.6.16}$$

Since $(\lambda I - A_\infty)D(A_\infty) = L^\infty(\Omega)$, (4.6.16) shows that[10] $\eta = R(\lambda; A_1')^\bullet z \in L^1(\Omega)$. It also follows from (4.6.15) that $\xi|_{D(A_\infty)} = \xi_z|_{D(A_\infty)}$ and we achieve (4.6.10) setting $\xi_s = \xi - \xi_z$. To show that (4.6.10) is a direct sum we must prove

$$\mathcal{R}_w(T)\cap \mathcal{S}_w(T) = \emptyset\,. \tag{4.6.17}$$

To this end, assume $\xi_z$ is a regular functional such that $\xi_z|_{D(A_\infty)} = 0$. Then it follows from the definition (4.6.1) that

$$\int_0^{T-\delta} \langle S_1'(T-\sigma)z\,,u(\sigma)\rangle d\sigma = \left\langle\!\!\left\langle \xi_z\,,\; S_c(\delta)\int_0^{T-\delta} S_\infty(T-\sigma-\delta)u(\sigma)d\sigma\right\rangle\!\!\right\rangle = 0$$

for $\delta > 0$ and $u(\cdot)\in L_w^\infty(0,T;L^\infty(\Omega))$; setting $u(\sigma)(x) = \operatorname{sign} S_1'(T-\sigma)z(x)$,

$$\int_0^{T-\delta} \|S_1'(T-\sigma)z\|_{L^1(\Omega)} = 0 \quad (\delta > 0)$$

so that $S_1'(T-\sigma)z = 0$ $(0\le\sigma<T) \implies z = 0 \Longrightarrow \xi_z = 0$ and (4.6.17) is proved. It remains to show the $L^1$ property (4.6.11) of the norm. If $\xi = \xi_z + \xi_s$, then

$$\|\xi\|_{R_w^\infty(T)^\star} \le \|\xi_z\|_{R_w^\infty(T)^\star} + \|\xi_s\|_{R_w^\infty(T)^\star}\,.$$

For the opposite inequality, let $\varepsilon > 0$,

$$y_z = \int_0^T S_\infty(T-\sigma)u(\sigma)d\sigma\,,\quad y_s = \int_0^T S_\infty(T-\sigma)u(\sigma)d\sigma$$

be elements of the unit ball $B_{w,1}^\infty$ with

$$\langle\!\langle \xi_z, y_z\rangle\!\rangle = \int_0^T \langle S_1'(T-\sigma)z\,,u(\sigma)\rangle \ge \|\xi_z\|_{R_w^\infty(T)^\star} - \varepsilon\,,$$
$$\langle\!\langle \xi_s, y_s\rangle\!\rangle \ge \|\xi_s\|_{R_w^\infty(T)^\star} - \varepsilon\,,$$

"elements of the unit ball" meaning we can assume that[11]

$$\|u(\cdot)\|_{L_w^\infty(0,T;L^\infty(\Omega))} \le 1+\varepsilon\,,\quad \|v(\cdot)\|_{L_w^\infty(0,T;L^\infty(\Omega))} \le 1+\varepsilon\,. \tag{4.6.18}$$

[10] A striking consequence of this argument is that the only functionals (4.6.12) on $D(A_\infty)$ we actually use are those with $\eta \in L^1(\Omega)$. The spaces $L^\infty(\Omega)^*$ and $D(A_\infty)^*$ play an ephemeral role and then are seen no more.

[11] We can actually take $\varepsilon = 0$ in (4.6.18); this requires an existence theorem for norm optimal controls (see **4.8**). The same applies to (4.6.26). This is irrelevant here.

Define

$$w_\delta(\sigma) = \begin{cases} u(\sigma) & (0 \le \sigma \le T-\delta) \\ v(\sigma) & (T-\delta < \sigma \le T)\,, \end{cases} \qquad y_\delta = \int_0^T S_\infty(T-\sigma) w_\delta(\sigma) d\sigma\,.$$

Since $\xi_s$ is localized,

$$\langle\!\langle \xi_s, y_\delta \rangle\!\rangle = \langle\!\langle \xi_s, y_s \rangle\!\rangle \ge \|\xi_s\|_{R_w^\infty(T)^\star} - \varepsilon\,.$$

On the other hand, we can choose $\delta$ so small that

$$\begin{aligned} \langle\!\langle \xi_z, y_\delta \rangle\!\rangle &= \int_0^T \langle S_1'(T-\sigma)z\,, w_\delta(\sigma)\rangle d\sigma \\ &= \int_0^{T-\delta} \langle S_1'(T-\sigma)z\,, u(\sigma)\rangle d\sigma + \int_{T-\delta}^T \langle S_1'(T-\sigma)z\,, v(\sigma)\rangle d\sigma \\ &\ge \langle\!\langle \xi_z, y_z \rangle\!\rangle - \varepsilon \ge \|\xi_z\|_{R_w^\infty(T)^\star} - 2\varepsilon\,, \end{aligned}$$

so that

$$\langle\!\langle \xi, y_\delta \rangle\!\rangle = \langle\!\langle \xi_z + \xi_s, y_\delta \rangle\!\rangle \ge \|\xi_z\|_{R_w^\infty(T)^\star} + \|\xi_s\|_{R_w^\infty(T)^\star} - 3\varepsilon\,,$$

which implies

$$\|\xi\|_{R_w^\infty(T)^\star} \ge \frac{1}{1+\varepsilon}\big(\|\xi_z\|_{R_w^\infty(T)^\star} + \|\xi_s\|_{R_w^\infty(T)^\star} - 3\varepsilon\big)$$

for arbitrary $\varepsilon > 0$. This ends the proof of (4.6.11). That $\mathcal{S}_w(T)$ is a closed subspace is obvious. As for $\mathcal{R}_w(T)$, if $\xi_{z_n}$ is a convergent sequence of regular functionals then $\xi_{z_n} + 0 \to \xi_z + \xi_s$ thus $0 \to \xi_s = 0$.

We call $\xi_z$ (resp. $\xi_s$) the *regular part* (resp. the *singular part*) of $\xi = \xi_z + \xi_s$. We denote by $B_{w,\rho}^\infty(T) \subset R_w^\infty(T)$ the ball of center 0 and radius $\rho$.

**Lemma 4.6.3.** Let $\bar{y} \in B_{w,\rho}^\infty(T)$ be such that

$$\langle\!\langle \xi, y \rangle\!\rangle \le \langle\!\langle \xi, \bar{y} \rangle\!\rangle \quad (y \in B_{w,\rho}^\infty(T)) \tag{4.6.19}$$

for a functional $\xi = \xi_z + \xi_s$. Then

$$\langle\!\langle \xi_z, y \rangle\!\rangle \le \langle\!\langle \xi_z, \bar{y} \rangle\!\rangle \quad (y \in B_{w,\rho}^\infty(T))\,. \tag{4.6.20}$$

*Proof.* (4.6.19) implies

$$\begin{aligned} &\left\langle\!\!\left\langle \xi_z + \xi_s\,, \int_0^T S_\infty(T-\sigma) u(\sigma) d\sigma \right\rangle\!\!\right\rangle \\ &\le \left\langle\!\!\left\langle \xi_z + \xi_s\,, \int_0^T S_\infty(T-\sigma) \bar{u}(\sigma) d\sigma \right\rangle\!\!\right\rangle \quad \big(\|u\|_{L_w^\infty(0,T;L^\infty(\Omega))} \le \rho\big)\,. \end{aligned} \tag{4.6.21}$$

Since $\xi_s$ is localized at $T$,

$$\left\langle\!\!\left\langle \xi_s\,, \int_0^T S_\infty(T-\sigma)u(\sigma)d\sigma \right\rangle\!\!\right\rangle = \left\langle\!\!\left\langle \xi_s\,, \int_0^T S_\infty(T-\sigma)\bar{u}(\sigma)d\sigma \right\rangle\!\!\right\rangle$$

if $u(\sigma) = \bar{u}(\sigma)$ for $T-\delta \le \sigma \le T$, so that (4.6.21) becomes

$$\left\langle\!\!\left\langle \xi_z\,, \int_0^{T-\delta} S_\infty(T-\sigma)u(\sigma)d\sigma \right\rangle\!\!\right\rangle \le \left\langle\!\!\left\langle \xi_z\,, \int_0^{T-\delta} S_\infty(T-\sigma)\bar{u}(\sigma)d\sigma \right\rangle\!\!\right\rangle$$
$$\big(\|u\|_{L_w^\infty(0,T-\delta;L^\infty(\Omega))} \le \rho,\ \delta > 0\big)\,, \tag{4.6.22}$$

or, after (4.6.1),

$$\int_0^{T-\delta} \langle S_1'(T-\sigma)z\,, u(\sigma)\rangle d\sigma \le \int_0^{T-\delta} \langle S_1'(T-\sigma)z\,, \bar{u}(\sigma)\rangle d\sigma$$
$$\big(\|u\|_{L_w^\infty(0,T;L^\infty(\Omega))} \le \rho,\ \delta > 0\big)\,. \tag{4.6.23}$$

This is equivalent to $\bar{u}(\sigma)(x) = \operatorname{sign} S_1'(T-\sigma)z(x)$ a. e. in $0 \le \sigma \le T-\delta$, $x \in \Omega$. However, $\delta > 0$ is arbitrary, so that

$$\bar{u}(\sigma)(x) = \operatorname{sign} S_1'(T-\sigma)z(x) \quad \text{a. e. in } 0 \le \sigma \le T,\ x \in \Omega\,. \tag{4.6.24}$$

Arguing backwards, this equality implies (4.6.22) $\Leftrightarrow$ (4.6.23) in $0 \le \sigma \le T$, thus we finally obtain

$$\left\langle\!\!\left\langle \xi_z\,, \int_0^T S_\infty(T-\sigma)u(\sigma)d\sigma \right\rangle\!\!\right\rangle \le \left\langle\!\!\left\langle \xi_z\,, \int_0^T S_\infty(T-\sigma)\bar{u}(\sigma)d\sigma \right\rangle\!\!\right\rangle$$
$$\big(\|u\|_{L_w^\infty(0,T;L^\infty(\Omega))} \le \rho\big)\,. \tag{4.6.25}$$

If $y \in B_{w,\rho}^\infty(T)$ then, for every $\varepsilon > 0$ there exists $u_\varepsilon(\cdot) \in L_w^\infty(0,T;L^\infty(\Omega))$ with

$$y = \int_0^T S_\infty(T-\sigma)u_\varepsilon(\sigma)d\sigma\,, \quad \|u(\cdot)\|_{L_w^\infty(0,T;L^\infty(\Omega))} \le \rho + \varepsilon\,, \tag{4.6.26}$$

so that (4.6.25) applied to $\rho u(\sigma)/(\rho+\varepsilon)$ gives $\rho\langle\!\langle \xi_z, y\rangle\!\rangle/(\rho+\varepsilon) \le \langle\!\langle \xi_z, \bar{y}\rangle\!\rangle$. Letting $\varepsilon \to 0$ (4.6.20) follows. The proof is complete.

**Remark 4.6.4.** We may apply the "strongly measurable controls" theory in Chapters 2 and 3 to the generator $A_c$ and the semigroup $S_c(t)$ in $C(\overline{\Omega})$. Theorem 2.4.1 gives the decomposition

$$R^\infty(T)^\star = \mathcal{R}(T) \oplus \mathcal{S}(T) \tag{4.6.27}$$

for the dual $R^\infty(T)^\star$ of the reachable space $R^\infty(T)$. In addition, the semigroup $S_c(t)$ is analytic, thus Theorem 2.4.3 says that the direct sum (4.6.27) is $L^1$.

On the other hand, we have proved in Theorem 4.4.4 that $R_w^\infty(T) = R^\infty(T)$ with equivalent norms; accordingly, $R_w^\infty(T)^\star = R^\infty(T)^\star$ with equivalent norms. Combining (4.6.27) and (4.6.10),

$$\mathcal{R}(T) \oplus \mathcal{S}(T) = \mathcal{R}_w(T) \oplus \mathcal{S}_w(T)\,. \tag{4.6.28}$$

Elements of $\mathcal{R}(T)$ are functionals of the form

$$\left\langle\!\!\left\langle \xi\,,\int_0^T S_c(T-\sigma)u(\sigma)d\sigma \right\rangle\!\!\right\rangle = \int_0^T \langle S'_\Sigma(T-\sigma)z\,,\,u(\sigma)\rangle d\sigma$$

where $\zeta \in Z(T)$, the subspace of $\Sigma(\overline{\Omega})_{-1}$ characterized by

$$\int_0^T \|S'_\Sigma(t)\zeta\|_{\Sigma(\overline{\Omega})}dt < \infty\,. \tag{4.6.29}$$

We have $L^1(\Omega) \subseteq \Sigma(\overline{\Omega})$ and $R(\mu; A'_1) \subseteq R(\mu; A'_\Sigma)$ so that $L^1(\Omega)_{-1} \subseteq \Sigma(\overline{\Omega})_{-1}$. Further, $S'_1(t) \subseteq S'_\Sigma(t)$, thus $Z^1(T) \subseteq Z(T)$. Define $\eta(t) = S'_\Sigma(t)\zeta$. It follows from (4.3.18) that $S'_\Sigma(t)\zeta \in L^1(\Omega)$ and that

$$\eta(s+t) = S'_\Sigma(s+t)\zeta = S'_1(t)S'_\Sigma(s)\zeta = S'_1(t)\eta(s)\,.$$

Combining this equality with (4.6.29), $\eta(t)$ satisfies the assumptions of Corollary 4.5.10 and there exists $z \in Z^1(T)$ such that $\eta(t) = S'_\Sigma(t)\zeta = S'_1(t)z$ $(t > 0)$. Accordingly, we have $Z(T) = Z_1(T)$ with equality of norms: $\|z\|_{Z(T)} = \|\zeta\|_{Z^1(T)}$. This implies the first equality in

$$\mathcal{R}_w(T) = \mathcal{R}(T)\,, \quad \mathcal{S}_w(T) = \mathcal{S}(T)\,. \tag{4.6.30}$$

The second equality follows from (4.6.28), and has dividends. Elements $\xi \in \mathcal{S}_w(T)$ are characterized by $\xi|_{D(A_\infty)} = 0$, whereas elements $\xi \in \mathcal{S}(T)$ are characterized by $\xi|_{D(A_c)} = 0$. This means

$$\xi|_{D(A_c)} = 0 \implies \xi|_{D(A_\infty)} = 0\,, \tag{4.6.31}$$

so that

$$\overline{D(A_c)} = \overline{D(A_\infty)} \quad (\text{closure in } R^\infty(T) = R_w^\infty(T))\,; \tag{4.6.32}$$

otherwise, by the Hahn - Banach theorem we would have a functional that vanishes in $D(A_c)$ but not in $D(A_\infty)$. This is an interesting piece of news, since (4.6.32) is not true if closure is taken in the norm of $D(A_\infty)$. To see this, consider the one-dimensional case $\Omega = (0,\pi)$, $A_c u(x) = u''(x)$, $u(0) = u(\pi) = 0$. Here, $D(A_c)$ is the set of all twice differentiable functions $u(x)$ with $u(\cdot), u''(\cdot)$ satisfying the Dirichlet boundary conditions. The operator $A_\infty$ is defined in the same way, but $D(A_\infty)$ consists of all continuously differentiable $u(\cdot)$ satisfying the boundary conditions, with absolutely continuous $u'(\cdot)$, and $u''(\cdot) \in L^\infty(0,\pi)$. We can take as norms of $D(A_c)$, $D(A_\infty)$ the graph norms

$$\|u(\cdot)\|_{D(A_c)} = \|u(\cdot)\|_{C_0[0,\pi]} + \|u''(\cdot)\|_{C_0[0,\pi]}\,,$$
$$\|u(\cdot)\|_{D(A_\infty)} = \|u(\cdot)\|_{L^\infty(0,\pi)} + \|u''(\cdot)\|_{L^\infty(0,\pi)}\,,$$

and we check that

$$\overline{D(A_c)} = D(A_c) \quad (\text{closure in } D(A_\infty))\,.$$

Since $S_c(t)$ is an analytic semigroup with unbounded infinitesimal generator $A_c$, Theorem 2.9.2 says that $\overline{D(A_c)} \neq R^\infty(T)$ (closure taken in $R^\infty(T)$) hence (4.6.32) shows that

$$\overline{D(A_\infty)} \neq R^\infty_w(T)\,.$$

Another dividend of (4.6.31). We have shown in Lemma 4.6.1 that all singular functionals are localized at $T$. The converse is also true: if $\xi \in R^\infty_w(T)^\star$ is localized at $T$, then $\xi$ is localized at $T$ as an element of $R^\infty(T)^\star$, thus, by virtue of Theorem 2.4.3 applied to the semigroup $S_c(t)$ and the generator $A_c$, $\xi$ is singular in $R^\infty(T)^\star$. Using the implication (4.6.31) we then see that $\xi$ is singular as an element of $R^\infty_w(T)^\star$. Summarizing,

**Theorem 4.6.5.** *A functional $\xi \in R^\infty_w(T)^\star$ is localized at $T$ if and only if it is singular.*

**Miscellaneous notes.** Some of the material in this section is taken from the author [2001:4] and [2002:2], but much of it has been considerably shortened and simplified. The identification of $Z^1(T)$ in [2001:4] is less precise than the one presented here. Some of the material is new.

**4.7. The maximum principle.** Let $\bar u(\cdot) \in L^\infty_w(0,T;L^\infty(\Omega))$ be a norm optimal control in the interval $0 \le t \le T$ under the target condition

$$y(t,\zeta,u) = \bar y\,.$$

Then

$$\bar y - S_\infty(T)\zeta = \int_0^T S_\infty(T-\sigma)\bar u(\sigma)d\sigma \tag{4.7.1}$$

belongs to the boundary of the ball $B^\infty_{w,\rho}(T)$ of center 0 and radius

$$\rho = \|\bar u(\cdot)\|_{L^\infty_w(0,T;L^\infty(\Omega))}$$

and it can be separated by a nonzero functional $\xi \in R^\infty_w(T)^\star$ from $B^\infty_{w,\rho}(T)$; this is

$$\langle\!\langle \xi, y\rangle\!\rangle \le \langle\!\langle \xi, \bar y - S_\infty(T)\zeta\rangle\!\rangle \quad (\|y\|_{R^\infty(T)} \le \rho)\,.$$

In view of (4.7.1), this implies

$$\left\langle\!\!\left\langle \xi\,,\int_0^T S_\infty(T-\sigma)u(\sigma)d\sigma \right\rangle\!\!\right\rangle \le \left\langle\!\!\left\langle \xi\,,\int_0^T S_\infty(T-\sigma)\bar{u}(\sigma)d\sigma \right\rangle\!\!\right\rangle \tag{4.7.2}$$

for $\|u(\cdot)\|_{L^\infty_w(0,T;L^\infty(\Omega))} \le \rho$. If $\xi = \xi_z$ is a regular functional, (4.7.2) implies

$$\int_0^T \langle S_1'(T-\sigma)z\,, u(\sigma)\rangle d\sigma \le \int_0^T \langle S_1'(T-\sigma)z\,, \bar{u}(\sigma)\rangle d\sigma$$

for $\|u(\cdot)\|_{L^\infty_w(0,T;L^\infty(\Omega))} \le \rho$, and it follows from this inequality that

$$\bar{u}(\sigma,x) = \bar{u}(\sigma)(x) = \rho\,\mathrm{sign}\, S_1'(T-\sigma)z(x) \quad \text{a. e. in } 0 \le \sigma \le T,\ x \in \Omega \tag{4.7.3}$$

with the qualifier "where $S_1'(T-\sigma)z(x) \neq 0$." We show later that if $z \neq 0$ the set where $S_1'(T-\sigma)z(x) = 0$ has Lebesgue measure zero, so that (4.7.3) determines the control $\bar{u}(\sigma,x)$ almost everywhere in $[0,T) \times \Omega$; however, the results in this section don't depend on this property.

With $\bar{u}(\cdot)$ given by (4.7.3) we have

$$\begin{aligned}\langle\!\langle \xi_z\,,\bar{y} - S_\infty(T)\zeta\rangle\!\rangle &= \int_0^T \langle S_1'(T-\sigma)z\,, \bar{u}(\sigma)\rangle d\sigma \\ &= \rho\int_0^T \|S_1'(T-\sigma)z\|_{L^1(\Omega)}d\sigma\,. \end{aligned}\tag{4.7.4}$$

Since $\langle S_1'(T-\sigma)z, \bar{u}(\sigma)\rangle \le \rho\|S_1'(T-\sigma)z\|_{L^1(\Omega)}$, this implies

$$\langle S_1'(T-\sigma)z\,, \bar{u}(\sigma)\rangle = \rho\|S_1'(T-\sigma)^*z\|_{L^1(\Omega)} \quad \text{a. e. in } 0 \le \sigma \le T\,, \tag{4.7.5}$$

which is equivalent to the maximum principle

$$\langle S_1'(T-\sigma)z\,, \bar{u}(t)\rangle = \max_{\|u\|_{L^\infty(\Omega)}\le\rho} \langle S_1'(T-\sigma)z\,, u\rangle \quad \text{a. e. in } 0 \le \sigma \le T\,. \tag{4.7.6}$$

**Theorem 4.7.1.** *(a) Assume $\bar{u}(t)$ is a norm optimal control and that*

$$\bar{y} \in \overline{D(A_\infty)}\,. \tag{4.7.7}$$

*Then there exists a regular functional separating $\bar{y} - S_\infty(T)\zeta$ from $B^\infty_{w,\rho}(T)$. (b) Assume $\bar{u}(t)$ is a time optimal control and that (4.7.7) is satisfied. Then there exists a regular functional separating $\bar{y} - S_\infty(T)\zeta$ from $B^\infty_{w,1}(T)$.*

*Proof.* Separating $\bar{y} - S_\infty(T)\zeta$ from the ball $B^\infty_{w,\rho}$ we may require

$$\langle\!\langle \xi, \bar{y} - S_\infty(T)\zeta\rangle\!\rangle = 1\,. \tag{4.7.8}$$

Since $S_\infty(T)L^\infty(\Omega) \subseteq D(A_\infty)$, (4.7.7) says $\bar{y} - S_\infty(T)\zeta \in \overline{D(A_\infty)}$, and we obtain from (4.7.8) that $\xi$ cannot be a singular functional, hence $\xi = \xi_z + \xi_s$, $\xi_z \neq 0$. It then follows from Lemma 4.6.3 that $\xi_z$ does the separation job as well, thus the argument after (4.7.2) achieves the proof for norm optimal controls. Time optimal controls are norm optimal; no separate argument is necessary.

**Theorem 4.7.2.** *Assume (4.7.7) holds. Then (a) If $\bar{u}(\cdot)$ is norm optimal it satisfies (4.7.3) $\Leftrightarrow$ (4.7.6) with $z \in Z^1(T)$, $z \neq 0$. (b) If $u(\cdot)$ is time optimal then (4.7.3) $\Leftrightarrow$ (4.7.6) holds with $\rho = 1$, $z \in Z^1(T)$, $z \neq 0$.*

*Proof.* Consequence of Theorem 4.7.1 and the equivalence of (4.7.3) and (4.7.2) for regular functionals.

**Theorem 4.7.3.** *Assume that $\bar{u}(\cdot) \in L^\infty_w(0, T, L^\infty(\Omega))$ satisfies (4.7.3) $\Leftrightarrow$ (4.7.6) with $z \in Z^1(T)$, $z \neq 0$, $\rho$ the minimum norm. Then $\bar{u}(t)$ is norm optimal.*

*Proof.* If $\bar{u}(\cdot)$ is not norm optimal there exists $u(\cdot) \in L^\infty_w(0, T; L^\infty(\Omega))$ with

$$\|u(\cdot)\|_{L^\infty_w(0,T;L^\infty(\Omega))} = \rho' < \rho, \quad y(T, \zeta, u) = y(T, \zeta, \bar{u}).$$

This means

$$\begin{aligned}\langle\!\langle \xi_z, \bar{y} - S_\infty(T)\zeta \rangle\!\rangle &= \int_0^T \langle S_1'(T-\sigma)z, u(\sigma)\rangle d\sigma \\ &\leq \rho' \int_0^T \|S_1'(T-\sigma)z\|_{L^1(\Omega)} d\sigma,\end{aligned}$$

which contradicts (4.7.4).

**Theorem 4.7.4.** *Assume the admissible control $\bar{u}(\cdot)$ satisfies (4.7.3) $\Leftrightarrow$ (4.7.6) with $\rho = 1$, $z \in Z^1(T)$, $z \neq 0$ and that, either*

$$\begin{aligned}&(a)\quad \bar{y} \in D(A_\infty), \ \|A_\infty \bar{y}\|_{L^\infty(\Omega)} < 1, \quad or \\ &(b)\quad \zeta \in D(A_\infty), \ \|A_\infty \zeta\|_{L^\infty(\Omega)} < 1.\end{aligned} \tag{4.7.9}$$

*Then $\bar{u}(t)$ is time optimal.*

*Proof.* The proof is about the same as that of Theorem 2.5.7 (and thus will only be sketched) but requires the equality

$$S_\infty(t)y - y = \int_0^t S_\infty(\sigma)A_\infty y d\sigma \quad (y \in D(A_\infty)), \tag{4.7.10}$$

which cannot be taken for granted, since $S_\infty(t)$ is not strongly continuous at $t = 0$. To show (4.7.10) we take $z \in D(A_1') \subseteq L^1(\Omega)$ and write

$$\begin{aligned}&\langle z, S_\infty(t)y - y\rangle = \langle S_1'(t)z - z, y\rangle \\ &= \left\langle \int_0^t S_1'(\sigma)A_1' z d\sigma, y \right\rangle = \int_0^t \langle S_1'(\sigma)A_1' z, y\rangle d\sigma \\ &= \int_0^t \langle z, S_\infty(\sigma)A_\infty y\rangle = \left\langle z, \int_0^t S_\infty(\sigma)A_\infty y d\sigma \right\rangle.\end{aligned}$$

Since $D(A_1')$ is dense in $L^1(\Omega)$, equality (4.7.10) follows.

Assume $\bar{u}(\cdot)$ is not time optimal. Then there exists $\delta > 0$ and an admissible control $\tilde{u}(\cdot) \in L_w^\infty(0, T-\delta; L^\infty(\Omega))$ that drives $\zeta$ to $\bar{y}$ in time $T-\delta$. In case $(a)$ we define

$$v(\sigma) = \begin{cases} \tilde{u}(\sigma) & (0 \le \sigma < T-\delta) \\ -A\bar{y} & (T-\delta \le \sigma \le T), \end{cases}$$

and note that, in view of (4.7.10),

$$\begin{aligned} \bar{y} - S_\infty(t-(T-\delta))\bar{y} &= -\int_0^{t-(T-\delta)} S_\infty(\sigma)A_\infty \bar{y}\,d\sigma \\ &= -\int_0^{t-(T-\delta)} S_\infty(t-(T-\delta)-\sigma)A_\infty \bar{y}\,d\sigma \\ &= -\int_{T-\delta}^{t} S_\infty(t-\sigma)A_\infty \bar{y}\,d\sigma\,, \end{aligned}$$

thus the trajectory $y(t,\zeta,v)$ starts at $\zeta$, reaches $\bar{y}$ at time $T-\delta$ and stays there until time $T$. This implies

$$\int_0^T S_\infty(T-\sigma)\bar{u}(\sigma)d\sigma = \int_0^T S_\infty(T-\sigma)v(\sigma)d\sigma\,. \tag{4.7.11}$$

By hypothesis, $\bar{u}(t)$ satisfies (4.7.6), which is equivalent to (4.7.2) with the regular functional $\xi = \xi_z$. In view of (4.7.11),

$$\left\langle\!\!\left\langle \xi_z\,, \int_0^T S_\infty(T-\sigma)u(\sigma)d\sigma \right\rangle\!\!\right\rangle \le \left\langle\!\!\left\langle \xi_z\,, \int_0^T S_\infty(T-\sigma)v(\sigma)d\sigma \right\rangle\!\!\right\rangle$$

for $\|u(\cdot)\|_{L_w^\infty(0,T;L^\infty(\Omega))} \le 1$, hence $v(t)$ satisfies (4.7.6) as well, which is impossible since $\|v(\sigma)\|_{L^\infty(t)} = \|A_\infty y\|_{L^\infty(\Omega)} < 1$ in $0 \le t \le \delta$ and, by analyticity, $S_1'(t)z \ne 0$ for all $t > 0$. In case $(b)$ the control $v(\sigma)$ is defined by

$$v(\sigma) = \begin{cases} -A\zeta & (0 \le \sigma \le \delta) \\ \tilde{u}(\sigma-\delta) & (\delta \le \sigma \le T), \end{cases}$$

and drives $\zeta$ to $\bar{y}$ in time $T$. The rest of argument is the same.

The next result establishes regularity of the function $z(t,x) = S_1'(T-t)z(x)$ appearing in the maximum principle. In fact, it does more: it shows regularity of

$$z(t,x) = S_1'(T-t)\psi(x)\,, \tag{4.7.12}$$

where $\psi$ is an arbitrary multiplier.

**Theorem 4.7.5.** *After eventual modification in a null set, the function $z(t,x)$ in (4.7.12) belongs to the space $C^\infty([0,T)\times\overline{\Omega})$ and satisfies the reverse adjoint equation*

$$\begin{aligned} \frac{\partial z(t,x)}{\partial t} &= -A'z(t,x) \quad (0\le t<T,\ x\in\Omega) \\ \beta' z(t,x) &= 0 \qquad\qquad (0\le t<T,\ x\in\Gamma) \end{aligned} \tag{4.7.13}$$

*in the classical sense.*

**Lemma 4.7.6.** *Let $n\ge 2$. We have*

$$D(A_1')\subseteq L^p(\Omega) \qquad 1\le p<\frac{n}{n-1}\,. \tag{4.7.14}$$

For a proof, see Amann - Escher [1996, Lemma 3.10, p. 52]. Actually, a lot more is shown there; in fact, Lemma 3.10 says that $D(A_1)\subseteq W_\beta^{p,1}(\Omega)$ for $p$ given by (4.7.14), which (via Sobolev imbeddings) makes it possible to extend the range of $p$. We don't need this added information.

*Proof of Theorem* 4.7.5. Let $\epsilon>0$. Then $S_1'(\epsilon/2)\psi\in L^1(\Omega)$ (by definition of multiplier), hence (4.7.14) implies $S_1'(\epsilon)\psi=S_1'(\epsilon/2)S_1'(\epsilon/2)\psi\in L^p(\Omega)$ for some $p>1$. Using the inclusion relation (3.3.12) for the semigroups we deduce that

$$S_1'(t)\psi=S_1'(t-\epsilon)S_1'(\epsilon)\psi=S_p'(t-\epsilon)S_1'(\epsilon)\psi\,,$$

and the smoothness claimed in Theorem 4.7.5, as well as the compliance with (4.7.13) follows from Theorem 3.3.1 in $t>\epsilon$; since $\epsilon$ is arbitrary, Theorem 4.7.5 stands proved.

We restate the results in this section "cleaning up" semigroup theory, that is, as theorems on optimal control of the distributed parameter system

$$\begin{aligned} \frac{\partial y(t,x)}{\partial t} &= Ay(t,x)+u(t,x) \quad (0\le t\le T,\ x\in\Omega) \\ y(0,x) &= \zeta(x) \qquad\qquad (x\in\Omega) \\ \beta y(t,x) &= 0 \qquad\qquad (0\le t\le T,\ x\in\Gamma) \end{aligned} \tag{4.7.15}$$

with controls in $L^\infty((0,T)\times\Omega)$. We note first that if $z(t,x)$ is a solution of the reverse adjoint equation (4.7.13) in the classical sense specified in Theorem 4.7.5 then $\theta(t,x)=z(T-t,x)$ satisfies (in the same sense)

$$\begin{aligned} \frac{\partial \theta(t,x)}{\partial t} &= A'\theta(t,x) \quad (0<t\le T,\ x\in\Omega) \\ \beta'\theta(t,x) &= 0 \qquad\qquad (0<t\le T,\ x\in\Gamma)\,. \end{aligned} \tag{4.7.16}$$

Define $\eta(t)=\theta(t,\cdot)$ $(t>0)$. Then $\eta(t)$ satisfies (4.5.31) and the argument in the proof of Corollary 4.5.10 provides a multiplier such that $\eta(t)=\theta(t,\cdot)=S_1'(t)\psi$, which is the same as (4.7.12). Hence, existence of a multiplier $\psi$ satisfying (4.7.12) is automatic for any solution $z(t, x)$ of the reverse adjoint equation (4.7.13) in $[0, T) \times \Omega$. The condition characterizing multipliers in $Z^1(T)$ is

$$\int_0^T \|S_1'(T-t)\psi\|_{L^1(\Omega)} dt < \infty \,, \tag{4.7.17}$$

or, in terms of (4.7.12),

$$\int_{(0,T)\times\Omega} |z(t,x)| dt dx < \infty \,. \tag{4.7.18}$$

**Theorem 4.7.7.** *(a) Let $\bar{u}(t, x)$ be a norm optimal control for the system (4.7.15) in the interval $0 \leq t \leq T$ with*

$$\bar{y}(\cdot) = y(T, \cdot) \in \overline{D(A_\infty)} \,. \tag{4.7.19}$$

*Then*

$$\bar{u}(t,x) = \rho \operatorname{sign} z(t,x) \,, \tag{4.7.20}$$

*where $z(t, x)$ is a nonzero solution of the reverse adjoint equation (4.7.13) satisfying the growth condition (4.7.18) and $\rho = \|\bar{u}(\cdot,\cdot)\|_{L^\infty((0,T)\times\Omega)}$. (b) If $\bar{u}(t, x)$ is time optimal and (4.7.19) holds then $\bar{u}(t, x)$ is given by (4.7.20) with $\rho = 1$ and $z(t, x)$ a nonzero solution of the reverse adjoint equation (4.7.13) satisfying (4.7.18).*

**Theorem 4.7.8.** *Assume $\bar{u}(t, x)$ satisfies (4.7.20) with $z(t, x)$ a nonzero solution of the reverse adjoint equation (4.7.13) satisfying the growth condition (4.7.18). Then $\bar{u}(t, x)$ is norm optimal. If $\rho = 1$ and*

$$\begin{aligned} &(a)\quad \bar{y}(\cdot) = y(T,\cdot) \in D(A_\infty)\,,\ \|A_\infty \bar{y}(\cdot)\|_{L^\infty(\Omega)} < 1\,, \quad \textit{or} \\ &(b)\quad \zeta(\cdot) = y(0,\cdot) \in D(A_\infty)\,,\ \|A_\infty \zeta(\cdot)\|_{L^\infty(\Omega)} < 1\,, \end{aligned} \tag{4.7.21}$$

*then $\bar{u}(t, x)$ is time optimal.*

Following the lead of **3.2**, we call (4.7.3) $\Leftrightarrow$ (4.7.6) with $z \in Z^1(T), z \neq 0$ the *strong* maximum principle; if $z$ is replaced by a multiplier $\psi$ not in $Z^1(T)$ we have the *weak* maximum principle. Extremals (that is, norm or time optimal controls) are classified according to the scheme in **3.2**:

*Strongly singular extremals:* do not satisfy the weak maximum principle.

*Weakly singular extremals:* satisfy the weak maximum principle but not the strong maximum principle,

*Regular extremals:* satisfy the strong maximum principle.

**Theorem 4.7.9.** *There exists a strongly singular norm optimal control, that is, a norm optimal control $\bar{u}(\cdot)$ of norm 1 such that (4.7.20) does not hold for any solution $z(t, x)$ of (4.7.13) (even if the growth condition (4.7.18) is given up).*

*Proof.* It suffices to apply the theory in **2.8** to the strongly continuous semigroup $S_c(t)$ in the space $C(\overline{\Omega})$. The control (2.8.28),

$$\bar{u}(\sigma) = \begin{cases} u_{2n-1}(\sigma) & (T - \delta_{2n-1} \leq \sigma < T - \delta_{2n}) \\ 0 & (T - \delta_{2n} \leq \sigma < T - \delta_{2n+1}) \end{cases}$$

constructed in Theorem 2.8.6 belongs to $L^\infty(0, T; C(\overline{\Omega}))$ and

$$\|\bar{u}(\cdot)\|_{L^\infty(0,T;C(\overline{\Omega}))} = \|\bar{u}(\cdot)\|_{L^\infty_w(0,T;L^\infty(\Omega))} = 1\,.$$

Assume that $v(\cdot) \in L^\infty(0, T; C(\overline{\Omega}))$ satisfies

$$\int_0^T S_\infty(T-\sigma)v(\sigma)d\sigma = \int_0^T S_\infty(T-\sigma)\bar{u}(\sigma)d\sigma\,. \tag{4.7.22}$$

For these controls, equality (4.7.22) is the same as

$$\int_0^T S_c(T-\sigma)v(\sigma)d\sigma = \int_0^T S_c(T-\sigma)\bar{u}(\sigma)d\sigma\,, \tag{4.7.23}$$

and it was proved in Theorem 2.8.6 that (4.7.23) implies $\|v(\cdot)\|_{L^\infty(0,T;C(\overline{\Omega}))} \geq 1$. However, to show that $\bar{u}(\cdot)$ is norm optimal in $R^\infty_w(T)$ we must show the same for $v(\cdot) \in L^\infty_w(0, T; L^\infty(\Omega))$; if (4.7.22) holds then $\|v(\cdot)\|_{L^\infty_w(0,T;L^\infty(\Omega))} \geq 1$. The argument is the same as that used in Theorem 2.8.6 and we omit it. Finally, since $\bar{u}(\sigma) = 0$ in $T - \delta_{2n} \leq \sigma < T - \delta_{2n+1}$, it is clear $\bar{u}(\cdot)$ cannot satisfy (4.7.20). This ends the proof of Theorem 4.7.9.

**Problem 4.7.10.** *Are there any weakly singular time or norm optimal controls for the system* (4.7.15)?

**Problem 4.7.11.** *Are there any strongly singular time optimal controls for the system* (4.7.15)?

The two problems are open. Although we are working with analytic semigroups the results in **3.3** on nonexistence of strongly singular time optimal controls require a space whose regularity far exceeds that of $C(\overline{\Omega})$ and the examples of weakly singular norm and time optimal controls in **3.4** and **3.5** are restricted to a specific Hilbert space and semigroup.

The *nodal set* $\mathcal{N}(z) \subseteq [0, T) \times \overline{\Omega}$ of a solution $z(t, x)$ of (4.7.13) is

$$\mathcal{N}(z) = \{(t, x) \in [0, t) \times \overline{\Omega};\ z(t, x) = 0\}\,. \tag{4.7.24}$$

The question of whether $\mathcal{N}(z)$ has Lebesgue measure zero is fundamental; in fact, (4.7.20) only determines $\bar{u}(t, x)$ outside[12] of the nodal set $\mathcal{N}(z)$. For instance,

[12] Whether or not the nodal set has measure zero doesn't affect the validity of Theorem 4.7.7 and Theorem 4.7.8.

(4.7.20) only implies the bang-bang property $|\bar{u}(t,x)| = \rho$ outside of $\mathcal{N}(z)$. Under the smoothness assumptions in vigor we have

**Theorem 4.7.12.** *The nodal set of a nonzero solution $z(t,x)$ of* (4.7.13) *has measure zero.*

The proof of Theorem 4.7.12 is a consequence of a result in Han - Lin [1994] (Theorem 4.7.14 below). Its statement needs the Hausdorff outer measure of sets $e$ in Euclidean space and several auxiliary definitions which we take from Falconer [1990]. The *diameter* $\delta(e)$ of a set $e \subseteq \mathbb{R}^m$ is $\delta(e) = \sup\{|x-y|; x, y \in e\}$. Given $\delta > 0$, a *countable $\delta$-cover* of a set $e$ is a cover by a countable collection $\{e_j\}$ of sets such that $\delta(e_j) \le \delta$. For $s > 0$ we define

$$\mathcal{H}^s_\delta(e) = \inf \sum_{j=1}^{\infty} \delta(e_j)^s \,,$$

the infimum taken over all countable $\delta$-covers of $e$. It is plain that $\mathcal{H}^s_\delta$ is an *outer measure* in the sense that

$$0 \le \mathcal{H}^s_\delta(e) \le \infty\,, \quad \mathcal{H}^s_\delta(\emptyset) = 0\,,$$
$$d \subseteq e \implies \mathcal{H}^s_\delta(d) \le \mathcal{H}^s_\delta(e)\,, \quad \mathcal{H}^s_\delta\Big(\bigcup_{n=1}^{\infty} e_n\Big) \le \sum_{n=1}^{\infty} \mathcal{H}^s_\delta(e_n)\,.$$

Both statements in the first line and the implication in the second line are trivial. So is the inequality in the second line when the sum is finite. If it is infinite, we select countable $\delta$-covers $\{e_{nj}; j = 1, 2, \dots\}$ of each $e_n$ such that

$$\sum_{j=1}^{\infty} \delta(e_{nj})^s \le \mathcal{H}^s_\delta(e_n) + \frac{\epsilon}{2^n}$$

and then use $\{e_{nj}; j, n = 1, 2, \dots\}$ as a cover of $e$. It is also true that

$$\delta \le \epsilon \implies \mathcal{H}^s_\delta(e) \ge \mathcal{H}^s_\epsilon(e)\,, \quad \delta \le 1,\ s \le t \implies \mathcal{H}^s_\delta(e) \ge \mathcal{H}^t_\delta(e)\,. \tag{4.7.25}$$

The first implication results from the fact that every $\delta$-cover is also an $\epsilon$-cover, thus the infimum defining $\mathcal{H}^s_\epsilon(e)$ is taken over more covers than the ones defining $\mathcal{H}^s_\delta(e)$. For the second implication, note that $\delta(e_j) \le \delta \le 1$ implies $\delta(e_j)^s \ge \delta(e_j)^t$.

The *Hausdorff $s$-dimensional outer measure* of $e$ is

$$\mathcal{H}^s(e) = \lim_{\delta \to 0} \mathcal{H}^s_\delta(e) = \sup_{\delta > 0} \mathcal{H}^s_\delta(e)\,, \tag{4.7.26}$$

the equality of lim and sup coming from the first implication (4.7.25). We have

$$0 \le \mathcal{H}^s(e) \le \infty\,, \quad \mathcal{H}^s(\emptyset) = 0\,,$$
$$d \subseteq e \implies \mathcal{H}^s(d) \le \mathcal{H}^s(e)\,, \quad \mathcal{H}^s\Big(\bigcup_{n=1}^{\infty} e_n\Big) \le \sum_{n=1}^{\infty} \mathcal{H}^s(e_n)\,,$$

thus $\mathcal{H}^s$ lives up to its "outer measure" name. Since we may assume that $\delta \le 1$ in all $\delta$-covers, the second implication (4.7.25) gives $s \le t \Rightarrow \mathcal{H}^s(e) \ge \mathcal{H}^t(e)$, although this inequality is true in a much stronger sense. To see this, let $s < t$, $\epsilon > 0$. Then there exists a countable $\delta$-cover $\{e_j\}$ of $e$ such that

$$\mathcal{H}^t_\delta(e) \le \sum_{j=1}^{\infty} \delta(e_j)^t = \sum_{j=1}^{\infty} \delta(e_j)^{t-s}\delta(e_j)^s \le \delta^{t-s}\sum_{j=1}^{\infty} \delta(e_j)^s \le \delta^{t-s}\mathcal{H}^s_\delta(e) + \epsilon\,,$$

hence $\mathcal{H}^t_\delta(e) \le \delta^{t-s}\mathcal{H}^s_\delta(e)$ for all $\delta > 0$. Taking infima on both sides,

$$\mathcal{H}^s(e) < \infty \Rightarrow \mathcal{H}^t(e) = 0 \quad (s < t)\,. \tag{4.7.27}$$

**Lemma 4.7.13.** *Let $e \subseteq \mathbb{R}^m$. Then we have*

$$\mathcal{H}^m(e) = \frac{\pi^{m/2}}{2^m\Gamma(1+m/2)}|e|\,,$$

*where $|\cdot|$ is Lebesgue outer measure.*

For a proof, see Falconer [1990, Theorem 1.12, p. 13].

The next result (Han - Lin [1994, Theorem 1.1, p. 1221]) happens in $\mathbb{R}\times\mathbb{R}^m$. We set [1994. p. 1220]

$$Q_\rho(t_0, x_0) = \left\{(t,x) \in \mathbb{R}\times\mathbb{R}^m;\ t_0 - 7\rho^2/8 \le t \le t_0 + \rho^2/8,\ |x - x_0| \le \rho\right\},$$

and consider a function $\eta(t,x)$ solving (classically) the parabolic equation

$$\frac{\partial\eta(t,x)}{\partial t} = \sum_{j=1}^{m}\sum_{k=1}^{m} a_{jk}(x)\frac{\partial^2\eta(t,x)}{\partial x_j\partial x_k} + \sum_{j=1}^{m} b_j(x)\frac{\partial\eta(t,x)}{\partial x_j} + c(x)\eta(t,x) \tag{4.7.28}$$

$(a_{jk}(x) = a_{kj}(x))$ in the domain

$$Q_1(0,0) = \left\{(t,x) \in \mathbb{R}\times\mathbb{R}^m;\ -7/8 \le t \le 1/8,\ |x| \le 1\right\}.$$

The partial differential operator in (4.7.28) is uniformly elliptic in $Q_1(0,0)$ :

$$\sum_{j=1}^{n}\sum_{k=1}^{n} a_{jk}(x)\xi_j\xi_k \ge c\sum_{j=1}^{n}\xi_j^2 \quad ((x,t) \in Q_1(0,0)) \tag{4.7.29}$$

for some $c > 0$. We assume the coefficients infinitely differentiable.

**Theorem 4.7.14.** *Assume that $\eta(0,0) = 0$ and let $\mathcal{N}(\eta)$ be the nodal set of $\eta(t,x)$ in $Q_1(0,0) \subset \mathbb{R}^{m+1} = \mathbb{R}\times\mathbb{R}^m$. Then there exists a finite constant $C$ such that*

$$\mathcal{H}^m(Q_{1/2}(0,0) \cap \mathcal{N}(\eta)) \le C\,. \tag{4.7.30}$$

Theorem 1.1 in Han - Lin [1994] is in fact considerably more general in hypotheses and conclusions: see Miscellaneous Notes below.

*Proof of Theorem* 4.7.12. We begin by setting $\theta(t,x) = z(T-t,x)$, so that $\theta(t,x)$ is a solution of

$$\frac{\partial\theta(t,x)}{\partial t} = A'\theta(t,x) \quad (t,x) \in (0,T)\times\Omega\,,$$

an equation of the form (4.7.28). We then take $(t_0,x_0) \in \mathcal{N}(\theta)\cap((0,T)\times\Omega)$, pick $\rho$ so small that $Q_\rho(t_0,x_0) \subseteq (0,T)\times\Omega$ and define $\eta(t,x) = \theta(t_0+\rho^2 t, x_0+\rho x)$. We have

$$(t,x)\in Q_1(0,0) \iff (t_0+\rho^2 t, x_0+\rho x) \in Q_\rho(t_0,x_0)\,, \tag{4.7.31}$$

and $\eta(t,x)$ satisfies in $Q_1(0,0)$ an equation of the same type as that satisfied by $\theta(t,x)$; if $a_{jk}(x), b_j(x), c(x)$ are the coefficients of (4.7.30), the coefficients of the new equation are

$$a_{jk}(x_0+\rho x), \qquad \frac{b_j(x_0+\rho x)}{\rho}, \qquad \frac{c(x_0+\rho x)}{\rho^2}\,,$$

thus (4.7.29) holds. Finally, $\eta(0,0) = \theta(t_0,x_0) = 0$, so we are in position to apply Theorem 4.7.14. The estimate (4.7.30) combined with (4.7.27) and Lemma 4.7.13 applied in the space $\mathbb{R}^{m+1} = \mathbb{R}\times\mathbb{R}^m$ imply

$$|Q_{1/2}(0,0)\cap\mathcal{N}(\eta))| = 0\,. \tag{4.7.32}$$

The "one-half" version of (4.7.31) is

$$(t,x)\in Q_{1/2}(0,0) \iff t_0+\rho^2 t, x_0+\rho x \in Q_{\rho/2}(t_0,x_0)\,,$$

thus we obtain from (4.7.32) that

$$|Q_{\rho/2}(t_0,x_0)\cap\mathcal{N}(\theta)| = 0\,.$$

At this point, we have proved that for every $(t,x) \in ((0,T)\times\Omega)\cap\mathcal{N}(\theta)$ there exists an open set $\mathcal{O}(t,x) = \mathrm{Int}(Q_{\rho/2}(t,x))$ such that $|\mathcal{O}(t,x)\cap\mathcal{N}(\theta)| = 0$. Using Lindelöf's theorem we may select a countable subcover $\{\mathcal{O}(t_n,x_n)\}$ from the cover $\{\mathcal{O}(t,x)\}$ so that we have

$$((0,T)\cap\Omega)\cap\mathcal{N}(\theta) = \bigcup_{n=1}^{\infty}\left(\mathcal{O}(t_n,x_n)\cap\mathcal{N}(\eta)\right).$$

This ends the proof of Theorem 4.7.12.

**Miscellaneous notes.** This section is based on the author [2001:4], although at that time it was not realized that the Han - Lin result applied to the to the measure of the nodal set in the general case; Theorem 4.7.12 is proved there only for self adjoint $A$ and $\beta$. Under the smoothness assumptions in vigor in this section, there is much additional information on the nodal set in Han - Lin [1994].

The results require a *doubling condition*, which is automatically satisfied for (4.7.28) due to smoothness and $t$-independence of the coefficients; see Remark 1.3 in Han - Lin [1994] and also Lin [1991].

**4.8. Existence, uniqueness and stability of optimal controls.** As in **3.1**, existence theory is lifted from the finite dimensional case. The equation is

$$y'(t) = A_c y(t) + u(t)\,, \quad y(0) = \zeta \tag{4.8.1}$$

with $\zeta \in L^\infty(\Omega)$ and $u(\cdot) \in L^\infty_w(0,T;L^\infty(\Omega)) = L^\infty((0,T)\times\Omega)$. For the norm optimal problem, a minimizing sequence $\{u_n(\cdot)\}$ satisfies

$$y(T,\zeta,u_n) = y_n \to \bar{y}\,, \quad \|u_n(\cdot)\|_{L^\infty_w(0,T;L^\infty(\Omega))} \to \rho\,, \tag{4.8.2}$$

where

$$\rho = \inf\{\|u(\cdot)\|_{L^\infty_w(0,T;L^\infty(\Omega))};\, y(T,\zeta,u) = \bar{y}\}\,. \tag{4.8.3}$$

For the time optimal problem, a minimizing sequence satisfies

$$y(t_n,\zeta,u_n) = y_n \to \bar{y}\,, \quad \|u_n(\cdot)\|_{L^\infty_w(0,t_n;L^\infty(\Omega))} \le 1\,, \quad t_n \to T \tag{4.8.4}$$

where $T$ is the infimum of all driving times from $\zeta$ to $\bar{y}$. For the norm optimal problem, if $y(T,\zeta,u_n) = \bar{y}$ we may require the norms $\|u_n(\cdot)\|_{L^\infty_w(0,T;L^\infty(\Omega))}$ to be decreasing, and for the time optimal problem, if $y(t_n,\zeta,u_n) = \bar{y}$ we may assume decreasing control intervals $[0,t_n]$. It is obvious from the definitions that, if we can drive $\zeta$ to $\bar{y}$ in time $T$ with a control $u(\cdot) \in L^\infty_w(0,T;L^\infty(\Omega))$ then there exists a minimizing sequence for the norm optimal problem in $0 \le t \le T$, and that if we can drive $\zeta$ to $\bar{y}$ in any time $t$ with a control $u(\cdot) \in L^\infty_w(0,t;L^\infty(\Omega))$ with $\|u(\cdot)\|_{L^\infty_w(0,t;L^\infty(\Omega))} \le 1$ then there exists a minimizing sequence for the time optimal problem. The existence results are based on the equality

$$L^\infty_w(0,T;L^\infty(\Omega)) = L^\infty((0,T)\times\Omega) = L^1((0,T)\times\Omega)^* = L^1(0,T;L^1(\Omega))\,,$$

which allows us to take weak limits of (subsequences of) minimizing sequences.[13] The first equality is Lemma 4.1.1 ($b$), the third is Lemma 4.1.1 ($a$), the middle one is standard $L^p$ duality theory.

**Theorem 4.8.1.** *If minimizing sequences exist, optimal controls exist.*

*Proof.* Let $u(\cdot) \in L^\infty_w(0,T;L^\infty(\Omega))$ drive $\zeta$ to $y$ in time $t$, and let $\mu \in C(\overline{\Omega})^* = \Sigma(\overline{\Omega})$. Then we have

$$\begin{aligned}\langle \mu\,, y - S_\infty(t)\zeta\rangle &= \left\langle \mu\,, \int_0^t S_\infty(t-\sigma)u(\sigma)d\sigma \right\rangle \\ &= \lim_{h\to 0}\left\langle \mu\,,\, S_c(h)\int_0^t S_\infty(t-\sigma)u(\sigma)d\sigma \right\rangle\end{aligned}$$

[13] $L^1(\Omega)$ is separable, thus $L^1(0,T;L^1(\Omega))$ is separable as well. This implies that the topology of the unit ball of $L^\infty_w(0,T;L^\infty(\Omega)) = L^1(0,T;L^1(\Omega))^*$ is defined by a metric, and we don't need to use generalized sequences. Sequences and subsequences are enough.

$$
\begin{aligned}
&= \lim_{h\to 0} \left\langle S'_\Sigma(h)\mu\,,\, \int_0^t S_\infty(t-\sigma)u(\sigma)d\sigma \right\rangle \\
&= \lim_{h\to 0} \int_0^t \langle S'_\Sigma(h)\mu\,,\, S_\infty(t-\sigma)u(\sigma)\rangle d\sigma \\
&= \lim_{h\to 0} \int_0^t \langle S'_1(t-\sigma)S'_\Sigma(h)\mu\,, u(\sigma)\rangle d\sigma \\
&= \lim_{h\to 0} \int_0^t \langle S'_\Sigma(t-\sigma+h)\mu\,, u(\sigma)\rangle d\sigma \\
&= \int_0^t \langle S'_\Sigma(t-\sigma)\mu\,, u(\sigma)\rangle d\sigma\,, \qquad (4.8.5)
\end{aligned}
$$

where the first three angled brackets indicate the duality of $C(\overline{\Omega})$ and $\Sigma(\overline{\Omega})$, the others the duality of $L^1(\Omega)$ and $L^\infty(\Omega)$. The limit in the last two lines is taken using (4.3.18), continuity of $S'_\Sigma(t)$ in $t > 0$ and the dominated convergence theorem.

If $\{u_n(\cdot)\} \subset L^\infty_w(0,T;L^\infty(\Omega))$ is a minimizing sequence for the norm optimal problem, we have

$$
y_n - S_\infty(T)\zeta = \int_0^T S_\infty(T-\sigma)u_n(\sigma)d\sigma\,,
$$

and $\|u_n(\cdot)\|_{L^\infty_w(0,T;L^\infty(\Omega))} \to \rho =$ optimal norm. Selecting a subsequence we may assume that $u_n(\cdot) \to \bar u(\cdot) \in L^\infty_w(0,T;L^\infty(\Omega))$ $L^1(0,T;L^1(\Omega))$-weakly, which in turn implies that $\|\bar u(\cdot)\|_{L^\infty_w(0,T;L^\infty(\Omega))} = \rho$. We write (4.8.5) for $\mu \in \Sigma(\overline{\Omega})$, $t = T$ and $u(\cdot) = u_n(\cdot)$,

$$
\langle \mu\,, y_n - S_\infty(T)\zeta\rangle = \int_0^T \langle S'_\Sigma(T-\sigma)\mu\,, u_n(\sigma)\rangle d\sigma\,,
$$

note that $S'_\Sigma(T-\,\cdot)\mu \in L^1(0,T;L^1(\Omega))$ and take limits. The result is, via (4.8.5),

$$
\begin{aligned}
\langle \mu\,, \bar y - S_\infty(T)\zeta\rangle &= \int_0^T \langle S'_\Sigma(T-\sigma)\mu\,, \bar u(\sigma)\rangle d\sigma \\
&= \left\langle \mu\,, \int_0^T S_\infty(T-\sigma)\bar u(\sigma)d\sigma \right\rangle.
\end{aligned}
$$

Since $\mu$ is arbitrary, this implies

$$
\bar y - S_\infty(T)\zeta = \int_0^T S_\infty(T-\sigma)\bar u(\sigma)d\sigma\,,
$$

and we conclude that $\bar u(\cdot)$ drives $\zeta$ to $\bar y$.

If $\{u_n(\cdot)\}$, $u_n(\cdot) \in L^\infty_w(0,t_n;L^\infty(\Omega))$ is a minimizing sequence for the time optimal problem we have

$$
y_n - S_\infty(t_n)\zeta = \int_0^{t_n} S_\infty(t_n-\sigma)u_n(\sigma)d\sigma\,, \qquad (4.8.6)
$$

and $t_n \to T =$ optimal time, $\|u(\cdot)\|_{L^\infty_w(0,t_n;L^\infty(\Omega))} \le 1$. Pick $\bar{t} >$ all $t_n$ and extend $u_n(\cdot)$ to $t_n \le \sigma \le \bar{t}$ setting $u_n(t) = 0$ there. Selecting a subsequence we may assume $u_n(\cdot) \to \bar{u}(\cdot) \in L^\infty_w(0,\bar{t}; L^\infty(\Omega))$ $L^1(0,t;L^1(\Omega))$-weakly. We may also take for granted that $T > 0$; in fact, if $T = 0$ we obtain taking limits directly in (4.8.6) that $\bar{y} = \lim_{n\to\infty} S_\infty(t_n)\zeta$; hence, for all $z \in L^1(\Omega)$,

$$\langle \bar{y}, z\rangle = \lim_{n\to\infty} \langle S_\infty(t_n)\zeta, z\rangle = \lim_{n\to\infty} \langle \zeta, S_1'(t_n)z\rangle = \langle \zeta, z\rangle ,$$

so that $\zeta = \bar{y}$ and we don't need to drive. With $T > 0$, we write (4.8.5) for $t = t_n$ :

$$\langle \mu\,, \bar{y} - S_\infty(t_n)\zeta\rangle = \int_0^{t_n} \langle S_\Sigma'(t_n - \sigma)\mu\,, u_n(\sigma)\rangle d\sigma\,. \tag{4.8.7}$$

To pass to the limit on the left side we use $C(\overline{\Omega})$-continuity of $S_\infty(t)$ in $t > 0$. For the limit on the right side, define

$$\chi_n(\sigma) = \begin{cases} S_\Sigma'(t_n - \sigma)\mu & 0 \le \sigma \le t_n \\ 0 & t_n < \sigma \le \bar{t}\,, \end{cases}$$

and

$$\chi(\sigma) = \begin{cases} S_\Sigma'(T - \sigma)\mu & 0 \le \sigma \le T \\ 0 & T < \sigma \le \bar{t}\,, \end{cases}$$

and note that, by continuity of $S_\Sigma'(t)$ for $t > 0$, $\chi_n(\cdot) \to \chi(\cdot)$ in $L^1(0,t;L^1(\Omega))$. We write (4.8.7) in the form

$$\langle \mu\,, y_n - S_\infty(t_n)\zeta\rangle = \int_0^{t} \langle \chi_n(\sigma)\,, u_n(\sigma)\rangle d\sigma\,,$$

take limits and use (4.8.5) again. The result is

$$\begin{aligned}\langle \mu\,, \bar{y} - S_\infty(T)\zeta\rangle &= \int_0^T \langle \chi(\sigma)\,, \bar{u}(\sigma)\rangle d\sigma \\ &= \int_0^T \langle S_\Sigma'(T-\sigma)\mu\,, u(\sigma) d\sigma = \left\langle \mu\,, \int_0^T S_\infty(T-\sigma)\bar{u}(\sigma)d\sigma \right\rangle,\end{aligned}$$

which again implies

$$\bar{y} - S_\infty(T)\zeta = \int_0^T S_\infty(\bar{t} - \sigma)\bar{u}(\sigma)d\sigma$$

and shows that $\bar{u}(t)$ drives $\zeta$ to $\bar{y}$ in optimal time $T$. This completes the proof.

**Corollary 4.8.2.** *Let $\bar{y} \in R^\infty_w(T)$. Then there exists $\bar{u}(\cdot) \in L^\infty_w(0,T;L^\infty(\Omega))$ with*

$$\bar{y} = \int_0^T S_\infty(T-\sigma)\bar{u}(\sigma)d\sigma\,, \quad \|\bar{y}\|_{R^\infty_w(T)} = \|\bar{u}(\cdot)\|_{L^\infty_w(0,T;L^\infty(\Omega))}\,. \tag{4.8.8}$$

Corollary 4.8.2 is just a restatement of the existence theorem for norm optimal controls.

Theorem 2.1.7 (uniqueness without intercession of the maximum principle) has no counterpart here. In fact, nothing at all seems to be known about time or norm optimal controls *without* the condition

$$\bar{y} \in \overline{D(A_\infty)}\,. \tag{4.8.9}$$

**Theorem 4.8.3.** *Let $\bar{u}(t,x)$ be a norm optimal control such that (4.8.9) holds for the target $\bar{y}$. Then*

$$|\bar{u}(t,x)| = \rho \qquad a.\ e.\ in\ (0,T)\times\Omega\,. \tag{4.8.10}$$

*where $\rho$ is the optimal norm. If $\bar{u}(t,x)$ is time optimal and (4.8.9) holds, then (4.8.10) is satisfied with $\rho = 1$.*

*Proof.* According to Theorem 4.7.7,

$$\bar{u}(t,x) = \rho\,\mathrm{sign}\, z(t,x)\,, \tag{4.8.11}$$

where $z(t,x)$ is a nonzero solution of the reverse adjoint equation (4.7.13). This shows that (4.8.10) holds except in the nodal set of $z(t,x)$, which has been shown to be of measure zero in Theorem 4.7.12. Same for time optimal controls.

**Corollary 4.8.4.** *Under (4.8.9) norm and time optimal controls are unique.*

*Proof.* Let $u_1(t,x))$, $u_2(t,x)$ be two time optimal controls, $T$ their common optimal time. Theorem 4.8.3 says that both controls satisfy (4.8.10) with $\rho = 1$,

$$|u_1(t,x)| = 1, \quad |u_2(t,x)| = 1 \quad \text{a. e. in } (0,T)\times\Omega\,.$$

The control

$$u(t,x) = \frac{1}{2}\big(u_1(t,x) + u_2(t,x)\big)$$

is also time optimal thus it also satisfies

$$|u(t,x)| = 1 \quad \text{a. e. in } (0,T)\times\Omega\,.$$

The last three equalities are compatible if and only if $u_1(t,x)$, $u_2(t,x)$ are both $= 1$ or both $-1$ except perhaps in a null set, which is what we wanted to prove. The argument is exactly the same for norm optimal controls, with $\rho$ instead of 1.

Without (4.8.9), uniqueness for the norm optimal problem collapses, as the following companion of Lemma 3.1.6 shows.

**Theorem 4.8.5.** *Given $T > 0$ there exists a target $\bar{y} \in C(\overline{\Omega})$ such that $0$ can be driven to $\bar{y}$ norm optimally in the interval $0 \le t \le T$ by different controls.*

*Proof.* The construction in Theorem 3.1.10, which can be performed with the space $E = C(\overline{\Omega})$ and the analytic semigroup $S_c(t)$ yields two different controls $\bar{u}_0(\cdot)$, $\bar{v}(\cdot) \in L^\infty(0,T;C(\overline{\Omega}))$ that drive $\zeta = 0$ to the target $y(T,\zeta,\bar{u}_0) = y(T,\zeta,\bar{v}) \in C(\overline{\Omega})$ and have optimal norm 1 *in the space* $L^\infty(0,T;C(\overline{\Omega}))$. Accordingly, we only have to show that these controls preserve their norm optimality if $L^\infty(0,T;C(\overline{\Omega}))$ is replaced by the larger space $L^\infty_w(0,T;L^\infty(\Omega))$. The first control $\bar{u}_0(t)$ coincides in an interval $[t_0,T]$ with the control in Theorem 4.7.9, thus we can lift the argument from there; if $\bar{u}_0(t)$ is norm optimal in $L^\infty_w(0,T;L^\infty(\Omega))$ then $\bar{v}(t)$, having the same norm, must be norm optimal too.

For the time optimal problem without (4.8.9) we don't know the answer to the following two problems.

**Problem 4.8.6.** *Does every time optimal control satisfy the bang-bang property*

$$|\bar{u}(t,x)| = 1 \quad \text{a. e. in } (0,T)\times\Omega\,?$$

**Problem 4.8.7.** *Are optimal controls unique without* (4.8.9)?

Of course, a "yes" answer to Problem 4.8.6 would produce a "yes" answer to Problem 4.8.7 via the argument in Corollary 4.8.4.

The bang-bang property (4.8.10) has convergence implications for the time and norm optimal problems. Generalizing slightly our previous definition, we call $\{u_n(t,x)\}$, $u_n(\cdot,\cdot) \in L^\infty((0,t_n)\times\Omega)$ a *minimizing sequence* for the time optimal problem if

$$\|u_n\|_{L^\infty((0,t_n)\times\Omega)} \le 1 + o(1)\,, \quad \|y(t_n,\zeta,u_n) - \bar{y}\|_{C(\overline{\Omega})} = o(1)\,, \\ t_n \to T = \text{optimal time}. \tag{4.8.12}$$

Existence results generalize trivially to the new definition.

**Theorem 4.8.8.** *Assume* $\bar{y} \in \overline{D(A_\infty)}$. *Let* $\{u_n(\cdot\,,\cdot)\}$ *be a minimizing sequence and let* $\bar{u}(\cdot\,,\cdot) \in L^\infty((0,T)\times\Omega)$ *be the control driving* $\zeta$ *time optimally to* $\bar{y}$. *Then, if* $1 \le p < \infty$ *we have*

$$\|u_n - \bar{u}\|_{L^p((0,T)\times\Omega)} \to 0 \quad \text{as } n\to\infty\,. \tag{4.8.13}$$

*Proof.* Pick $t >$ all $t_n$ and extend the $u_n(\sigma,x)$ and $u(\sigma,x)$ to $0 \le \sigma \le t$ by

$$u_n(\sigma,x) = 0 \;\; (t_n \le \sigma \le t)\,, \quad u(\sigma,x) = 0 \;\; (T \le \sigma \le t)\,.$$

Select an arbitrary subsequence of $\{u_n(\cdot,\cdot)\}$ and then select a further subsequence (both equally named) such that

$$u_n(\cdot\,,\cdot) \to \tilde{u}(\cdot\,,\cdot) \in L^\infty((0,T)\times\Omega) \quad L^1((0,T)\times\Omega)\text{-weakly}\,. \tag{4.8.14}$$

The argument in Theorem 4.8.1 (very slightly generalized to handle the first condition (4.8.12)) shows that $\tilde{u}(\cdot\,,\cdot)$ is time optimal. Since $\bar{y} \in D(A_\infty)$, Corollary 4.8.4 applies and $\tilde{u}(t,x) = \bar{u}(t,x)$; on the other hand, Theorem 4.8.3 says that the bang-bang property $|\bar{u}(t,x)| = 1$ a. e. in $(0,T)\times\Omega$ is satisfied. Noticing that (4.8.14) implies $u_n(\cdot\,,\cdot) \to \tilde{u}(\cdot\,,\cdot) = \bar{u}(\cdot\,,\cdot) \in L^2((0,T)\times\Omega)$ $L^2((0,T)\times\Omega)$-weakly, the $L^2$ part of Theorem 4.8.8 follows from

$$\begin{aligned}
&\|u_n - \bar{u}\|^2_{L^2((0,T)\times\Omega)} = (u_n - \bar{u}, u_n - \bar{u}) \\
&= \|u_n\|^2_{L^2((0,T)\times\Omega)} + \|\bar{u}\|^2_{L^2((0,T)\times\Omega)} - 2(u_n, \bar{u}) \\
&\le T|\Omega| + o(1) + T|\Omega| - 2(u_n, \bar{u}) \\
&\to 2T|\Omega| - 2\|\bar{u}\|^2_{L^2((0,T)\times\Omega)} = 0 \quad \text{as } n \to \infty\,.
\end{aligned}$$

Since $u(\cdot\,,\cdot)$ is bounded a. e. and the $u_n(\cdot\,,\cdot)$ are uniformly bounded a. e. it follows that $L^2$ convergence implies $L^p$ convergence for any $p < \infty$ (the one-dimensional argument in Remark 3.6.6 applies just as well in dimension $m$).

All the definitions and results above have counterparts for the norm optimal problem; a minimizing sequence satisfies

$$\|u_n\|_{L^\infty((0,T)\times\Omega)} \le \rho + o(1)\,, \quad \|y(T,\zeta,u_n) - \bar{y}\|_{C(\overline{\Omega})} = o(1) \tag{4.8.15}$$

where $\rho$ is the minimum norm. The conclusions are those of Theorem 4.8.8, that is, $L^p$ convergence of minimizing sequences for $1 \le p < \infty$.

Theorem 4.8.8 does not extend to the $L^\infty$ norm; counterexamples can be constructed as in the comments before and after (3.6.13). We also note that, in either the time or norm optimal problem, is not necessary to take the initial condition $\zeta$ fixed; just a sequence $\{\zeta_n\}$ with $\|\zeta_n - \zeta\|_{C(\overline{\Omega})} = o(1)$.

More precise convergence results hold under stronger assumptions. Assume $\bar{u}(\sigma,x)$ drives $\bar{\zeta}$ to $y(T,\bar{\zeta},\bar{u})$ in optimal time $T$ and that it satisfies the maximum principle (4.7.6). Let $y(t,\zeta,u)$ be another trajectory with $\|u(\cdot\,,\cdot)\|_{L^\infty((0,T)\times\Omega} \le 1$. Formula (4.8.11) with $\rho = 1$ implies

$$z(t,x)(\bar{u}(t,x) - u(t,x)) \ge 0\,, \tag{4.8.16}$$

so that

$$\begin{aligned}
&\int_{(0,T)\times\Omega} |z(\sigma,x)||\bar{u}(\sigma,x) - u(\sigma,x)|d\sigma dx \\
&\int_{(0,T)\times\Omega} z(\sigma,x)(\bar{u}(\sigma,x) - u(\sigma,x))d\sigma dx \\
&= \int_0^T \langle S_1'(T-\sigma)z\,, \bar{u}(\sigma) - u(\sigma)\rangle d\sigma \\
&= \left\langle\!\!\left\langle \xi_z\,, \int_0^T S_\infty(T-\sigma)\bar{u}(\sigma)d\sigma - \int_0^T S_\infty(T-\sigma)u(\sigma)d\sigma \right\rangle\!\!\right\rangle \\
&= \langle\!\langle \xi_z\,, (y(T,\bar{\zeta},\bar{u}) - S_\infty(T)\bar{\zeta}) - (y(T,\zeta,u) - S_\infty(T)\zeta)\rangle\!\rangle \\
&= \langle\!\langle \xi_z\,, y(T,\bar{\zeta},\bar{u}) - y(T,\zeta,u)\rangle\!\rangle - \langle\!\langle \xi_z\,, S_\infty(T)\bar{\zeta} - S_\infty(T)\zeta\rangle\!\rangle\,.
\end{aligned}$$

Accordingly,

$$\begin{aligned}\int_{(0,T)\times\Omega} &|z(t,x)||\bar u(t,x)-u(t,x)|dtdx \\ &\le \|\xi_z\|_{R^\infty_w(T)^\star}\|y(T,\bar\zeta,\bar u)-y(T,\zeta,u)\|_{R^\infty_w(T)} \\ &+ \|\xi_z\|_{R^\infty_w(T)^\star}\|S_\infty(T)\bar\zeta-S_\infty(T)\zeta\|_{R^\infty_w(T)} \\ &= \|z\|_{Z^1(T)}\|y(T,\bar\zeta,\bar u)-y(T,\zeta,u)\|_{R^\infty_w(T)} \\ &+ \|z\|_{Z^1(T)}\|S_\infty(T)\bar\zeta-S_\infty(T)\zeta\|_{R^\infty_w(T)}\,, \end{aligned} \tag{4.8.17}$$

which is much more explicit than the convergence relation (4.8.13). Since the norm of $R^\infty_w(T)$ is not amenable to practical computation, the estimate is especially useful when $y(T,\zeta,u)$, $y(T,\bar\zeta,\bar u)\in D(A_\infty)$; in this case (see (4.4.13)), the right hand of (4.8.17) is bounded by (a constant times)

$$\|y(T,\bar\zeta,\bar u)-y(T,\zeta,u)\|_{D(A_\infty)}+\|S_\infty(T)\bar\zeta-S_\infty(T)\zeta\|_{D(A_\infty)}\,.$$

Note, however, that if (4.8.17) is put to work to show convergence of a suboptimal sequence for the time optimal problem, the second requirement in(4.8.12) must be upgraded to $\|y(T,\zeta,u_n)-\bar y\|_{D(A_\infty)}=o(1)$, which uses a stronger norm than (4.8.12) and also presupposes knowledge of the optimal time $T$. For the norm optimal problem, we change $\|u(\cdot\,,\cdot)\|_{L^\infty((0,T)\times\Omega)}\le 1$ by $\|u(\cdot\,,\cdot)\|_{L^\infty((0,T)\times\Omega)}\le\rho$ in (4.8.16) and (4.8.18), and the conclusion is the same; here there is no $T$ to know since $T$ is fixed in advance.

**Miscellaneous notes.** This section is more or less taken from the author [2002:2] although many of the results are of older vintage (see the author [1999:2]).

**4.9. Examples and applications.** We include in this section some examples of optimal controls for the system

$$\begin{aligned}\frac{\partial y(t,x)}{\partial t} &= Ay(t,x)+u(t,x) \quad (0\le t\le T,\ x\in\Omega)\\ y(0,x) &= \zeta(x) \quad (x\in\Omega)\\ \beta y(t,x) &= 0 \quad (0\le t\le T,\ x\in\Gamma)\end{aligned} \tag{4.9.1}$$

produced by means of Theorems 4.7.3 and 4.7.4. These controls are given by

$$\bar u(t,x)=\rho\,\mathrm{sign}\,z(t,x)\,, \tag{4.9.2}$$

where $z(t,x)$ is a solution of the reverse adjoint equation

$$\begin{aligned}\frac{\partial z(t,x)}{\partial t} &= -A'z(t,x) \quad (0\le t< T,\ x\in\Omega)\\ \beta' z(t,x) &= 0 \quad (0\le t< T,\ x\in\Gamma)\end{aligned} \tag{4.9.3}$$

satisfying

$$\int_{(0,T)\times\Omega} |z(t,x)|dtdx < \infty\,. \tag{4.9.4}$$

To avoid repetition, we agree that "a control $\bar{u}(t)$ is norm optimal in the interval $0 \le t \le T$" means that it drives any initial condition $\zeta$ to $y(T, \zeta, u)$ norm optimally; that "it is time optimal" means that it drives any initial condition $\zeta$ to $y(T, \zeta, \bar{u})$ in optimal time $T$ under conditions (4.7.9) in Theorem 4.7.4 on the initial condition or the target.

When the space dimension is $m = 1$, a lot more is known about the nodal set $\mathcal{N}(z)$ of solutions of (4.9.3). Here, $\Omega = (a, b)$ and

$$Ay(x) = (a(x)y'(x))' + b(x)y'(x) + c(x)y(x)\,,$$

with the formal adjoint given by

$$\begin{aligned} A'y(x) &= (a(x)y'(x))' - (b(x)y(x))' + c(x)y(x) \\ &= a(x)y''(x) + (a'(x) - b(x))y'(x) + (c(x) - b'(x))y(x) \\ &= A(x)y''(x) + B(x)y'(x) + C(x)y(x)\,, \end{aligned}$$

so that the equation satisfied by $\theta(t,x) = z(T-t,x)$ is

$$\frac{\partial\theta(t,x)}{\partial t} = A(x)\frac{\partial^2\theta(t,x)}{\partial x^2} + B(x)\frac{\partial\theta(t,x)}{\partial x} + C(x)\theta(t,x)$$
$$(0 < t \le T,\ a \le x \le b)\,.$$

Let $\eta(t)$, the *zero counting function,* be the number of zeros of $\theta(t,x)$ in $a < x < b$. Applying Theorem C in Angenent [1988] in the rectangles $[\epsilon, T] \times [a, b]$ we deduce that, if the boundary condition is Dirichlet[14] ($\theta(t, 0) = \theta(t, \pi) = 0$ for $t > 0$) then $\eta(t)$ is finite and nonincreasing in $t \ge \epsilon$, thus it is finite and nonincreasing in $t > 0$; it may tend to infinity as $t \to 0$, since $\theta(t,x)$ is smooth only in the open rectangle $(0, T] \times [a, b]$. Translating this to the costate $z(t,x) = \theta(T-t,x)$ we obtain

**Theorem 4.9.1.** *Assume $m = 1$ and that the boundary condition is Dirichlet, and let $\eta(t)$ be the number of zeros of $z(t,x)$ in $a < x < b$. Then $\eta(t) < \infty$ $(0 \le t < T)$ and $\eta(t)$ is nondecreasing.*

**Example 4.9.2.** For the heat equation

$$\frac{\partial y(t,x)}{\partial t} = \frac{\partial^2 y(t,x)}{\partial x^2} + u(t,x)\,, \quad y(t,0) = y(t,\pi) = 0 \tag{4.9.5}$$

[14] The result in Angenent [1988] admits time dependent coefficients, gives more information on the nodal sets and works (under additional restrictions) with other boundary conditions

the reverse adjoint equation is

$$\frac{\partial z(t,x)}{\partial t} = -\frac{\partial^2 z(t,x)}{\partial x^2}, \quad z(t,0) = z(t,\pi) = 0 . \tag{4.9.6}$$

Condition (4.9.4) for a solution $z(t,x)$ of (4.9.6) in $(0,T] \times [0,\pi]$ is

$$\int_0^\pi \int_0^T |z(t,x)| dt dx < \infty . \tag{4.9.7}$$

The solution of (4.9.6) with "final condition"

$$z(T,x) = z(x) = \sum_{n=1}^\infty c_n \sin nx$$

is given by

$$z(t,x) = \sum_{n=1}^\infty e^{-n^2(T-t)} c_n \sin nx = \sum_{n=1}^\infty a_n e^{n^2 t} \sin nx ,$$

where the choice of the $c_n$ (or of the $a_n$) must heed (4.9.7). This is automatic if the sum is finite. Figure 4.9.1 below shows the nodal set for

$$z(t,x) = e^{t-6} \sin x - e^{4t-6} \sin 2x + e^{16t-34} \sin 4x - e^{25t-34} \sin 5x$$

in the rectangle $(0,6] \times [0,\pi]$. The curves are the *switching curves* where $\bar{u}(t,x)$ switches from $\rho$ to $-\rho$ and back. The control (4.9.2) is norm optimal (and time optimal if $\rho = 1$). Of course, so is $-\bar{u}(t,x) = \text{sign}(-z(t,x))$, thus we may exchange the roles of $\rho$ and $-\rho$.

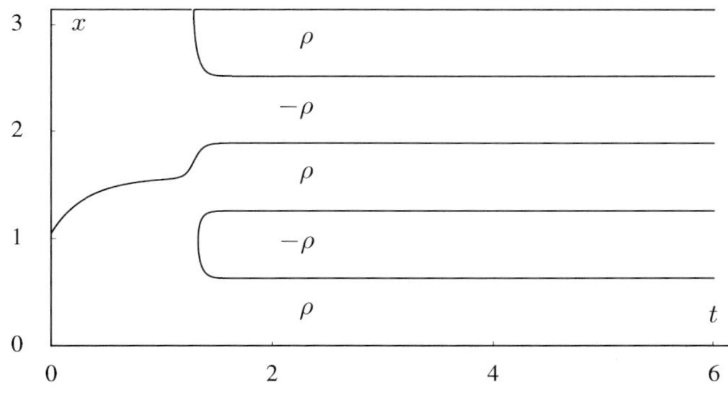


Figure 4.9.1

Theorem 4.9.1 is seen in action here. The number $\eta(t)$ of zeros in $(0,\pi)$ is

$$\eta(t) = \begin{cases} 1 & 0 \le t < 1.2902 \\ 2 & 1.2902 \le t < 1.3297 \\ 3 & t = 1.3297 \\ 4 & t > 1.3297 \end{cases}$$

The successive bifurcations from 1 to 2 to 3 to 4 zeros are shown with snapshots of $z(t,x)$ for $t = 1.29, 1.31, 1.33, 1.34$ in Figure 4.9.2.

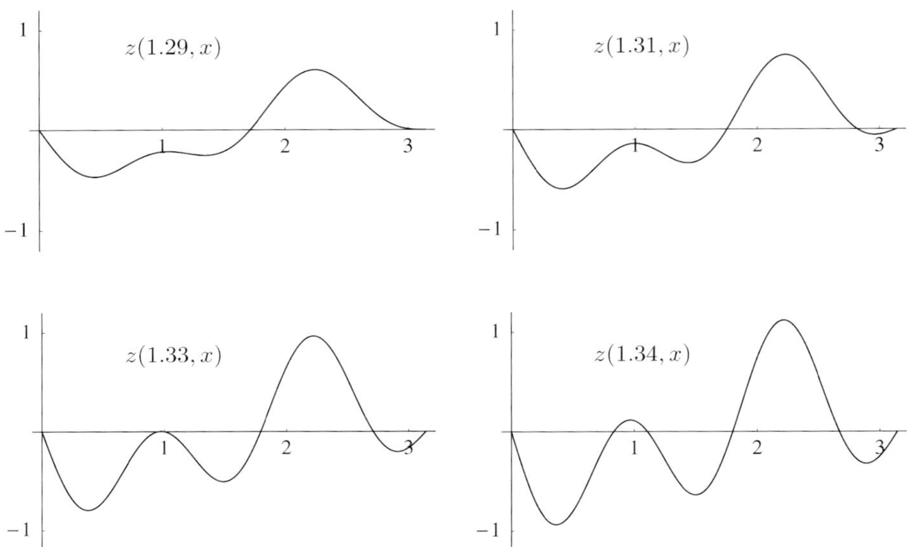


Figure 4.9.2

**Remark 4.9.3.** We can produce optimal controls where the number of switchings tends to infinity as $t \to T$. To this end, we need solutions of (4.9.6) that satisfy the summability condition (4.9.7) in $[0, T) \times (0, \pi)$ and have a zero counting function $\eta(t)$ such that

$$\lim_{t \to T} \eta(t) = \infty\,.$$

Let $e \subseteq (0, \pi)$ be a measurable set such that

$$|(a,b) \cap e| > 0\,, \quad |(a,b) \cap e^c| > 0 \quad (0 \le a < b \le \pi)\,,$$

where $^c$ denotes complement (Natanson [1955, Exercise 8, p. 88]). Define

$$z(x) = \begin{cases} 1 & x \in e\,, \\ -1 & x \notin e\,. \end{cases}$$

We obtain a solution $\theta(t,x)$ of the heat equation continuing $z(x)$ as an odd $2\pi$-periodic function in $(-\infty, \infty)$ and using Weierstrass' formula

$$\theta(t,x) = \frac{1}{2\sqrt{\pi t}} \int_{\infty}^{\infty} e^{-(x-\xi)^2/4t} z(\xi)d\xi\,, \tag{4.9.8}$$

the Dirichlet boundary conditions automatically satisfied due to the way $z(x)$ is continued. It follows from properties of the Weierstrass kernel that

$$\|\theta(t,\cdot)\|_{L^\infty(0,\pi)} \le \|z(\cdot)\|_{L^\infty(0,\pi)} = 1 \quad (t>0)\,, \qquad \lim_{t\to 0+} \theta(t,x) = z(x) \quad \text{a. e.}$$

thus $z(t,x) = \theta(T-t,x)$, satisfies the reverse adjoint equation (4.9.6) and it is uniformly bounded (so that (4.9.7) is automatic). Moreover,

$$\|z(t,\cdot)\|_{L^\infty(0,\pi)} \le \|z(\cdot)\|_{L^\infty(0,\pi)} = 1 \quad (t<T)\,, \qquad \lim_{t\to T-} z(t,x) = z(x) \quad \text{a. e.}$$

In view of the definition of $z(x)$, given $N$ and a finite set of nonempty disjoint open intervals $\{(a_n,b_n);\, n=1,2,\dots,N\}$ we can select $x_n, y_n$ in each interval with

$$\lim_{t\to T-} z(t,x_n) = 1, \qquad \lim_{t\to T-} z(t,y_n) \to -1$$

so that, for $t<T$ sufficiently near $T$, $z(t,x)$ will have (at least) $N$ zeros.

**Remark 4.9.4.** The control $\bar{u}(t,x)$ in (4.9.2) may be associated with very different costates $z(t,x)$. For an example, consider the system 4.9.5 and the costate

$$z_0(t,x) = e^{T-t}\sin x\,,$$

for which

$$\bar{u}(t,x) = \operatorname{sign} z_0(t,x) = 1\,.$$

For a (numerous) family of costates associated with the same control, consider a solution $z_1(t,x)$ of (4.9.6) with final condition $z_1(T,x) = \mu$, where $\mu$ is an arbitrary nonzero finite positive measure in $[0,\pi]$; we have $z(t,x)>0$ in $[0,T)\times\Omega$, so that $\bar{u}(t,x) = \operatorname{sign} z_1(t,x) = 1$. This indeterminacy of $z(t,x)$ contrasts with results for the equation

$$y'(t) = Ay(t) + u(t)$$

with $A$ the infinitesimal generator of, say, an analytic semigroup in a Hilbert space $E$. In this case, the analog of (4.9.2) is

$$\bar{u}(t) = \rho\,\frac{S(T-t)^*z}{\|S(T-t)^*z\|}$$

with $z \in Z(T)$, which characterizes $z$ within a multiplicative constant. In fact, freezing the equality

$$\bar{u}(t) = \frac{S(T-t)^*z_1}{\|S(T-t)^*z_1\|} = \frac{S(T-t)^*z_2}{\|S(T-t)^*z_2\|}$$

at any $t$ we get

$$S(T-t)^*z_2 = S(T-t)^*\alpha z_1\,.$$

We apply $R(\lambda; A^*)$ to both sides of this equality for $\lambda$ large enough and use the facts that $R(\lambda; A^*)Z(T) \subseteq E^* = E$ and that $S(t)^*$ is one-to-one, obtaining $z_2 = \alpha z_1$.

To show how the necessary and sufficient conditions mesh we solve the problem (of no particular importance) of identifying all norm and time optimal controls for the one dimensional heat equation (4.9.5) that are *time-independent,*[15] that is, $\bar{u}(t,x) = \bar{u}(x)$.

**Theorem 4.9.5.** *$\bar{u}(t,x) = \bar{u}(x)$ is norm optimal (time optimal if $\rho = 1$) if and only if it switches from $\pm\rho$ to $\mp\rho$ across the equispaced lines $x = x_j$, where*

$$\{x_0, x_1, x_2, \ldots, x_n\} = \mathcal{N}_n = \left\{0, \frac{\pi}{n}, \frac{2\pi}{n}, \ldots, \frac{(n-1)\pi}{n}, \pi\right\}.$$

*Proof.* $\mathcal{N}_n$ is the set of zeros of $\sin nx$. The function

$$z_n(t,x) = e^{-n^2(T-t)} \sin nx$$

is a solution of the reverse adjoint equation (4.9.6) in $[0,T) \times [0,\pi]$, and (4.9.6) is trivially satisfied. We have

$$\bar{u}(t,x) = \rho \operatorname{sign} z(t,x) \quad \text{or} \quad u(t,x) = \rho \operatorname{sign}(-z(t,x)),$$

depending on wether $\bar{u}(t,x) = \rho$ or $\bar{u}(t,x) = -\rho$ in the initial strip $0 \le x \le \pi/n$. Hence, $\bar{u}(t,x)$ is norm optimal, time optimal if $\rho = 1$.

Conversely, assume $\bar{u}(t,x) = \bar{u}(x)$ is time or norm optimal in an interval $0 \le t \le T$. To achieve the necessary condition (4.7.19) in Theorem 4.7.7 it is enough to show that

$$\int_0^T S_\infty(T-\sigma)u d\sigma = \int_0^T S_\infty(\sigma)u d\sigma \in D(A_\infty) \tag{4.9.9}$$

for every $u \in L^\infty(\Omega)$. To do this, take $y \in D(A_1')$. We have

$$\left\langle \int_0^T S_\infty(\sigma)ud\sigma\,,\, A_1'y \right\rangle = \int_0^T \langle S_\infty(\sigma)u\,,\, A_1'y\rangle d\sigma$$
$$= \int_0^T \langle u\,,\, S_1'(\sigma)A_1'y\rangle d\sigma = \left\langle u\,, \int_0^T S_1'(\sigma)A_1'y d\sigma \right\rangle = \langle u\,,\, S_1'(T)y - y\rangle,$$

which shows that the first angled bracket is a bounded functional of $y$ in the $L^1(\Omega)$ norm. This is equivalent to (4.9.9).

Assume $\bar{u}(t,x) = \bar{u}(x)$ is (time or norm) optimal. By Theorem 4.7.7,

$$\bar{u}(x) = \rho \operatorname{sign} z(t,x)$$

[15] This does not mean these controls are optimal among time independent controls; they are optimal in competition with all controls doing the same drive.

where $z(t,x)$ satisfies (4.9.6) and (4.9.7). If $\epsilon > 0$ then the costate $z(t,x)$ is regular in $0 \leq t \leq T - \epsilon$ and

$$z(t,x) = \sum_{n=1}^{\infty} e^{-n^2(T-\epsilon-t)} a_n(\epsilon) \sin nx \,, \tag{4.9.10}$$

where the $a_n(\epsilon)$ come from the expansion $z(T-\epsilon, x) = \sum_{n=1}^{\infty} a_n(\epsilon) \sin nx$ and can thus be assumed bounded (in fact, due to smoothness of $z(T-\epsilon, T)$ the $a_n(\epsilon)$ converge rapidly to zero). Since $\bar{u}(x)$ is time independent, the nodal set of $z(t,x)$ must be independent of $t$, thus it consists of a finite set of lines $x = x_j$. By the uniqueness theorem for Dirichlet series (Bernstein [1933 p. 12]) we obtain from (4.9.10) that

$$\sin nx_j = 0 \quad j = 0, 1, \ldots, n \quad \text{whenever } a_n(\epsilon) \neq 0 \tag{4.9.11}$$

($a_n$ the coefficients in (4.9.10)) and, conversely, (4.9.11) implies that the line $x = x_j$ belongs to the nodal set. Hence

$$\mathcal{N}(z) = [0, T) \times \bigcap_{a_n(\epsilon) \neq 0} \mathcal{N}_n \,,$$

and, as each $\mathcal{N}_n$ is equispaced in $[0, \pi]$, $\mathcal{N}(z)$ consists of a finite set of equispaced lines. The theorem can be reformulated as

**Theorem 4.9.6.** *A time independent control $\bar{u}(t,x) = \bar{u}(x)$ is norm optimal (time optimal if $\rho = 1$) if and only if*

$$\bar{u}(x) = \rho \, \mathrm{sign} \, \phi_n(x) \tag{4.9.12}$$

*where $\phi_n(x)$ $(= \pm \sin nx)$ is an eigenfunction of $d^2/dx^2$.*

**Remark 4.9.7.** Fragments of Theorem 4.9.6 extend to the $m$-dimensional system (4.9.1). In fact, let $\bar{u}(x)$ be given by (4.9.12), where $\phi_n(x)$ is an eigenfunction of $A'$ (boundary condition $\beta'$) and $-\lambda_n$ is the corresponding eigenvalue. Then $\bar{u}(x)$ is a norm optimal control (time optimal if $\rho = 1$). To see this, note that

$$z(t,x) = e^{-\lambda_n(T-t)} \phi_n(x)$$

is a solution of the reverse adjoint equation and apply Theorems 4.7.3 and 4.7.4.

As for the necessity part, assume we are in the self adjoint case $A = A'$, $\beta = \beta'$. Here, $A_2' = A_2$ is selfadjoint with pure point spectrum, that is, the eigenfunctions $\{\phi_n\}$ (corresponding to eigenvalues $\{-\lambda_n\}$, $\lambda_n \to \infty$) are a complete orthonormal system in $L^2(\Omega)$. Solutions of the reverse adjoint equation (4.9.3) are regular in $0 \leq t \leq T - \epsilon$, and can be written

$$z(t,x) = \sum_{n=1}^{\infty} a_n(\epsilon) e^{-\lambda_n(T-\epsilon-t)} \phi_n(x) \,, \tag{4.9.13}$$

where the $a_n(\epsilon)$ come from the expansion $z(T-\epsilon, x) = \sum_{n=1}^{\infty} a_n(\epsilon)\phi_n(x)$ and can thus be assumed bounded (as in the one dimensional case, they actually die down rapidly). The series (4.9.13) converges in the norm of $L^2(\Omega)$. Using Theorem 3.3.2 and its consequence (3.3.18) we obtain

$$\begin{aligned}\|\phi_n\|_{W^{2k,2}(\Omega)} &\leq C_k \|(\mu I - A_2)^k \phi_n\|_{L^2(\Omega)} \\ &= C_k(\mu - \lambda_n)^k \|\phi_n\|_{L^2(\Omega)} = C_k(\mu - \lambda_n)^k\end{aligned}$$

for arbitrary $k$, so that, taking $k$ large enough and using Theorem 3.3.3 we obtain

$$\|\phi_n(\cdot)\|_{C(\overline{\Omega})} \leq C\lambda_n^k ,$$

which, together with the asymptotic formula for the eigenvalues

$$\lambda_n = \frac{C}{n^{m/2}} + o(1)$$

in Gårding [1953] shows that (4.9.13) is uniformly convergent in $\overline{\Omega}$ for $t > 0$. Now, if $\bar{u}(x)$ is a an optimal control and $T$ is the optimal time, then by (4.9.9) (which holds in arbitrary dimension), $y(T, \zeta, \bar{u}) \in D(A_\infty)$ so that

$$\bar{u}(x) = \rho \operatorname{sign} z(t, x)$$

with $z(t, x)$ a smooth solution of (4.9.3). The nodal set $\mathcal{N}(z)$ must be independent of $t$, so that arguing with (4.9.13) on the basis of the uniqueness argument for Dirichlet series in the same way as with (4.9.10) we obtain

$$\mathcal{N}(z) = [0, T) \times \bigcap_{a_n(\epsilon)\neq 0} \mathcal{N}_n .$$

where $\mathcal{N}_n$ is the nodal set of $\phi_n(x)$. However, unlike in dimension 1, $\mathcal{N}(z)$ doesn't have to be the nodal set of a single eigenfunction.

As a byproduct of the results in this section, we support the observation after (4.4.22) that the control space $L^\infty(0, T; C(\overline{\Omega}))$ is a bad space for existence. The result below gives an example of nonexistence a of a norm optimal control in this control space and thus justifies the introduction of weakly measurable controls.

**Corollary 4.9.8.** *There exists $y \in R^\infty(T)$ such that*

$$y = \int_0^T S_c(T-\sigma)u(\sigma)d\sigma , \quad u(\cdot) \in L^\infty(0, T; C(\overline{\Omega})) \tag{4.9.14}$$

*implies*

$$\|u(\cdot)\|_{L^\infty(0,T;C(\overline{\Omega}))} > \|y\|_{R^\infty(T)}. \tag{4.9.15}$$

*Proof.* We use the equation (4.9.5). The semigroup $S_\infty(t)$ corresponding to the operator $A = d^2/dx^2$ in $[0,\pi]$ with Dirichlet boundary conditions can be expressed by means of Weierstrass' formula (4.9.8),

$$S_\infty(t)y(x) = \frac{1}{2\sqrt{\pi t}} \int_\infty^\infty e^{-(x-\xi)^2/4t} y(\xi)d\xi$$

with $y(x)$ extended as in (4.9.8). It follows from this representation that

$$\|S_\infty(t)\|_{\mathcal{L}(L^\infty(\Omega),L^\infty(\Omega))} \le 1 \quad (t \ge 0)\,,$$

thus Theorem 4.4.4 says

$$R_w^\infty(T) = R^\infty(T)\,, \quad \|y\|_{R_w^\infty(T)} = \|y\|_{R^\infty(T)}\,. \tag{4.9.16}$$

Define a control in $L_w^\infty(0,T;L^\infty(\Omega))$ by $\bar{u}(t,x) = \bar{u}(x) = \operatorname{sign}\sin nx$ for any $n \ge 2$. In view of the first equality (4.9.16) there exists $u(\cdot) \in L^\infty(0,T;C(\overline{\Omega}))$ such that

$$\begin{aligned} y &= \int_0^T S_\infty(T-\sigma)\bar{u}(\sigma)d\sigma \\ &= \int_0^T S_c(T-\sigma)u(\sigma)d\sigma = \int_0^T S_\infty(T-\sigma)u(\sigma)d\sigma\,. \end{aligned} \tag{4.9.17}$$

According to Theorem 4.9.6 the time invariant control $\bar{u}(t) = \bar{u}$ is norm optimal, hence we have $\|y\|_{R_w^\infty(T)} = \|\bar{u}(\cdot)\|_{L_w^\infty(0,T;L^\infty(\Omega))}$. Assume (4.9.15) is false for $u(t)$, that is, assume $\|u(\cdot)\|_{L^\infty(0,T;C(\overline{\Omega}))} = \|y\|_{R^\infty(T)}$. Then, using (4.9.16),

$$\begin{aligned} \|u(\cdot)\|_{L^\infty(0,T;L^\infty(\Omega))} &= \|u(\cdot)\|_{L^\infty(0,T;C(\overline{\Omega}))} \\ &= \|y\|_{R^\infty(T)} = \|y\|_{R_w^\infty(T)} = \|\bar{u}\|_{L_w^\infty(0,T;L^\infty(\Omega))}\,, \end{aligned}$$

so that $u(\cdot)$ is norm optimal as well. In view of (4.9.9), the element $y$ defined in (4.9.17) belongs to $D(A^\infty)$, thus the uniqueness Corollary 4.8.4 yields $\bar{u}(\cdot) = u(\cdot)$, absurd since $\bar{u}(t) \notin C(\overline{\Omega})$.

**Example 4.9.9.** We do the two dimensional controlled heat equation in the unit circle $\Omega \subset \mathbb{R}^2$ with Dirichlet boundary condition,

$$\frac{\partial y(t,x)}{\partial t} = \Delta y(t,x) + u(t,x) \ \ (x \in \Omega)\,, \quad y(t,x) = 0 \ \ (x \in \Gamma)\,. \tag{4.9.18}$$

The reverse adjoint equation is

$$\frac{\partial z(t,x)}{\partial t} = -\Delta z(t,x) \ \ (x \in \Omega)\,, \quad z(t,x) = 0 \ \ (x \in \Gamma)\,. \tag{4.9.19}$$

and we can construct optimal controls using the formula (4.9.2) for costates that satisfy the growth condition (4.9.4). The eigenfunctions of the Laplacian in the unit circle are

$$J_m(\gamma_{mn}r)\sin m\theta, \quad J_m(\gamma_{mn}r)\cos m\theta, \qquad m, n = 0, 1, 2, \dots$$

corresponding to the eigenvalue $-\gamma_{mn}^2$, where $(r, \theta)$ are the polar coordinates in the plane and, for each $m$, $0 < \gamma_{m1} < \gamma_{m2} < \dots$ are the zeros of the Bessel function $J_m(r)$. Figure 4.9.3 below shows the cross section of the (radially symmetric) nodal set $\mathcal{N}(z)$ of the radially symmetric costate

$$z(t, r) = 10\, e^{\gamma_{01}^2 t} J_0(\gamma_{01}r) - e^{\gamma_{02}^2 t} J_0(\gamma_{02}r) + 0.001\, e^{\gamma_{03}^2 t} J_0(\gamma_{03}r)$$

in the interval $0 \le t \le T = 2$. The nodal set consists of of two disjoint surfaces dividing the cylinder

$$\{(t, r, \theta); 0 \le t \le 2,\ 0 \le r \le 1,\ 0 \le \theta \le 2\pi\}$$

in three disjoint regions, shown in cross section in Figure 4.9.3 with the values that the optimal control takes in each region. Of course, we can switch $\rho$ and $-\rho$ since $\bar{u}(t, x) = \rho\, \mathrm{sign}\,(-z(t, x))$ is optimal as well.

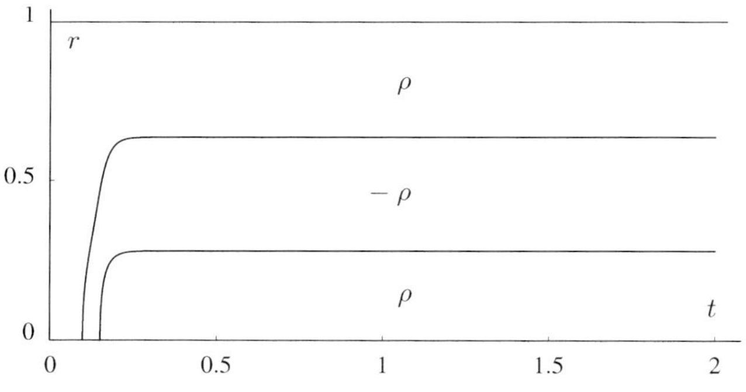


Figure 4.9.3

The intersection of the upper curve in Figure 4.9.3 (the bottom of the outer surface) is $t = 0.09630$, whereas the intersection of the lower curve in Figure 4.9.3 (the bottom of the inner surface) is $t = 0.14896$.

Another way to describe the nodal set is: the intersection $\mathcal{N}(t, z)$ of the nodal set $\mathcal{N}(z)$ with the plane parallel to the $(r, \theta)$ plane cutting the $t$-axis at $t$ is

$$\mathcal{N}(t, z) = \begin{cases} \emptyset & 0 \le t < 0.09630 \\ \text{a single point} & t = 0.09630 \\ \text{an expanding circle} & 0.09630 < t < 0.14896 \\ \text{a circle plus its center} & t = 0.14896 \\ \text{two concentric expanding circles} & t > 0.14896 \end{cases}$$

Figure 4.9.4 below illustrates the successive bifurcations with snapshots of $z(t,x)$ for $t = 0.09, 0.105, 0.155, 0.2$.

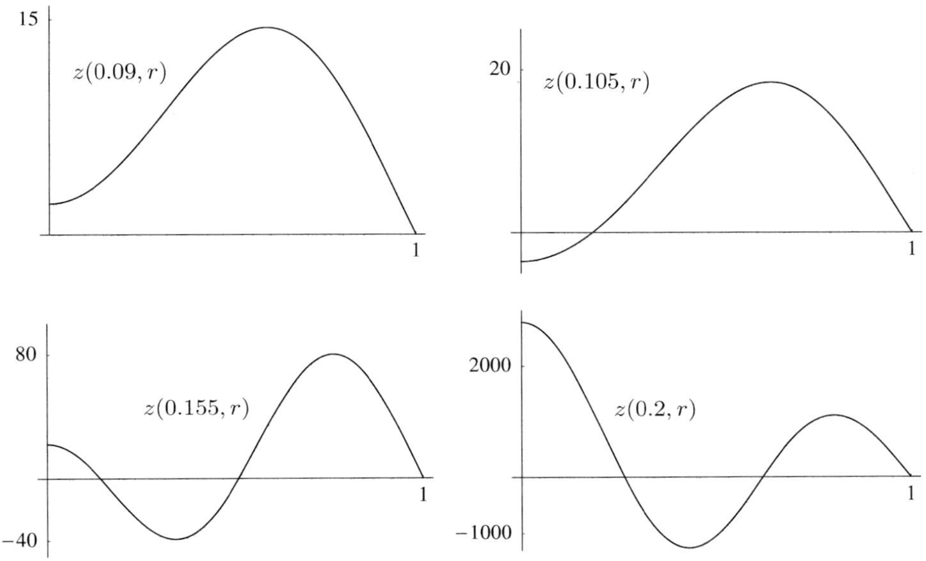


Figure 4.9.4

**Miscellaneous notes.** Most of the material in this section is taken from the author [2002:2], with various improvements and additions.

We note as a curiosity that Theorem 4.9.6 is a $L^\infty$ version of the the following (much simpler) Hilbert space result.

**Theorem 4.9.10.** *Let $A$ be the infinitesimal generator of a strongly continuous semigroup $S(t)$ in the Hilbert space $E$. A constant control $\bar{u}(t) = \bar{u}$ is norm optimal (resp. time optimal) in $0 \le t \le T$ if and only if $\bar{u} = \phi$ is an eigenvector of the operator $A^*$ (resp. an eigenvector of $A^*$ with $\|\phi\| = 1$).*

*Proof.* Any control of the form

$$\bar{u}(t) = \rho \frac{S(T-t)^* z}{\|S(T-t)^* z\|} \tag{4.9.20}$$

with $z \in Z(T)$ and nonzero denominator is norm optimal, and also time optimal if $\rho = 1$ (Theorems 2.5.5 and 2.5.7). As a particular case, if $\bar{u} = \phi$ is an eigenvector of $A^*$ with eigenvalue $-\lambda$ (we may assume $\|\phi\| = 1$) the control

$$\bar{u}(t) = \rho \frac{S(T-t)^* \phi}{\|S(T-t)^* \phi\|} = \rho \frac{e^{-\lambda(T-t)}\phi}{\|e^{-\lambda(T-t)}\phi\|} = \rho \frac{\phi}{\|\phi\|} = \rho\phi$$

is norm optimal. It is also time optimal if $\rho = 1$.

Conversely, let $\bar{u}(t) = \bar{u}$ be a time invariant norm optimal control. We have

$$\int_0^T S(T-\sigma)ud\sigma \in D(A)$$

for all $u \in E$, so that Theorem 2.5.3 applies and there exists $z \in Z(T)$, $z \neq 0$ with

$$\langle S(T-t)^* z, \bar{u}\rangle = \max_{\|u\| \le \rho} \langle S(T-t)^* z, u\rangle ,$$

which is the same as (4.9.20) in the interval $T - \delta < t < T$ where $S(T-t)^* z \neq 0$. Taking $\mu$ large enough and applying $R(\mu; A^*)$ to both sides of (4.9.20),

$$\|S(T-t)^* z\| R(\mu; A^*)\bar{u} = S(T-t)^* R(\mu; A^*)z \quad (T - \delta < t < T) .$$

Now, $R(\mu; A^*)Z(T) \subseteq E^* = E$, thus letting $t \to T$, $R(\mu; A^*)z = R(\mu; A^*)u$, which implies $z = \bar{u}$ and we can rewrite (4.9.20) as

$$\bar{u} = \frac{S(T-t)^* \bar{u}}{\|S(T-t)^* \bar{u}\|} .$$

This implies $S(t)^*\bar{u} = \eta(t)\bar{u}$ in $0 \le t < \delta$. Writing $S(t)^*\bar{u} = S(t/n)^* \cdots S(t/n)^*\bar{u}$ ($n$ factors) for $n$ large enough we deduce that $S(t)^*$ is a scalar multiple of $\bar{u}$ for any $t$, or $S(t)^*\bar{u} = \eta(t)\bar{u}$ for $t \ge 0$. Integrating,

$$R(\mu; A^*)\bar{u} = \int_0^\infty e^{-\mu t} S(t)^* \bar{u} dt = \left( \int_0^\infty e^{-\mu t} \eta(t) dt \right) \bar{u} = c(\mu)\bar{u} ,$$

and applying $\mu I - A^*$ to both sides we get $\bar{u} = c(\mu)(\mu I - A^*)\bar{u}$. This ends the proof for the norm optimal case; when $\bar{u}(t)$ is time optimal it is also norm optimal thus we don't need a separate proof.

## CHAPTER 5

# OPTIMAL CONTROL OF DIFFUSIONS

**5.1. Modeling of parabolic equations.** We retake the study of the parabolic equation

$$\begin{aligned} \frac{\partial y(t,x)}{\partial t} &= Ay(t,x) + u(t,x) \quad && (0 \le t \le T,\ x \in \Omega) \\ y(0,x) &= \zeta(x) && (x \in \Omega) \\ \beta y(t,x) &= 0 && (0 \le t \le T,\ x \in \Omega) \end{aligned} \tag{5.1.1}$$

in a bounded $C^\infty$ domain $\Omega \subset \mathbb{R}^m$ with boundary $\Gamma$, where $A$ is the operator

$$Ay(x) = \sum_{j=1}^m \sum_{k=1}^m \frac{\partial}{\partial x_j}\Big(a_{jk}(x)\frac{\partial}{\partial x_k}y(x)\Big) + \sum_{j=1}^m b_j(x)\frac{\partial}{\partial x_j}y(x) + c(x)y(x)\,. \tag{5.1.2}$$

The assumptions are the same in **3.3** and **4.1**: $a_{jk}(x) = a_{kj}(x)$ and $A$ is uniformly elliptic. The boundary condition $\beta$ on the boundary $\Gamma$ is either Dirichlet, $\beta y(x) = y(x) = 0$ or variational, $\beta y(x) = \partial^\nu y(x) - \gamma(x)y(x) = 0$ $(x \in \Gamma)$ where $\partial^\nu$ is the conormal derivative at the boundary $\Gamma$ (see **3.3**). We assume $C^{(\infty)}$ coefficients for the operator and boundary condition.

In this chapter, we use (5.1.1) to model a diffusion in $\Omega$ (Kolmogorov [1931], Stroock - Varadhan [1979]).[1] If $y(t,x)$ is a nonnegative solution of (5.1.1) then $y(t,x)$ is the concentration at $x$ (at time $t$) of the diffusing matter, so that

$$\int_\Omega y(t,x)dx$$

is the total amount of diffusing matter at time $t$. The inhomogeneous term $u(t,x)$ represents matter being pumped in or absorbed out of the process, with $u(t,x)$ the concentration at $x$ at time $t$. Two natural bounds for $u(t,x)$ are

$$\int_\Omega |u(t,x)|dx \le C\,, \qquad \int_0^T \int_\Omega |u(t,x)|dxdt \le C\,. \tag{5.1.3}$$

[1] In the modeling of diffusions, $c(x) = 0$; however, the treatment is the same for $c(x) \ne 0$.

The first puts a limit on the total amount of matter being pumped in or absorbed out at time $t$, while the second limits the total amount for the duration of the process.

We consider the time and norm optimal problems corresponding to the first bound (5.1.3). The modeling of (5.1.1) ignores the obvious restriction $y(t,x) \geq 0$ (the concentration cannot be negative) but experience with this type of *pointwise state constraint* indicates it would complicate considerably the treatment.

The choice of state space is clearly $E = L^1(\Omega)$. The control space associated with the first bound (5.1.3) is $L^\infty(0,T;L^1(\Omega))$, but this space puts us in a bad existence situation. We use instead spaces of measure valued controls,

$$L^\infty_w(0,T;\Sigma(\overline{\Omega})) \quad \text{or} \quad L^\infty_w(0,T;\Sigma_0(\overline{\Omega}))$$

depending on whether the boundary condition is variational or Dirichlet. The abstract model for (5.1.1) is

$$y'(t) = A_1 y(t) + \mu(t)\,, \quad y(0) = \zeta \tag{5.1.4}$$

where $A_1$ is the semigroup generator in $L^1(\Omega)$ determined by $A$ and $\beta$. We also make use of the semigroup generator $A_p$ in $L^p(\Omega)$ ($1 < p < \infty$) determined by $A$ and $\beta$. For the definitions, see **3.3** and **4.1**. The operators $A_p$, $p > 1$ can be defined in strong or weak form. $A_1$ is only defined in weak form:

*Weak definition of* $A_1$ : $y \in D(A_1)$ if and only if there exists $z$ $(= A_1 y)$ in $L^1(\Omega)$ such that

$$\int_\Omega y(x)(A'v)(x)dx = \int_\Omega z(x)v(x)dx$$

for every $v \in C^{(2)}_{\beta'}(\overline{\Omega})$.

We recall from **4.1** that if $X$ is a Banach space, a function $t \to u(t) \in X^*$ is $X$*-weakly measurable* if $t \to \langle u(t), v\rangle$ is measurable for every $v \in X$. The space $L^\infty_w(0,T;X^*)$ consists of all $X^*$-valued $X$-weakly measurable functions such that

$$\langle u(t), v\rangle \leq C\|v\| \quad \text{a. e.} \tag{5.1.5}$$

"a. e." depending on $v$. The norm of $u(\cdot) \in L^\infty_w(0,T;X^*)$ is the infimum of all the constants $C$ satisfying (5.1.5).

**Theorem 5.1.1.** *Let $X$ be an arbitrary Banach space. Then*

$$L^1(0,T;X)^* = L^\infty_w(0,T;X^*)\,,$$

*the duality of both spaces given by*

$$\langle u(\cdot), v(\cdot)\rangle = \int_0^T \langle u(t), v(t)\rangle dt$$

*for $u(\cdot) \in L^\infty_w(0,T;X^*), v(\cdot) \in L^1(0,T;X)$.*

For a proof see Ionescu Tulcea [1969, Theorem 7, p. 94]; it is reproduced in the author [1999:2, Theorem 12.2.11, p. 625]. The interpretation of the integral is explained in the comments after (4.1.11).

**Lemma 5.1.2.** *Assume $X$ is separable, and let $u(\cdot) \in L^\infty_w(0,T;X^*)$. Then the function $t \to \|u(t)\|$ is measurable, and*

$$\|u(\cdot)\|_{L^\infty_w(0,T;X^*)} = \operatorname*{ess.\,sup}_{0\le t\le T} \|u(t)\| \,. \tag{5.1.6}$$

*Proof.* Let $\{v_n\}$ be a countable dense set in the unit ball of $X$. Then the norm $\|u(t)\| = \sup_n \langle u(t), v_n\rangle$ is measurable. The inequality

$$\|u(\cdot)\|_{L^\infty_w(0,T;X^*)} \le \operatorname*{ess.\,sup}_{0\le t\le T} \|u(t)\|$$

is obvious from the definitions. On the other hand,

$$\langle u(t), v_n\rangle \le \|u(\cdot)\|_{L^\infty_w(0,T;X^*)} \quad (t \in e_n)$$

where $e_n \in [0,T]$ is a set of full measure in $[0,T]$. Accordingly,

$$\|u(t)\| = \sup_n \langle u(t), v_n\rangle \le \|u(\cdot)\|_{L^\infty_w(0,T;X^*)} \quad (t \in e = \cap_n e_n)$$

where $e \in [0,T]$ has full measure in $[0,T]$. This gives the opposite inequality and ends the proof.

We use Theorem 5.1.1 for $X = C(\overline{\Omega}), C_0(\overline{\Omega})$, obtaining

$$\begin{aligned} L^\infty_w(0,T;\Sigma(\overline{\Omega})) &= L^1(0,T;C(\overline{\Omega}))^* \,, \\ L^\infty_w(0,T;\Sigma_0(\overline{\Omega})) &= L^1(0,T;C_0(\overline{\Omega}))^* \,. \end{aligned} \tag{5.1.7}$$

We need a variation-of-constants formula for the control spaces (5.1.7). To this end, we use the semigroup generator $A'_c$ in $C(\overline{\Omega})$ constructed with the formal adjoint $A'$ and the adjoint boundary condition $\beta'$ and the semigroup $S'_c(t)$ generated by $A'_c$. We bring back

$$A_\Sigma = {A'}^*_c, \qquad S_\Sigma(t) = S'_c(t)^*$$

defined in (4.2.18) and (4.3.16) and

$$A'_\infty = A^*_1 \,, \qquad S'_\infty(t) = S_1(t)^*$$

defined in (4.3.1) and (4.3.9). Applying the inclusion relations (4.3.17) and (4.3.18) to the formal adjoint $A'$ and the adjoint boundary condition $\beta'$ we obtain

$$S_\Sigma(t)\Sigma(\overline{\Omega}) \subseteq D(A_\Sigma) \subseteq L^1(\Omega) \quad (t > 0) \tag{5.1.8}$$

and

$$S_1(t) \subseteq S_\Sigma(t)\,, \quad S_\Sigma(t)|_{L^1(\Omega)} = S_1(t) \quad (t > 0)\,. \tag{5.1.9}$$

Likewise, using the inclusion relations (4.3.10) and (4.3.11) for $A'$ and $\beta'$ we obtain

$$S'_\infty(t)L^\infty(\Omega) \subseteq D(A'_\infty) \subseteq C(\overline{\Omega}) \tag{5.1.10}$$

and

$$S'_c(t) \subseteq S'_\infty(t)\,, \quad S'_\infty(t)|_{C(\overline{\Omega})} = S'_c(t) \quad (t > 0)\,. \tag{5.1.11}$$

The variation-of-constants formula is

$$y(t, \zeta, \mu) = S_\Sigma(t)\zeta + \int_0^t S_\Sigma(t-\sigma)\mu(\sigma)d\sigma\,. \tag{5.1.12}$$

**Theorem 5.1.3.** *Let $\zeta \in \Sigma(\overline{\Omega})$, $\mu(\cdot) \in L^\infty_w(0, T; \Sigma(\overline{\Omega}))$. Then the integrand in (5.1.12) takes values in $L^1(\Omega)$ in $0 \le \sigma < t$; moreover*

$$S_\Sigma(t - \,\cdot\,)\mu(\cdot) \in L^\infty(0, T; L^1(\Omega))\,,$$

*$t \to y(t, \zeta, \mu)$ is a continuous, $L^1(\Omega)$-valued function in $0 < t \le T$ and*

$$\lim_{t\to 0} y(t, \zeta, \mu) = \zeta \quad C(\overline{\Omega})\text{-weakly in } \Sigma(\overline{\Omega})\,.$$

*If $\zeta \in L^1(\Omega)$ then $y(t, \zeta, \mu)$ is continuous in $0 \le t \le T$ and $y(0, \zeta, u) = \zeta$.*

The proof is a mirror image of that of Theorem 4.3.3 so that we only check the various steps. Formulas (5.1.8) and (5.1.9) show that $S_\Sigma(t-\sigma)\mu(\sigma) \in L^1(\Omega)$ in $0 \le \sigma < t$ and

$$\begin{aligned} S_\Sigma(t-\sigma)\mu(\sigma) &= S_\Sigma(h)S_\Sigma(t-\sigma-2h)S_\Sigma(h)\mu(\sigma) \\ &= S_1(h)S_1(t-\sigma-2h)S_\Sigma(h)\mu(\sigma) \end{aligned}$$

if $h > 0$, $0 \le \sigma \le t - 2h$, so that, if $u \in L^\infty(\Omega)$,

$$\begin{aligned} \langle u\,,\, S_\Sigma(t-\sigma)\mu(\sigma)\rangle &= \langle u\,,\, S_1(h)S_1(t-\sigma-2h)S_\Sigma(h)\mu(\sigma)\rangle \\ &= \langle S'_\infty(t-\sigma-2h)S'_\infty(h)u\,,\, S_\Sigma(h)\mu(\sigma)\rangle \\ &= \langle S'_c(t-\sigma-2h)S'_\infty(h)u\,,\, S_\Sigma(h)\mu(\sigma)\rangle \\ &= \langle S'_c(h)S'_c(t-\sigma-2h)S'_\infty(h)u\,,\, \mu(\sigma)\rangle \\ &= \langle S'_c(t-\sigma-h)S'_\infty(h)u\,,\, \mu(\sigma)\rangle\,. \end{aligned}$$

The left side of the last duality bracket, as a function of $\sigma$, is $C(\overline{\Omega})$-strongly measurable and the right side is $C(\overline{\Omega})$-weakly measurable, thus the bracket is measurable (in $\sigma > 0$ by arbitrariness of $h > 0$). Since $u$ is arbitrary, the function

$\sigma \to S_\Sigma(t-\sigma)\mu(\sigma)$ is $L^\infty(\Omega)$-weakly measurable; by separability of $L^1(\Omega)$, it is strongly measurable. In view of (5.1.6),

$$\|\mu(\sigma)\|_{\Sigma(\overline{\Omega})} \le \|\mu(\cdot)\|_{L^\infty_w(0,T;\Sigma(\overline{\Omega}))} \quad \text{a. e.}$$

so that, if $C$ is a bound for the semigroup,

$$\|S_\Sigma(t-\sigma)\mu(\sigma)\|_{L^1(\Omega)} \le C\|\mu(\cdot)\|_{L^\infty_w(0,T;\Sigma(\overline{\Omega}))} \quad \text{a. e.}$$

From here on, the proof is just the same as that of Theorem 4.3.3 and we omit it.

Given $t > 0$, the *reachable space* $R^\infty_w(t)$ (at time $t$) consists of all elements $y \in L^1(\Omega)$ of the form

$$y = y(t,0,u) = \int_0^t S_\Sigma(t-\sigma)\mu(\sigma)d\sigma\ , \quad \mu(\cdot) \in L^\infty_w(0,t;\Sigma(\overline{\Omega}))\,, \tag{5.1.13}$$

and is equipped with the norm

$$\|y\|_{R^\infty_w(t)} = \inf\left\{\|\mu\|_{L^\infty_w(0,t;\Sigma(\overline{\Omega}))};\ \int_0^t S_\Sigma(t-\sigma)\mu(\sigma)d\sigma = y\right\}. \tag{5.1.14}$$

With

$$M(t) = \max_{0\le\sigma\le t}\|S_\Sigma(\sigma)\|_{\mathcal{L}(\Sigma(\overline{\Omega}),\Sigma(\overline{\Omega}))} \tag{5.1.15}$$

we have

$$\|y\|_{L^1(\Omega)} \le tM(t)\|y\|_{R^\infty_w(t)}\,, \tag{5.1.16}$$

so that $R^\infty_w(T) \hookrightarrow L^1(\Omega)$. The space $R^\infty_w(T)$ is a Banach space; this is shown as in Lemma 4.4.1. All results in **4.4** generalize to the system (5.1.4) using the "replacement chart"

$$\begin{aligned} C(\overline{\Omega}) &\quad\longrightarrow\quad L^1(\Omega)\,, \\ A_\infty &\quad\longrightarrow\quad A_\Sigma\,, \\ S_\infty(t) &\quad\longrightarrow\quad S_\Sigma(t)\,, \\ u(\cdot) \in L^\infty_w(0,T;L^\infty(\Omega)) &\quad\longrightarrow\quad \mu(\cdot) \in L^\infty_w(0,T;\Sigma(\overline{\Omega}))\,. \end{aligned} \tag{5.1.17}$$

To illustrate this principle, consider the mirror image of formula (4.4.11):

$$y = \int_0^T S_\Sigma(T-\sigma)\frac{1}{T}(y - \sigma A_\Sigma y)d\sigma \quad (y \in D(A_\Sigma))\,. \tag{5.1.18}$$

Using (2.1.12) for $A_1$ and $S_1(t)$ with $S_\Sigma(h)y$ instead of $y$ we obtain

$$\begin{aligned} S_\Sigma(h)y &= \int_0^T S_1(T-\sigma)\frac{1}{T}(S_\Sigma(h)y - \sigma A_1 S_\Sigma(h)y)d\sigma \\ &= \int_0^T S_\Sigma(T-\sigma)\frac{1}{T}(S_\Sigma(h)y - \sigma A_1 S_\Sigma(h)y)d\sigma\,. \end{aligned}$$

We apply the element of $L^1(\Omega) \subseteq \Sigma(\overline{\Omega})$ on both sides of this equality to $z \in C(\overline{\Omega})$, obtaining

$$\begin{aligned}\langle S_c'(h)z, y\rangle &= \langle z, S_\Sigma(h)y\rangle \\ &= \left\langle z\,,\, \int_0^T S_\Sigma(T-\sigma)\frac{1}{T}(S_\Sigma(h)y - \sigma A_1 S_\Sigma(h)y)d\sigma\right\rangle \\ &= \frac{1}{T}\int_0^T \langle S_c'(T-\sigma)z\,,\, S_\Sigma(h)y - \sigma A_1 S_\Sigma(h)y\rangle d\sigma \\ &= \frac{1}{T}\int_0^T \langle S_c'(T-\sigma)z\,,\, S_\Sigma(h)y - \sigma S_\Sigma(h)A_\Sigma y\rangle d\sigma \\ &= \frac{1}{T}\int_0^T \langle S_c'(T+h-\sigma)z\,,\, y - \sigma A_\Sigma y\rangle d\sigma\,,\end{aligned}$$

and take limits as $h \to 0$; the result is

$$\begin{aligned}\langle z, y\rangle &= \frac{1}{T}\int_0^T \langle S_c'(T-\sigma)z\,,\, y - \sigma A_\Sigma y\rangle d\sigma\,, \\ &= \left\langle z\,,\, \int_0^T S_\Sigma(T-\sigma)\frac{1}{T}(y - \sigma A_\Sigma y)d\sigma\right\rangle\end{aligned}$$

which, since $z \in C(\overline{\Omega})$ is arbitrary, is the same as (5.1.18).

The domain $D(A_\Sigma) \subseteq L^1(\Omega)$ of $A_\Sigma$ is equipped with the graph norm

$$\|y\|_{\Sigma(\overline{\Omega})} = \|y\|_{L^1(\Omega)} + \|A_\Sigma y\|_{\Sigma(\overline{\Omega})}.$$

Using (5.1.18) we obtain $D(A_\Sigma) \subseteq R_w^\infty(T)$, and

$$\|y\|_{R_w^\infty(T)} \le \frac{\|y\|_{L^1(\Omega)} + T\|A_\Sigma y\|_{\Sigma(\overline{\Omega})}}{T} \le \frac{\max(1,T)}{T}\|y\|_{D(A_\Sigma)}\,, \tag{5.1.19}$$

hence

$$D(A_\Sigma) \hookrightarrow R_w^\infty(T) \hookrightarrow L^1(\Omega)\,. \tag{5.1.20}$$

We prove the analog of Theorem 4.4.2 using the chart (5.1.17).

**Theorem 5.1.4.** *Let the control $\nu(\cdot) \in L_w^\infty(0,t;\Sigma(\overline{\Omega}))$, $\|\nu(\cdot)\|_{L_w^\infty(0,t;\Sigma(\overline{\Omega}))} = 1$ drive $\zeta \in \Sigma(\overline{\Omega})$ to $\bar{y} \in L^1(\Omega)$ in the interval $0 \le \sigma \le t$. Assume $\nu(\cdot)$ is not norm optimal. Then it is not time optimal.*

*Proof.* Assume first that $\zeta \in L^1(\Omega)$. That the control $\nu(\cdot)$ is not norm optimal means there is another control $\mu(\cdot) \in L_w^\infty(0,t;\Sigma(\overline{\Omega}))$ driving $\zeta$ to $\bar{y}$ in the same interval,

$$\bar{y} = S_1(t)\zeta + \int_0^t S_\Sigma(t-\sigma)\mu(\sigma)d\sigma\,, \tag{5.1.21}$$

and such that $\|\mu(\cdot)\|_{L^\infty_w(0,t;\Sigma(\overline{\Omega}))} = \rho < 1$. If $s < t$ we have

$$\begin{aligned}\bar{y} &= S_1(s)\zeta + (S_1(t) - S_1(s))\zeta + \int_0^t S_\Sigma(t-\sigma)\mu(\sigma)d\sigma \\ &= S_1(s)\zeta + S_1(s)(S_1(t-s) - I)\zeta + \int_0^t S_\Sigma(t-\sigma)\mu(\sigma)d\sigma \\ &= S_1(s)\zeta + \int_0^s S_1(s-\sigma)\frac{S_1(\sigma)(S_1(t-s)-I)\zeta}{s}\,d\sigma \\ &\quad + \int_0^s S_1(s-\sigma)\Big(\frac{S_1(\sigma)}{s}\int_0^{t-s} S_\Sigma(t-s-\tau)\mu(\tau)d\tau\Big)d\sigma \\ &\quad + \int_0^s S_\Sigma(s-\sigma)\mu(\sigma+t-s)d\sigma \\ &= S_1(s)\zeta + \int_0^s S_\Sigma(s-\sigma)\omega(\sigma)d\sigma\,, \end{aligned} \tag{5.1.22}$$

where

$$\begin{aligned}\omega(\sigma) &= \frac{S_1(\sigma)(S_1(t-s)-I)\zeta}{s} \\ &\quad + \frac{S_1(\sigma)}{s}\int_0^{t-s} S_\Sigma(t-s-\tau)\mu(\tau)d\tau + \mu(\sigma+t-s)\end{aligned}$$

and

$$\begin{aligned}\|\omega(\sigma)\|_{L^1(\Omega)} &\le \frac{M(s)}{s}\|S_1(t-s)\zeta - \zeta\|_{L^1(\Omega)} \\ &\quad + \Big(1 + \frac{t-s}{s}M(s)\,M(t-s)\Big)\rho\,. \end{aligned} \tag{5.1.23}$$

According to the last line of (5.1.22) the control $\omega(\cdot)$ drives $\zeta$ to $\bar{y}$ in time $s$ and it follows from (5.1.23) that, if $s$ is sufficiently close to $t$ then $\|\omega(\cdot)\|_{L^\infty_w(0,T;\Sigma(\overline{\Omega}))} \le 1$, which shows that $t$ is not the optimal driving time.

Assume finally that $\zeta \in \Sigma(\overline{\Omega})$, and let $0 < h < t$. Using the fact that $y(h,\zeta,u) \in L^1(\Omega)$ we can construct a control $\omega(\cdot)$ that improves the driving time in $h \le \sigma \le t$. This, by the optimality principle, implies that the driving time can be improved in the whole interval and ends the proof.

**Theorem 5.1.5.** *Let $\bar{\mu}(t)$ be a time optimal control. Then*

$$\|\bar{\mu}(t)\|_{\Sigma(\overline{\Omega})} = 1 \quad \text{a. e. in } 0 \le t \le T. \tag{5.1.24}$$

Theorem 5.1.5 (as Theorem 4.4.3) does not imply uniqueness for optimal controls since $\Sigma(\overline{\Omega})$ is not strictly convex. However, and very much unlike Theorem 4.4.3, there are situations where Theorem 5.1.5 is essentially sufficient as an optimality condition (Theorem 5.4.12), thus the proof is worth sketching. It is a generalization (with no surprises) of that of Theorem 2.1.3.

Given a measurable set $e \subseteq I\!R$ we denote by $R_w^\infty(t;e)$ the subspace of $R_w^\infty(t)$ corresponding to controls $\mu(\cdot) \in L_w^\infty((0,t) \cap e; \Sigma(\overline{\Omega}))$ (or, rather, controls in $L_w^\infty(0,t;\Sigma(\overline{\Omega}))$ with support in $e$). The space $R_w^\infty(t;e)$ is equipped with the norm

$$\|y\|_{R_w^\infty(t;e)} = \inf \|u(\cdot)\|_{L^\infty((0,t)\cap e;\Sigma(\overline{\Omega}))},$$

the infimum taken over all $u(\cdot) \in L_w^\infty((0,t)\cap e;\Sigma(\overline{\Omega}))$ such that

$$y = \int_0^t S_\Sigma(t-\sigma)\mu(\sigma)d\sigma .$$

Obviously $R_w^\infty(t;e) \hookrightarrow R_w^\infty(t)$ and, if $|e| > 0$,

$$R_w^\infty(t;e) = R_w^\infty(t) \tag{5.1.25}$$

(with equivalent norms) for almost all $t \in e$. To prove this, we recall (Lemma 2.1.5) that, given $\rho$ with $0 < \rho < 1$ there exists $c > 0$ such that for almost all $t \in e$ there exists a sequence $\{t_n\}$, $t_1 < t_2 < \ldots < t$, $t_n \to t$ such that

$$|(t_n, t_{n+1}) \cap e| \geq \rho(t_{n+1} - t_n), \quad \frac{t_n - t_{n-1}}{t_{n+1} - t_n} \leq c \quad (n = 1, 2, \ldots). \tag{5.1.26}$$

Writing

$$\begin{aligned} y &= \int_0^t S_\Sigma(t-\sigma)\mu(\sigma)d\sigma = S_1(t-t_1)\int_0^{t_1} S_\Sigma(t_1-\tau)\mu(\tau)d\tau \\ &\quad + \int_{t_1}^t S_\Sigma(t-\sigma)\mu(\sigma)d\sigma \\ &= \int_{t_1}^t S_1(t-\sigma)\left(\frac{S_1(\sigma-t_1)}{t-t_1}\int_0^{t_1} S_\Sigma(t_1-\tau)\mu(\tau)d\tau\right)d\sigma \\ &\quad + \int_{t_1}^t S_\Sigma(t-\sigma)\mu(\sigma)d\sigma \end{aligned}$$

we deduce that $R_w^\infty(t) = R_w^\infty(t;(t_1,t))$ with equivalent norms, and we can prove (5.1.25) with $R_w^\infty(t;(t_1,t))$ on the right side. We have

$$\begin{aligned} y &= \int_{t_1}^t S_\Sigma(t-\sigma)\mu(\sigma)d\sigma \\ &= \sum_{n=1}^\infty S_1(t-t_{n+1})\int_{t_n}^{t_{n+1}} S_\Sigma(t_{n+1}-\tau)\mu(\tau)d\tau \\ &= \sum_{n=1}^\infty \int_{(t_{n+1},t_{n+2})\cap e} S_1(t-\sigma) \\ &\quad \times \left(\frac{S_1(\sigma-t_{n+1})}{|(t_{n+1},t_{n+2})\cap e)|}\int_{t_n}^{t_{n+1}} S_\Sigma(t_{n+1}-\tau)\mu(\tau)d\tau\right)d\sigma \\ &= \int_{t_1}^t S_\Sigma(t-\sigma)\nu(\sigma)d\sigma , \end{aligned}$$

where

$$\nu(\sigma) = \frac{S_1(\sigma - t_{n+1})}{|(t_{n+1}, t_{n+2}) \cap e)|} \int_{t_n}^{t_{n+1}} S_\Sigma(t_{n+1} - \tau)\mu(\tau)d\tau$$
$$(\sigma \in (t_{n+1}, t_{n+2}) \cap e,\ n = 1, 2, \ldots),$$
$$\nu(\sigma) = 0 \qquad \text{elsewhere},$$

and

$$\begin{aligned}\|\nu(\sigma)\|_{L^1(\Omega)} &\leq \frac{M(t_{n+2} - t_{n+1})M(t_{n+1} - t_n)(t_{n+1} - t_n)}{|(t_{n+1}, t_{n+2}) \cap e)|} \|\mu(\cdot)\|_{L_w^\infty(0,t;\Sigma(\overline{\Omega}))} \\ &\leq \frac{M(t_{n+2} - t_{n+1})M(t_{n+1} - t_n)(t_{n+1} - t_n)}{\rho(t_{n+2} - t_{n+1})} \|\mu(\cdot)\|_{L_w^\infty(0,t;\Sigma(\overline{\Omega}))} \\ &\leq \frac{cM(t_{n+2} - t_{n+1})M(t_{n+1} - t_n)}{\rho} \|\mu(\cdot)\|_{L_w^\infty(0,t;\Sigma(\overline{\Omega}))}\end{aligned}$$

in $t_{n+1} \leq \sigma \leq t_{n+2}$ which ends the proof of (5.1.25).

To complete the proof of Theorem 5.1.5 we assume that the set $\subseteq (0, t)$ where $\|\mu(t)\|_{\Sigma(\overline{\Omega})} < 1$ has positive measure. Then there exists $\epsilon > 0$ and a set $e$ of positive measure where

$$\|\mu(t)\|_{\Sigma(\overline{\Omega})} \leq 1 - \epsilon \quad (t \in e).$$

Using equality (5.1.25) there exists $s \in e$ such that we can construct a control $\nu(\cdot) \in L_w^\infty((0, s) \cap e; \Sigma(\overline{\Omega}))$ with

$$\begin{aligned}y(s, \zeta, \nu) &= S(s)\zeta + \int_0^s S_\Sigma(s - \sigma)\nu(\sigma)d\sigma \\ &= S(s)\zeta + \int_0^s S_\Sigma(s - \sigma)\mu(\sigma)d\sigma = y(s, \zeta, \mu).\end{aligned}$$

Let $0 < \delta < 1$, $\nu_\delta(\sigma) = (1 - \delta)\mu(\sigma) + \delta\nu(\sigma)$. Then we have $y(s, \zeta, \nu_\delta) = y(s, \zeta, \mu)$ for any $\delta$. If $C = \|\nu(\cdot)\|_{L_w^\infty(0,T;\Sigma(\overline{\Omega}))}$,

$$\begin{aligned}\|\nu_\delta(\sigma)\|_{\Sigma(\overline{\Omega})} &\leq (1 - \delta)(1 - \epsilon) + C\delta \\ &\leq 1 - \epsilon + C\delta \leq 1 - \frac{\epsilon}{2} \quad \text{a. e. in } (t_1, t) \cap e),\end{aligned}$$

the last inequality for $\delta > 0$ sufficiently small; on the other hand, outside of $e$ we have $\|\nu_\delta(\sigma)\| \leq 1 - \delta$, so that $\|\nu_\delta(\cdot)\|_{L_w^\infty(0,s;\Sigma(\overline{\Omega}))} < 1$ and $\nu_\delta(\cdot)$ is not norm optimal in $(0, s)$, thus not time optimal in the same interval (Theorem 5.1.4). Since $\mu(\sigma)$ drives from $\zeta$ to the same target as $\nu_\delta(\sigma)$ in the interval $[0, s]$ it follows that $\mu(\sigma)$ itself is not time optimal, thus by the optimality principle (see **1.3**) it is not time optimal in the entire interval $(0, t)$. This completes the proof of Theorem 5.1.5.

Define $R^\infty(t)$ as the space of all elements $y \in L^1(\Omega)$ of the form

$$y = y(t, 0, u) = \int_0^t S_1(t-\sigma)u(\sigma)d\sigma\,, \quad u(\cdot) \in L^\infty(0, t; L^1(\Omega)) \tag{5.1.27}$$

equipped with the norm

$$\|y\|_{R^\infty(t)} = \inf\left\{\|u\|_{L^\infty(0,t;L^1(\Omega))};\ \int_0^t S_1(t-\sigma)u(\sigma)d\sigma = y\right\}. \tag{5.1.28}$$

**Theorem 5.1.6.** *We have $R^\infty(t) = R_w^\infty(t)$ with equivalent norms. If $\|S_\Sigma(t)\| \le 1$ for $t \ge 0$ then*

$$\|y\|_{R_w^\infty(t)} = \|y\|_{R^\infty(t)} \quad (t > 0)\,.$$

The proof is a mirror image of that of Theorem 4.4.4 and is omitted.

**Miscellaneous notes.** Some of the results in this section may be new, others follow from the unified treatment in the author [1999:2] of semilinear equations in nonreflexive spaces using Phillips' $^\odot$-adjoint theory, which encompasses the theory of (4.1.4) and (5.1.4).

**5.2. The reachable space and its dual, I.** Necessary optimality conditions for the system

$$y'(t) = A_1y(t) + \mu(t), \quad y(0) = \zeta \tag{5.2.1}$$

(controls in $L^\infty(0, T; \Sigma(\overline{\Omega}))$) can be obtained under the condition $\bar{y} \in \overline{D(A_1)}$ on the target, where the bar indicates closure in $R^\infty(T)$. Closures in $R^\infty(T)$ are difficult to compute, thus in practice one would use the condition $\bar{y} \in D(A_1)$. This condition turns out to be too strong, thus we work under the weaker condition

$$\bar{y} \in D(A_\Sigma)\,. \tag{5.2.2}$$

Example 5.2.1 below is the companion of Example 4.5.1, where the system is modelled in $C(\overline{\Omega})$ and the conditions on the target are $\bar{y} \in D(A_c)$ and $\bar{y} \in D(A_\infty)$.

**Example 5.2.1.** The system is

$$\frac{\partial y(t,x)}{\partial t} = \frac{\partial^2 y(t,x)}{\partial x^2} + \mu(t)\,, \quad y(t,0) = y(t,\pi) = 0$$

with initial condition $u(0, x) = \zeta(x) = 0$ and time independent control

$$\mu_a(t, x) = \delta(x - a) \quad (0 \le x \le \pi)\,.$$

The domain of $A_1y(x) = y''(x)$ consists of all $y(x)$ continuously differentiable in $0 \le x \le \pi$ having absolutely continuous derivative $y'(x)$ with $y''(\cdot) \in L^1(0, \pi)$ and satisfying the boundary conditions $y(0) = y(\pi) = 0$. Equivalently, $D(A_1)$ is the set of all $y(x)$ such that the second derivative $y''(x)$ in the sense of distributions belongs to $L^1(0, \pi)$ and $y(0) = y(\pi) = 0$.

The domain of $A_\Sigma y(x) = y''(x)$ is described in a similar way. It is the set of all $y(x)$ such that the second derivative $y''$ in the sense of distributions belongs to $\Sigma_0[0, \pi]$; the function satisfies $y(0) = y(\pi) = 0$. The taking of boundary conditions needs justification; it is enough to show that any $y(\cdot)$ with second derivative $y'' = \mu$ in $\Sigma[0, \pi]$ is continuous. We have

$$y(x) = \int_0^x (x - \xi)\mu(d\xi) + ax + b = y_0(x) + ax + b\,.$$

If $x' > x$ then

$$\begin{aligned} y_0(x') - y_0(x) &= \int_0^{x'} (x' - \xi)\mu(d\xi) - \int_0^x (x - \xi)\mu(d\xi) \\ &= \int_x^{x'} (x' - \xi)\mu(d\xi) + \int_0^x ((x' - \xi) - (x - \xi))\mu(d\xi)\,, \end{aligned}$$

so that $|y_0(x') - y_0(x)| \le 2\pi |x' - x| \|\mu\|_{\Sigma[0,\pi]}$; $y(x)$ is actually Lipschitz continuous.

Expanding in sine Fourier series we have

$$\mu_a(t, x) = \delta(x - a) = \frac{1}{\pi} \sum_{n=1}^{\infty} \sin na \sin nx$$

(the series convergent in the sense of distributions, or more precisely, in $H^{-1}(0, \pi)$). The solution of

$$\frac{\partial y(t,x)}{\partial t} = \frac{\partial^2 y(t,x)}{\partial x^2} + \delta(x - a)\,, \quad y(t, 0) = y(t, \pi) = 0\,,\ y(0, x) = 0$$

is

$$y(t, x) = \frac{1}{\pi} \sum_{n=1}^{\infty} \frac{1 - e^{-n^2 t}}{n^2} \sin na \sin nx\,,$$

so that

$$\begin{aligned} \frac{\partial^2 y(t,x)}{\partial x^2} &= -\frac{1}{\pi} \sum_{n=1}^{\infty} \sin na \sin nx + \frac{1}{\pi} \sum_{n=1}^{\infty} e^{-n^2 t} \sin na \sin nx \\ &= -\delta(x - a) + \phi(t, x)\,, \end{aligned}$$

where the function $\phi(t, x)$ is infinitely differentiable in $t$ and $x$ for $t > 0$. It follows that $y(t, \cdot) \notin D(A_1)$ for $t > 0$. On the other hand, $y(t, \cdot) \in D(A_\Sigma)$ for $t > 0$, thus a theory based on (5.2.2) will be a lot more inclusive than one that uses $\bar{y} \in D(A_1)$.

We begin with the characterization of the dual of $D(A_\Sigma) \subseteq L^1(\Omega)$ equipped with its graph norm or with the equivalent norm

$$\|y\|_{D(A_\Sigma)} = \|R(\lambda; A_\Sigma)y\|_{\Sigma(\overline{\Omega})} = \|(\lambda I - A_\Sigma)^{-1} y\|_{\Sigma(\overline{\Omega})} \tag{5.2.3}$$

with $\lambda > \omega$ ($\omega$ the constant in (4.1.9)); $(\lambda I - A'_c)^{-1}$ is everywhere defined and bounded, thus the same is true of $(\lambda I - A_\Sigma)^{-1} = ((\lambda I - A'_c)^{-1})^*$ (see (4.2.12)). Since the space $D(A_\Sigma)$ is "based on $\Sigma(\overline{\Omega})$", we need the dual space $\Sigma(\overline{\Omega})^*$ of $\Sigma(\overline{\Omega})$. Characterizations of $\Sigma(\overline{\Omega})^*$ exist (see Kaplan [1985]) but we will avoid any specific construction of the dual.

The domain $D(A_\Sigma)$ of $A_\Sigma$ is not dense in $\Sigma(\overline{\Omega})$ (see (4.2.22)) thus the operator $A_\Sigma^*$ in $\Sigma(\overline{\Omega})^*$ is multivalued. In view of (4.2.3), if $\eta \in D(A_\Sigma^*)$ the set $[A_\Sigma^*\eta]$ of all values of $A_\Sigma^*\eta$ satisfies

$$[A_\Sigma^*\eta] = \zeta + \mathcal{N}_{A_\Sigma},$$

where $\zeta$ is an arbitrary element of $[A_\Sigma^*\eta]$ and

$$\mathcal{N}_{A_\Sigma} = \{\eta \in \Sigma(\overline{\Omega})^*; \langle \eta, \mu\rangle = 0,\ \mu \in D(A_\Sigma)\}.$$

We have shown in (4.2.25) that $D(A_\Sigma)$ is dense in $L^1(\Omega)$, hence

$$\mathcal{N}_{A_\Sigma} = \mathcal{N}_1(\overline{\Omega}) = \{\eta \in \Sigma(\overline{\Omega})^*; \langle \eta, f\rangle = 0,\ f \in L^1(\Omega)\}. \tag{5.2.4}$$

Due to (4.2.12) we have $(\lambda I - A_\Sigma^*)^{-1} = ((\lambda I - A_\Sigma)^{-1})^*$ but $A_\Sigma$ is not densely defined, so this equality must be interpreted as

$$((\lambda I - A_\Sigma)^{-1})^*(\lambda I - A_\Sigma^*) \subseteq I,$$
$$[(\lambda I - A_\Sigma^*)((\lambda I - A_\Sigma)^{-1})^*\eta] = \eta + \mathcal{N}_1(\Omega) \quad (\eta \in \Sigma(\overline{\Omega})),$$

where $[\ \cdot\ ]$ denotes the set obtained applying a multivalued operator (see **4.2**). According to (4.2.11), $\mathcal{N}_1(\overline{\Omega})$ is the nullspace of $((\lambda I - A_\Sigma)^{-1})^*$ :

$$\mathcal{N}_1(\overline{\Omega}) = \{\eta \in \Sigma(\overline{\Omega})^*; (\lambda I - A_\Sigma^*)^{-1}\eta = 0\}.$$

**Lemma 5.2.2.**

$$A_c \subseteq A_\Sigma^*, \quad D(A_\Sigma^*) \subseteq C(\overline{\Omega}). \tag{5.2.5}$$

*Proof.* The first inclusion follows from general duality theory, with $E$ a Banach space and $A$ a linear operator in $E$ (not necessarily densely defined); we have $E \overset{i}{\hookrightarrow} E^{**}$ under the canonical isometric imbedding and, if $y \in D(A)$, $\langle y, A^*y^*\rangle = \langle Ay, y^*\rangle$ $(y^* \in D(A^*))$ so that $y \in D(A^{**})$ and $A^{**}y = Ay$. Taking multivaluedness into account, $[Ay] \subseteq [A^{**}y]$ $(y \in D(A))$ and this the way $A_c \subseteq A_\Sigma^*$ in (5.2.5) is understood; the left side is single-valued, the right side multivalued.

To show the second inclusion in (5.2.5) we select $\eta \in D(A_\Sigma^*)$, and define $\xi = (\lambda I - A_\Sigma^*)\eta \in \Sigma(\overline{\Omega})^*$. We have

$$\langle (\lambda I - A_\Sigma)\mu, \eta\rangle = \langle \mu, (\lambda I - A_\Sigma^*)\eta\rangle = \langle \mu, \xi\rangle \quad (\mu \in D(A_\Sigma)). \tag{5.2.6}$$

In view of (4.2.22) and (4.2.25), $D(A_\Sigma) \subseteq L^1(\Omega)$ is dense in $L^1(\Omega)$, thus (5.2.6) defines a bounded linear functional of $\mu \in L^1(\Omega)$. Accordingly, there exists $u \in L^\infty(\Omega)$ such that

$$\langle (\lambda I - A_\Sigma)\mu, \eta\rangle = \langle \mu, u\rangle \quad (\mu \in D(A_\Sigma)).$$

Restricting $\mu = f$ to $D(A_1)$ and using $A_1 \subseteq A_\Sigma$ (see (4.2.19)) we obtain

$$\langle(\lambda I - A_1)f, \eta\rangle = \langle f, u\rangle \quad (f \in D(A_1)) . \tag{5.2.7}$$

Using (4.3.3) the operator $\lambda I - A'_\infty = (\lambda I - A_1)^*$ is invertible, thus we can write $u = (\lambda I - A'_\infty)z$ for some $z \in D(A'_\infty) \subset C(\overline{\Omega})$. Doing this, (5.2.7) becomes

$$\begin{aligned}&\langle(\lambda I - A_\Sigma)f, \eta\rangle = \langle(\lambda I - A_1)f, \eta\rangle = \langle f, u\rangle \\ &= \langle f, (\lambda I - A'_\infty)z\rangle = \langle(\lambda I - A_1)f, z\rangle = \langle(\lambda I - A_\Sigma)f, z\rangle \quad (f \in D(A_1)) ,\end{aligned}$$

or, equivalently,

$$\begin{aligned}&\langle(\lambda I - A_\Sigma)f, \eta\rangle = \langle f, (\lambda I - A_\Sigma^*)\eta\rangle \\ &= \int_\Omega z(x)((\lambda I - A_\Sigma)f)(x)dx \quad (f \in D(A_1)) .\end{aligned} \tag{5.2.8}$$

The next step is to extend (5.2.8) from $f \in D(A_1)$ to $\mu \in D(A_\Sigma) \subseteq L^1(\Omega)$. For this, we take $\mu \in D(A_\Sigma)$ arbitrary and use Lemma 4.2.1 to select a sequence $\{g_n(\cdot)\} \subset L^1(\Omega)$ such that

$$\begin{aligned}&\langle y, g_n\rangle \to \langle y, (\lambda I - A_\Sigma)\mu\rangle \quad (y(\cdot) \in C(\overline{\Omega})) , \\ &\|g_n(\cdot)\|_{L^1(\Omega)} \le C\|(\lambda I - A_\Sigma)\mu\|_{\Sigma(\overline{\Omega})} .\end{aligned} \tag{5.2.9}$$

Let $f_n = (\lambda I - A_1)^{-1}g_n = (\lambda I - A_\Sigma)^{-1}g_n$. Since $\|g_n\|_{L^1(\Omega)}$ is bounded and the operator $(\lambda I - A_1)^{-1}$ is compact, it follows that (a subsequence of) the sequence $\{f_n(\cdot)\}$ is convergent to $f \in L^1(\Omega)$ in $L^1(\Omega)$, hence, passing to the subsequence,

$$\begin{aligned}\langle y, f\rangle &= \lim_{n\to\infty}\langle y, f_n\rangle = \lim_{n\to\infty}\langle y, (\lambda I - A_\Sigma)^{-1}g_n\rangle \\ &= \lim_{n\to\infty}\langle(\lambda I - A'_c)^{-1}y, g_n\rangle = \langle(\lambda I - A'_c)^{-1}y, (\lambda I - A_\Sigma)\mu\rangle \\ &= \langle(\lambda I - A'_c)(\lambda I - A'_c)^{-1}y, \mu\rangle = \langle y, \mu\rangle \quad (y(\cdot) \in C(\overline{\Omega})) ,\end{aligned}$$

and we deduce that $f = \mu$. We write (5.2.8) for $f_n = (\lambda I - A_1)^{-1}g_n$,

$$\begin{aligned}\langle f_n, (\lambda I - A_\Sigma^*)\eta\rangle &= \int_\Omega z(x)((\lambda I - A_\Sigma)f_n)(x)dx \\ &= \int_\Omega z(x)g_n(x)dx\end{aligned}$$

and take limits; on the right we use (5.2.9), on the left $L^1(\Omega)$-convergence (thus $\Sigma(\overline{\Omega})$-convergence) of $f_n$ to $f = \mu$. The result is

$$\langle\mu, (\lambda I - A_\Sigma^*)\eta\rangle = \int_{\overline{\Omega}} z(x)(\lambda I - A_\Sigma)\mu(dx)$$

or

$$\langle(\lambda I - A_\Sigma)\mu, \eta\rangle = \langle(\lambda I - A_\Sigma)\mu, z\rangle\,.$$

Since $(\lambda I - A_\Sigma)D(A_\Sigma) = \Sigma(\overline{\Omega})$, this equality says that $\eta = z$ as elements of $\Sigma(\overline{\Omega})^*$ and thus ends the proof of Lemma 5.2.2.

Since $S'_c(t)$ is analytic in $\mathcal{L}(C(\overline{\Omega}), C(\overline{\Omega}))$ in $\mathrm{Re}\, t > 0$, the semigroup

$$S_\Sigma(t)^* = S'_c(t)^{**} \tag{5.2.10}$$

in $\Sigma(\overline{\Omega})^*$ is analytic in $\mathcal{L}(\Sigma(\overline{\Omega})^*, \Sigma(\overline{\Omega})^*)$ in $\mathrm{Re}\, t > 0$ but it is not strongly continuous at $t = 0$. Using the fact that $S_\Sigma(t)A_\Sigma$ (domain $D(A_\Sigma)$) is bounded, we use (4.2.9) to deduce that $(S_\Sigma(t)A_\Sigma)^* \supseteq A_\Sigma^* S_\Sigma(t)^*$ (the operators on both sides are multivalued) which shows that $A_\Sigma^* S_\Sigma(t)^*$ is bounded and

$$S_\Sigma(t)^*\Sigma(\overline{\Omega})^* \subseteq D(A_\Sigma^*) \subseteq C(\overline{\Omega}) \tag{5.2.11}$$

(the second inclusion after Lemma 5.2.2).

**Lemma 5.2.3.** *We have*

$$S'_c(t) \subseteq S_\Sigma(t)^*, \quad S_\Sigma(t)^*|_{C(\overline{\Omega})} = S'_c(t) \quad (t > 0)\,. \tag{5.2.12}$$

The proof of Lemma 5.2.3 comes from duality theory; if $B$ is bounded and everywhere defined in $E$, then everything is single-valued and $B \subseteq B^{**}$, $B^{**}|_E = B$.

The space $C(\overline{\Omega})_{-1}(\lambda)$ is the completion of $C(\overline{\Omega})$ under the norm

$$\|y\|_{C(\overline{\Omega})_{-1},\lambda} = \|R(\lambda; A'_c)y\|_{C(\overline{\Omega})} = \|(\lambda I - A'_c)^{-1}y\|_{C(\overline{\Omega})}\,.$$

Construction of $C(\overline{\Omega})_{-1}(\lambda)$ from $A'_c$ follows the same lines as that of $L^1(\Omega)_{-1}(\lambda)$ from $A'_1$ (the latter done in **4.5**) thus we only outline the main steps. All norms $\|\cdot\|_{C(\overline{\Omega})_{-1},\lambda}$, $\lambda > \omega$ define the same space $C(\overline{\Omega})_{-1} = C(\overline{\Omega})_{-1}(\lambda)$ and we have the imbedding

$$C(\overline{\Omega}) \hookrightarrow C(\overline{\Omega})_{-1}\,. \tag{5.2.13}$$

The operator $R(\lambda; A'_c)$ is extended to $C(\overline{\Omega})_{-1}$ by

$$R(\lambda; A'_c)^\bullet z = \lim_{n\to\infty} R(\lambda; A'_c)y_n$$

where $\{y_n\} \subset C(\overline{\Omega})$ is a sequence such that $\|y_n - z\|_{C(\overline{\Omega})_{-1}} \to 0$. This definition implies that the operator

$$R(\lambda; A'_c)^\bullet : (C(\overline{\Omega})_{-1}(\lambda), \|\cdot\|_{C(\overline{\Omega})_{-1},\lambda}) \to (C(\overline{\Omega}), \|\cdot\|_{C(\overline{\Omega})}) \tag{5.2.14}$$

is an isometry onto (a bounded operator with a bounded inverse if $\lambda > \omega$ in the norm is changed). We extend then $A'_c$ to $C(\overline{\Omega})$ setting

$$A_c'^{\bullet} = \lambda I - (R(\lambda; A'_c)^{\bullet})^{-1},$$

and we prove as in **4.5** that $A_c'^{\bullet}$ does not depend on $\lambda > \omega$ and that

$$A_c'^{\bullet}|_{D(A'_c)} = A'_c, \quad R(\lambda; A_c'^{\bullet})|_{C(\overline{\Omega})} = R(\lambda; A'_c). \tag{5.2.15}$$

We extend $S'_c(t)$ to $C(\overline{\Omega})_{-1}$ by

$$S'_c(t)^{\bullet}\psi = (\lambda I - A'_c)S'_c(t)R(\lambda; A'_c)^{\bullet}\psi,$$

and we have

$$S'_c(t)^{\bullet}C(\overline{\Omega})_{-1} \subseteq C(\overline{\Omega}) \quad (t > 0), \quad S'_c(t) = S'_c(t)^{\bullet}|_{C(\overline{\Omega})}. \tag{5.2.16}$$

The inclusion in (5.2.16) justifies simplifying $S'_c(t)^{\bullet}z$ ($z \in C(\overline{\Omega})_{-1}$) to $S'_c(t)z$. The space $Z_c(T)$ is the subspace of $C(\overline{\Omega})_{-1}$ defined by the summability condition

$$\|z\|_{Z_c(T)} = \int_0^T \|S'_c(t)z\|_{C(\overline{\Omega})}dt < \infty. \tag{5.2.17}$$

Membership in $Z_c(T)$ doesn't depend on $T > 0$, that is, the spaces $Z_c = Z_c(T)$ coincide for all $T > 0$ and all the norms $\|\cdot\|_{Z_c}$ are equivalent.

Multiplier spaces for $S'_c(t)$ are defined in the same way as for $S'_1(t)$ and we have this companion of Lemma 4.5.9:

**Lemma 5.2.4.** *Let $\psi$ be a multiplier such that*

$$\int_0^T \|S'_c(t)\psi\|_{C(\overline{\Omega})}dt < \infty.$$

*Then there exists a unique $z \in Z_c(T)$ such that*

$$S'_c(t)\psi = S'_c(t)z \quad (t > 0).$$

The proof is very similar to that of Lemma 4.5.9 and is omitted.

**Corollary 5.2.5.** *Let $\eta(t)$ $(t > 0)$ be a $C(\overline{\Omega})$-valued function such that*

$$\eta(s + t) = S'_c(t)\eta(s) \quad (s, t > 0)$$

*and*

$$\int_0^T \|\eta(t)\|_{C(\overline{\Omega})}dt < \infty.$$

*Then there exists $z \in Z_c(T)$ such that*

$$\eta(t) = S_c'(t)z \quad (t > 0) .$$

The proof follows the lines of Corollary 4.5.10.

In the following result, $[\eta]$ denotes the equivalence class of $\eta \in \Sigma(\overline{\Omega})^*$ in the quotient space $\Sigma(\overline{\Omega})^*/\mathcal{N}_1(\overline{\Omega})$ equipped with its standard quotient norm. The subspace $\mathcal{N}_1(\overline{\Omega})$ is defined in (5.2.4).

**Theorem 5.2.6.** *Every bounded linear functional $\Phi$ in $D(A_\Sigma)$ is given by*

$$\Phi(y) = \langle [\eta], (\lambda I - A_\Sigma)y \rangle = \langle \eta, (\lambda I - A_\Sigma)y \rangle \quad (y \in D(A_\Sigma)) \tag{5.2.18}$$

*for some $[\eta] \in \Sigma(\overline{\Omega})^*/\mathcal{N}_1(\overline{\Omega})$. Moreover, if the space $D(A_\Sigma)$ is given the norm (5.2.3) we have*

$$\|\Phi\|_{D(A_\Sigma)^*} = \|[\eta]\|_{\Sigma(\overline{\Omega})^*/\mathcal{N}_1(\overline{\Omega})} \, . \tag{5.2.19}$$

*Proof:* Let $\eta \in \Sigma(\overline{\Omega})^*$. Then (5.2.18) defines a bounded linear functional in $D(A_\Sigma)$ and, since $\mathcal{N}_{A_\Sigma} = \mathcal{N}_1(\overline{\Omega})$, (5.2.18) only depends on the equivalence class of $[\eta]$. That (5.2.19) holds follows from the facts that

$$(\lambda I - A_\Sigma)D(A_\Sigma) = \Sigma(\overline{\Omega}) \, , \qquad \|(\lambda I - A_\Sigma)y\|_{\Sigma(\overline{\Omega})} = \|y\|_{D(A_\Sigma)} \, .$$

Conversely, let $\Phi$ be a bounded linear functional in $D(A_\Sigma)$. Then $\Phi((\lambda I - A_\Sigma)^{-1}y)$ is a bounded linear functional in $\Sigma(\overline{\Omega})$, so that $\Phi((\lambda I - A_\Sigma)^{-1}y) = \langle \eta, y \rangle$, which is equivalent to (5.2.18).

**Miscellaneous notes.** Most of the material in this section is taken (or in some cases generalized) from the author [2001:4] and [2002:1].

**5.3. The reachable space and its dual, II.** An element $z \in Z_c(T)$ defines a functional $\xi_z \in R_w^\infty(T)^\star$ according to the formula

$$\left\langle\!\!\left\langle \xi_z \, , \int_0^T S_\Sigma(T-\sigma)\mu(\sigma)d\sigma \right\rangle\!\!\right\rangle = \int_0^T \langle S_c'(T-\sigma)z \, , \, \mu(\sigma) \rangle d\sigma \, , \tag{5.3.1}$$

the integral interpreted as in Theorem 5.1.1. We check that (5.3.1) is compatible with the equivalence relation in $R_w^\infty(T)$; this means

$$\int_0^T S_\Sigma(T-\sigma)\mu(\sigma) = 0 \implies \left\langle\!\!\left\langle \xi_z \, , \int_0^T S_\Sigma(T-\sigma)\mu(\sigma)d\sigma \right\rangle\!\!\right\rangle = 0 \, . \tag{5.3.2}$$

To prove this implication, note that

$$
\begin{aligned}
&\int_0^T \langle S_c'(T-\sigma)z\,,\,\mu(\sigma)\rangle d\sigma \\
&\quad = \lim_{h\to 0}\int_0^T \langle S_c'(T-\sigma+h)z\,,\,\mu(\sigma)\rangle d\sigma \\
&\quad = \lim_{h\to 0}\int_0^T \langle S_c'(T-\sigma)S_c'(h)z\,,\,\mu(\sigma)\rangle d\sigma \\
&\quad = \lim_{h\to 0}\int_0^T \langle S_c'(h)z\,,\,S_\Sigma(T-\sigma)\mu(\sigma)\rangle d\sigma \\
&\quad = \lim_{h\to 0}\left\langle S_c'(h)z\,,\,\int_0^T S_\Sigma(T-\sigma)\mu(\sigma)\right\rangle d\sigma\,,
\end{aligned}
\tag{5.3.3}
$$

where the fact that the first integral equals the first limit is justified using the dominated convergence theorem. Equation (5.3.3) has another dividend. Applied to an element of $R_w^\infty(T)$ satisfying

$$
\int_0^T S_\Sigma(T-\sigma)\mu(\sigma)d\sigma \in D(A_\Sigma)
\tag{5.3.4}
$$

it gives, with $\lambda > \omega$, $\omega$ the constant in (4.1.9),

$$
\begin{aligned}
&\left\langle\!\!\left\langle \xi_z\,,\,\int_0^T S_\Sigma(T-\sigma)\mu(\sigma)d\sigma\right\rangle\!\!\right\rangle \\
&\quad = \lim_{h\to 0}\left\langle S_1'(h)z\,,\,\int_0^T S_\Sigma(T-\sigma)\mu(\sigma)d\sigma\right\rangle \\
&\quad = \lim_{h\to 0}\left\langle (\lambda I - A_c')S_c'(h)R(\lambda;A_c')^\bullet z\,,\,\int_0^T S_\Sigma(T-\sigma)\mu(\sigma)d\sigma\right\rangle \\
&\quad = \lim_{h\to 0}\left\langle S_c'(h)R(\lambda;A_c')^\bullet z\,,\,(\lambda I - A_\Sigma)\int_0^T S_\Sigma(T-\sigma)\mu(\sigma)d\sigma\right\rangle \\
&\quad = \left\langle R(\lambda;A_c')^\bullet z\,,\,(\lambda I - A_\Sigma)\int_0^T S_\Sigma(T-\sigma)\mu(\sigma)d\sigma\right\rangle.
\end{aligned}
\tag{5.3.5}
$$

It follows from (5.3.1) that

$$
\begin{aligned}
\left\langle\!\!\left\langle \xi_z\,,\,\int_0^T S_\Sigma(T-\sigma)\mu(\sigma)d\sigma\right\rangle\!\!\right\rangle &\le \|\mu(\cdot)\|_{L_w^\infty(0,T;\Sigma(\overline{\Omega}))}\int_0^T \|S_c'(T-\sigma)z\|_{C(\overline{\Omega})}d\sigma \\
&= \|z\|_{Z_c(T)}\|\mu(\cdot)\|_{L_w^\infty(0,T;\Sigma(\overline{\Omega}))}\,,
\end{aligned}
$$

hence $\|\xi_z\|_{R_w^\infty(T)^\star} \le \|z\|_{Z_c(T)}$. In fact, we have

$$
\|\xi_z\|_{R_w^\infty(T)^\star} = \|z\|_{Z_c(T)}\,.
\tag{5.3.6}
$$

To see this we recall (Theorem 5.1.1) that $L_w^\infty(0,T;\Sigma(\overline{\Omega})) = L^1(0,T;C(\overline{\Omega}))^*$, duality of both spaces given by

$$\langle \mu(\cdot), y(\cdot)\rangle = \int_0^T \langle y(t), \mu(t)\rangle dt\,,$$

so that, applying the Hahn-Banach theorem we can pick $\bar{\mu}(\cdot) \in L_w^\infty(0,T;\Sigma(\overline{\Omega}))$, $\|\bar{\mu}(\cdot)\|_{L_w^\infty(0,T;\Sigma(\overline{\Omega}))} = 1$ with

$$\begin{aligned}\left\langle\!\!\left\langle \xi_z\,, \int_0^T S_\Sigma(T-\sigma)\bar{\mu}(\sigma)d\sigma \right\rangle\!\!\right\rangle &= \int_0^T \langle S_c'(T-\sigma)z\,,\, \bar{\mu}(\sigma)\rangle d\sigma \\ &= \int_0^T \|S_c'(T-\sigma)\|_{C(\overline{\Omega})} d\sigma = \|z\|_{Z_c(T)}\,. \end{aligned} \tag{5.3.7}$$

A functional of the form (5.3.1) is called *regular* and $\mathcal{R}_w(T)$ is the space of all regular functionals, $\mathcal{R}_w(T) \subseteq R_w^\infty(T)^\star$. According to (5.1.20),

$$D(A_\Sigma) \hookrightarrow R_w^\infty(T)\,, \tag{5.3.8}$$

and a functional $\xi_s \in R_w^\infty(T)^\star$ is *singular* if $\xi_s|_{D(A_\Sigma)} = 0$. The space of all singular functionals is $\mathcal{S}_w(T) \subseteq R^\infty(T)^\star$.

**Lemma 5.3.1.** *Singular functionals $\xi_s$ are localized*[2] *at $T$ in the sense that*

$$\left\langle\!\!\left\langle \xi_s\,, \int_0^T S_\Sigma(T-\sigma)\mu(\sigma)d\sigma \right\rangle\!\!\right\rangle = \left\langle\!\!\left\langle \xi_s\,, \int_0^T S_\Sigma(T-\sigma)\nu(\sigma)d\sigma \right\rangle\!\!\right\rangle$$

*if $\mu(\sigma) = \nu(\sigma)$ in $T-\delta \le \sigma \le T$ for some $\delta > 0$.*

*Proof.* Assume $\xi_s \in R_w^\infty(T)^\star$ is singular. We have

$$\begin{aligned}&\int_0^T S_\Sigma(T-\sigma)\mu(\sigma)d\sigma \\ &= S_c(\delta)\int_0^{T-\delta} S_\Sigma(T-\delta-\sigma)\mu(\sigma)d\sigma + \int_{T-\delta}^T S_\Sigma(T-\sigma)\mu(\sigma)d\sigma\,,\end{aligned}$$

where the first term belongs to $D(A_1) \subset D(A_\Sigma)$, hence

$$\left\langle\!\!\left\langle \xi_s\,, \int_0^T S_\Sigma(T-\sigma)\mu(\sigma)d\sigma \right\rangle\!\!\right\rangle = \left\langle\!\!\left\langle \xi_s\,, \int_{T-\delta}^T S_\Sigma(T-\sigma)\mu(\sigma)d\sigma \right\rangle\!\!\right\rangle,$$

and a similar equality holds for $\nu(\cdot)$. This ends the proof.

[2] The converse is also true: localized functionals are singular. See Theorem 5.3.7 below.

**Theorem 5.3.2.** *The subspaces $\mathcal{R}_w(T)$, $\mathcal{S}_w(T)$ are closed. We have*

$$R_w^\infty(T)^\star = \mathcal{R}_w(T) \oplus \mathcal{S}_w(T) \tag{5.3.9}$$

*with $L^1$ direct sum: if $\xi = \xi_z + \xi_s$ then*

$$\|\xi\|_{R_w^\infty(T)^\star} = \|\xi_z\|_{R_w^\infty(T)^\star} + \|\xi_s\|_{R_w^\infty(T)^\star} . \tag{5.3.10}$$

*Proof.* In view of (5.3.8), a bounded functional $\xi \in R_w^\infty(T)^\star$ is also a bounded functional in $D(A_\Sigma)$. By Theorem 5.2.6 there exists $\eta \in \Sigma(\overline{\Omega})^*$ such that

$$\left\langle\!\!\left\langle \xi , \int_0^T S_\Sigma(T-\sigma)\mu(\sigma)d\sigma \right\rangle\!\!\right\rangle = \left\langle \eta , (\lambda I - A_\Sigma)\int_0^T S_\Sigma(T-\sigma)\mu(\sigma)d\sigma \right\rangle \tag{5.3.11}$$

whenever (5.3.4) holds. Among controls satisfying (5.3.4) are those $\mu(\cdot)$ with $\mu(\sigma) = 0$ $(T-\delta \le t \le T)$ and for them we may introduce $\lambda I - A_\Sigma$ inside the integral on the right of (5.3.11). Using Lemma 5.2.2 and Lemma 5.2.3 we obtain

$$\begin{aligned}
\langle \eta , (\lambda I - A_\Sigma)S_\Sigma(t)y\rangle &= \langle \eta , S_\Sigma(t/2)(\lambda I - A_\Sigma)S_\Sigma(t/2)y\rangle \\
&= \langle S_\Sigma(t/2)^*\eta , (\lambda I - A_\Sigma)S_\Sigma(t/2)y\rangle \\
&= \langle (\lambda I - A_c')S_\Sigma(t/2)^*\eta , S_\Sigma(t/2)y\rangle \\
&= \langle S_\Sigma(t/2)^*(\lambda I - A_c')S_\Sigma(t/2)^*\eta , y\rangle \\
&= \langle (\lambda I - A_c')S_\Sigma(t)^*\eta , y\rangle \quad (t > 0) ,
\end{aligned}$$

so that (5.3.11) and the fact that $\xi$ is bounded in $R_w^\infty(T)$ give

$$\begin{aligned}
&\int_0^{T-\delta} \langle (\lambda I - A_c')S_\Sigma(T-\sigma)^*\eta , \mu(\sigma)\rangle d\sigma \\
&\quad = \left\langle\!\!\left\langle \xi , \int_0^{T-\delta} S_\Sigma(T-\sigma)\mu(\sigma)d\sigma \right\rangle\!\!\right\rangle \le \|\xi\|_{R_w^\infty(T)^\star}\|\mu(\cdot)\|_{L_w^\infty(0,T;\Sigma(\overline{\Omega}))} .
\end{aligned}$$

Taking into account that $\mu(\cdot)$ is arbitrary,

$$\int_0^{T-\delta} \|(\lambda I - A_c')S_\Sigma(T-\sigma)^*\eta\|_{C(\overline{\Omega})}d\sigma \le \|\xi\|_{R_w^\infty(T)^\star} .$$

This estimate must hold for every $\delta > 0$, hence

$$\begin{aligned}
&\int_0^T \|(\lambda I - A_c')S_\Sigma(\sigma)^*\eta\|_{C(\overline{\Omega})}d\sigma \\
&\quad = \int_0^T \|(\lambda I - A_c')S_\Sigma(T-\sigma)^*\eta\|_{C(\overline{\Omega})}d\sigma \le \|\xi\|_{R_w^\infty(T)^\star} .
\end{aligned}$$

By virtue of (5.2.11) and of the second relation (5.2.12) we have

$$\begin{aligned}(\lambda I - A_c')S_\Sigma(t+s)^*\eta &= (\lambda I - A_c')S_\Sigma(t)^*S_\Sigma(s)^*\eta \\ &= (\lambda I - A_c')S_c'(t)S_\Sigma(s)^*\eta = S_c'(t)(\lambda I - A_c')S_\Sigma(s)^*\eta\,,\end{aligned}$$

thus we may apply Corollary 5.2.5 to the function $\eta(\sigma) = (\lambda I - A_\Sigma)S_\Sigma(\sigma)^*\eta$ and prove there exists $z \in Z_c(T)$ such that

$$\eta(\sigma) = (\lambda I - A_c')S_\Sigma(\sigma)^*\eta = S_c'(t)z\,.$$

We define a regular functional $\xi_z$ in all of $R_w^\infty(T)$ by

$$\left\langle\!\!\left\langle \xi_z\,, \int_0^T S_\Sigma(T-\sigma)\mu(\sigma)d\sigma \right\rangle\!\!\right\rangle = \int_0^T \langle S_c'(T-\sigma)z\,, \mu(\sigma)\rangle d\sigma\,.$$

Under (5.3.4) we can use (5.3.5) and (5.3.11) together, obtaining

$$\begin{aligned}&\left\langle R(\lambda; A_c')^\bullet z\,,\ (\lambda I - A_\Sigma)\int_0^T S_\Sigma(T-\sigma)\mu(\sigma)d\sigma\right\rangle \\ &\quad = \left\langle \eta\,, (\lambda I - A_\Sigma)\int_0^T S_\Sigma(T-\sigma)\mu(\sigma)d\sigma\right\rangle \qquad (5.3.12)\end{aligned}$$

for all $\mu(\cdot)$ such that (5.3.4) is satisfied. Equality (5.3.12) is

$$\begin{aligned}&\langle R(\lambda; A_c')^\bullet z\,, (\lambda I - A_\Sigma)y\rangle = \langle \eta\,, (\lambda I - A_\Sigma)y\rangle \\ &\qquad (y \in R_w^\infty(T) \cap D(A_\Sigma) = D(A_\Sigma))\,. \qquad (5.3.13)\end{aligned}$$

Since $(\lambda I - A_\Sigma)D(A_\Sigma) = \Sigma(\overline{\Omega})$, (5.3.13) shows that[3] $\eta = R(\lambda; A_c')^\bullet z \in C(\overline{\Omega})$. It also follows from (5.3.13) that $\xi|_{D(A_\Sigma)} = \xi_z|_{D(A_\Sigma)}$; accordingly, we achieve the decomposition (5.3.9) setting $\xi_s = \xi - \xi_z$. To show that the sum is direct we must prove that $\mathcal{R}_w(T) \cap \mathcal{S}_w(T) = \{0\}$. To this end, assume $\xi_z$ is a regular functional such that $\xi_z|_{D(A_\Sigma)} = 0$. It follows from the definition (5.3.1) that

$$\int_0^{T-\delta} \langle S_c'(T-\sigma)z\,, \mu(\sigma)\rangle d\sigma = \left\langle\!\!\left\langle \xi_z\,, S_1(\delta)\int_0^{T-\delta} S_\Sigma(T-\sigma-\delta)\mu(\sigma)d\sigma \right\rangle\!\!\right\rangle = 0$$

for $\delta > 0$ arbitrary. Using the $\bar{\mu}(\cdot)$ in (5.3.7) and letting $\delta \to 0$ we obtain

$$\int_0^T \|S_c'(T-\sigma)z\|_{C(\overline{\Omega})} d\sigma = 0\,,$$

[3] This excuses our lack of interest in characterizing $\Sigma(\overline{\Omega})^*$; this space plays a purely auxiliary role and is no longer needed after (5.3.13).

so that $S'_c(T-\sigma)z = 0$ $(0 \le \sigma \le T) \implies z = 0 \implies \xi_z = 0$.

It remains to show that the $L^1$ property (5.3.10) of the norm is in force. Let $\xi = \xi_z + \xi_s$. We have

$$\|\xi\|_{R^\infty_w(T)^\star} \le \|\xi_z\|_{R^\infty_w(T)^\star} + \|\xi_s\|_{R^\infty_w(T)^\star}\,.$$

For the opposite inequality, let $\varepsilon > 0$ arbitrary and let

$$y_z = \int_0^T S_\Sigma(T-\sigma)\mu(\sigma)d\sigma\,, \quad y_s = \int_0^T S_\Sigma(T-\sigma)\nu(\sigma)d\sigma$$

be elements of the unit ball $B^\infty_{w,1}(T)$ with

$$\langle\!\langle \xi_z, y_z\rangle\!\rangle = \int_0^T \langle S'_c(T-\sigma)z, \mu(\sigma)\rangle d\sigma \ge \|\xi_z\|_{R^\infty_w(T)^\star} - \varepsilon\,,$$
$$\langle\!\langle \xi_s, y_s\rangle\!\rangle \ge \|\xi_s\|_{R^\infty_w(T)^\star} - \varepsilon\,,$$

"elements of the unit ball" meaning we can assume that

$$\|\mu(\cdot)\|_{L^\infty_w(0,T;\Sigma(\overline{\Omega}))} \le 1+\varepsilon\,, \qquad \|\nu(\cdot)\|_{L^\infty_w(0,T;\Sigma(\overline{\Omega}))} \le 1+\varepsilon\,.$$

We define

$$\omega_\delta(\sigma) = \begin{cases} \mu(\sigma) & (0 \le \sigma \le T-\delta) \\ \nu(\sigma) & (T-\delta < \sigma \le T)\,, \end{cases} \qquad y_\delta = \int_0^T S_\Sigma(T-\sigma)\omega_\delta(\sigma)d\sigma\,.$$

Since $\xi_s$ is localized,

$$\langle\!\langle \xi_s, y_\delta\rangle\!\rangle = \langle\!\langle \xi_s, y_s\rangle\!\rangle \ge \|\xi_s\|_{R^\infty_w(T)^\star} - \varepsilon\,.$$

On the other hand, we can choose $\delta$ so small that

$$\begin{aligned}\langle\!\langle \xi_z, y_\delta\rangle\!\rangle &= \int_0^T \langle S'_c(T-\sigma)z\,,\, \omega_\delta(\sigma)\rangle d\sigma \\ &= \int_0^{T-\delta} \langle S'_c(T-\sigma)z\,,\, \mu(\sigma)\rangle d\sigma + \int_{T-\delta}^T \langle S'_c(T-\sigma)z\,,\, \nu(\sigma)\rangle d\sigma \\ &\ge \langle\!\langle \xi_z, y_z\rangle\!\rangle - \varepsilon \ge \|\xi_z\|_{R^\infty_w(T)^\star} - 2\varepsilon\,,\end{aligned}$$

so that

$$\langle\!\langle \xi, y_\delta\rangle\!\rangle = \langle\!\langle \xi_z + \xi_s, y_\delta\rangle\!\rangle \ge \|\xi_z\|_{R^\infty_w(T)^\star} + \|\xi_s\|_{R^\infty_w(T)^\star} - 3\varepsilon\,,$$

thus

$$\|\xi\|_{R^\infty_w(T)^\star} \ge \frac{1}{1+\varepsilon}\big(\|\xi_z\|_{R^\infty_w(T)^\star} + \|\xi_s\|_{R^\infty_w(T)^\star} - 3\varepsilon\big)$$

for arbitrary $\varepsilon > 0$. This ends the proof of (5.3.10).

That $\mathcal{S}_w(T)$ is a closed subspace is obvious. On the other hand, if $\{\xi_{z_n}\}$ is a convergent sequence of regular functionals then $\xi_{z_n}+0 \to \xi_z+\xi_s$, thus $0 \to \xi_s = 0$. This shows the limit is a regular functional, and ends the proof of Theorem 5.3.2.

We call $\xi_z$ (resp. $\xi_s$) the *regular part* (resp. the *singular part*) of $\xi = \xi_z + \xi_s$. We denote by $B^\infty_{w,\rho}(T) \subset R^\infty_w(T)$ the ball of center 0 and radius $\rho$.

**Lemma 5.3.3.** *Let $\bar{y} \in B^\infty_{w,\rho}(T)$, $\xi = \xi_z + \xi_s \in R^\infty_w(T)^\star$ be such that*

$$\langle\!\langle \xi, y \rangle\!\rangle \le \langle\!\langle \xi, \bar{y} \rangle\!\rangle \quad (y \in B^\infty_{w,\rho}(T)). \tag{5.3.14}$$

*Then*

$$\langle\!\langle \xi_z, y \rangle\!\rangle \le \langle\!\langle \xi_z, \bar{y} \rangle\!\rangle \quad (y \in B^\infty_{w,\rho}(T)). \tag{5.3.15}$$

*Proof.* Essentially the same as that of Lemma 4.6.3. Taking advantage of the fact that $\xi_s$ is localized we show that (5.3.14) implies, for a control $\mu(\sigma) = \bar{\mu}(\sigma)$ in $T-\delta \le \sigma \le T$,

$$\left\langle\!\!\left\langle \xi_z\,, \int_0^{T-\delta} S_\Sigma(T-\sigma)\mu(\sigma)d\sigma \right\rangle\!\!\right\rangle \le \left\langle\!\!\left\langle \xi_z\,, \int_0^{T-\delta} S_\Sigma(T-\sigma)\bar{\mu}(\sigma)d\sigma \right\rangle\!\!\right\rangle$$
$$\big(\|\mu\|_{L^\infty_w(0,T;\Sigma(\overline{\Omega}))} \le \rho,\ \delta > 0\big)$$

or, according to (5.3.1),

$$\int_0^{T-\delta} \langle S'_c(T-\sigma)z\,,\, \mu(\sigma)\rangle d\sigma \le \int_0^{T-\delta} \langle S'_c(T-\sigma)z\,,\, \bar{\mu}(\sigma)\rangle d\sigma$$
$$\big(\|\mu\|_{L^\infty_w(0,T;\Sigma(\overline{\Omega}))} \le \rho,\ \delta > 0\big).$$

We take limits as $\delta \to 0$ using the dominated convergence theorem and obtain

$$\int_0^{T} \langle S'_c(T-\sigma)z\,,\, \mu(\sigma)\rangle d\sigma \le \int_0^{T} \langle S'_c(T-\sigma)z\,,\, \bar{\mu}(\sigma)\rangle d\sigma$$
$$\big(\|\mu\|_{L^\infty_w(0,T;\Sigma(\overline{\Omega}))} \le \rho\big)$$

which is (5.3.15). This ends the proof of Lemma 5.3.3.

The final statements in **4.6** have a mirror image here, and we only sketch the details. Applying the theory in Chapter 2 (in particular, **2.3** and **2.4**) to the generator $A_1$ and the semigroup $S_1(t)$ in the space $L^1(\Omega)$, Theorem 2.4.1 gives the decomposition

$$R^\infty(T)^\star = \mathcal{R}(T) \oplus \mathcal{S}(T)$$

where $\mathcal{R}(T)$ (resp. $\mathcal{S}(T)$) is the subspace of regular (resp. singular) functionals in the dual $R^\infty(T)^\star$. On the other hand, we have shown in Theorem 5.1.6 that

$R^\infty(T) = R^\infty_w(T)$ with equivalent norms, hence $R^\infty(T)^\star = R^\infty_w(T)^\star$ with equivalent norms. This implies

$$\mathcal{R}(T) \oplus \mathcal{S}(T) = \mathcal{R}_w(T) \oplus \mathcal{S}_w(T)\,. \tag{5.3.16}$$

Elements of $\mathcal{R}(T)$ are functionals of the form

$$\left\langle\!\!\left\langle \xi\,, \int_0^T S_1(T-\sigma)u(\sigma)d\sigma \right\rangle\!\!\right\rangle = \int_0^T \langle S_\infty(T-\sigma)\zeta\,, u(\sigma)\rangle d\sigma$$

where $\zeta \in Z_w(T)$, the subspace of $L^\infty(\Omega)_{-1}$ such that

$$\int_0^T \|S'_\infty(t)\zeta\|_{L^\infty(\Omega)}dt < \infty\,.$$

We prove exactly as in **4.6** that, given $\zeta \in Z_w(T)$ there exists $z \in Z_c(T)$ such that

$$S'_\infty(t)\zeta = S'_c(t)z \quad (t > 0)\,,$$

thus $Z_w(T) = Z_c(T)$ and it follows from the definition that the norms in both spaces are equivalent. This implies the first equality in

$$\mathcal{R}_w(T) = \mathcal{R}(T)\,, \quad \mathcal{S}_w(T) = \mathcal{S}(T)\,. \tag{5.3.17}$$

The second equality is a consequence of (5.3.16), and implies

$$\xi|_{D(A_1)} = 0 \implies \xi|_{D(A_\Sigma)} = 0$$

so that

$$\overline{D(A_1)} = \overline{D(A_\Sigma)} \quad (\text{closure in } R^\infty(T) = R^\infty_w(T))\,. \tag{5.3.18}$$

Since $S_1(t)$ is an analytic semigroup, Theorem 2.9.2 says that $\overline{D(A_1)} \neq R^\infty(T)$ (closure in $R^\infty(T)$) hence (5.3.18) shows that

$$\overline{D(A_\Sigma)} \neq R^\infty(T) \quad (\text{closure in } R^\infty(T) = R^\infty_w(T))\,. \tag{5.3.19}$$

In particular, (5.3.19) says that nontrivial singular functionals exist. Equality (5.3.18) is not true if closure is taken in the norm of $D(A_\Sigma)$; in fact, since $A_1 \subseteq A_\Sigma$,

$$\overline{D(A_1)} = D(A_1) \quad (\text{closure in } D(A_\Sigma)) \tag{5.3.20}$$

The following result is proved as Theorem 4.6.5:

**Theorem 5.3.4.** *A functional $\xi \in R^\infty_w(T)^\star$ is localized at $T$ if and only if it is singular.*

**Miscellaneous notes.** The material in this section expands the author [2002:1].

**5.4. The maximum principle.** Let $\bar{\mu}(\cdot) \in L^\infty_w(0, T; \Sigma(\overline{\Omega}))$ be a norm optimal control in $0 \le t \le T$ under the target condition

$$y(t, \zeta, u) = \bar{y}\,.$$

Then

$$\bar{y} - S_\Sigma(T)\zeta = \int_0^T S_\Sigma(T-\sigma)\bar{\mu}(\sigma)d\sigma \tag{5.4.1}$$

belongs to the boundary of the ball $B^\infty_{w,\rho}(T)$ of center 0 and radius

$$\rho = \text{optimal norm} = \|\bar{\mu}(\cdot)\|_{L^\infty_w(0,T;\Sigma(\overline{\Omega}))}\,,$$

and it can be separated from $B^\infty_{w,\rho}(T)$ by a nonzero functional $\xi \in R^\infty_w(T)^\star$; the separation inequality is

$$\langle\langle \xi, y\rangle\rangle \le \langle\langle \xi, \bar{y} - S_\Sigma(T)\zeta\rangle\rangle \quad (\|y\|_{R^\infty_w(T)} \le \rho)$$

which, in view of (5.4.1), implies

$$\left\langle\!\!\left\langle \xi\,, \int_0^T S_\Sigma(T-\sigma)\mu(\sigma)d\sigma \right\rangle\!\!\right\rangle \le \left\langle\!\!\left\langle \xi\,, \int_0^T S_\Sigma(T-\sigma)\bar{\mu}(\sigma)d\sigma \right\rangle\!\!\right\rangle \tag{5.4.2}$$

for $\|\mu(\cdot)\|_{L^\infty_w(0,T;\Sigma(\overline{\Omega}))} \le \rho$. If $\xi = \xi_z$ is a regular functional, (5.4.2) is

$$\int_0^T \langle S'_c(T-\sigma)z\,, \mu(\sigma)\rangle d\sigma \le \int_0^T \langle S'_c(T-\sigma)z\,, \bar{\mu}(\sigma)\rangle d\sigma \tag{5.4.3}$$

for $\|\mu(\cdot)\|_{L^\infty_w(0,T;L^\infty(\Omega))} \le \rho$. The $C(\overline{\Omega})$-valued function $\sigma \to S'_c(t-\sigma)z$ is continuous in $0 \le \sigma < T$, thus it is just as continuous as a $\Sigma(\overline{\Omega})^* = C(\overline{\Omega})^{**}$-valued function, and it follows from Lemma 2.2.10 (the nonseparable version of Lemma 2.2.1) that

$$\sup_{\|\mu(\cdot)\|_{L^\infty(0,T;\Sigma(\overline{\Omega}))} \le \rho} \int_0^T \langle S'_c(T-\sigma)z\,, \mu(\sigma)\rangle d\sigma = \rho \int_0^T \|S'_c(T-\sigma)z\|_{C(\overline{\Omega})} d\sigma\,.$$

In view of (5.4.3), the supremum doesn't change if we take $\mu(\cdot) \in L^\infty_w(0, T; \Sigma(\overline{\Omega}))$ instead of $L^\infty(0, T; \Sigma(\overline{\Omega}))$. It then follows that (5.4.3) is equivalent to

$$\begin{aligned}\langle\langle \xi, \bar{y} - S_\Sigma(T)\zeta\rangle\rangle &= \int_0^T \langle S'_c(T-\sigma)z\,, \bar{\mu}(\sigma)\rangle d\sigma \\ &= \rho \int_0^T \|S'_c(T-\sigma)z\|_{C(\overline{\Omega})} d\sigma\,,\end{aligned}$$

which, since $\langle S_c'(T-\sigma)z, \bar\mu(\sigma)\rangle \le \rho\|S_c'(T-\sigma)z\|_{C(\overline{\Omega})}$ is equivalent to

$$\langle S_c'(T-\sigma)z, \bar\mu(\sigma)\rangle = \rho\|S_c'(T-\sigma)z\|_{C(\overline{\Omega})} \quad \text{a. e. in } 0 \le \sigma \le T\,.$$

This, in turn, is equivalent to the maximum principle

$$\langle S_c'(T-t)z, \bar\mu(t)\rangle = \max_{\|\mu\|_{\Sigma(\overline{\Omega})}\le\rho} \langle S_c'(T-t)z, \mu\rangle \quad \text{a. e. in } 0 \le t \le T\,. \tag{5.4.4}$$

**Theorem 5.4.1.** (*a*) *Assume $\bar\mu(t)$ is a norm optimal control and that*

$$\bar y \in \overline{D(A_\Sigma)}\,. \tag{5.4.5}$$

*Then there exists a regular functional separating $\bar y - S_\Sigma(T)\zeta$ from $B^\infty_{w,\rho}(T)$.* (*b*) *Assume $\bar\mu(t)$ is a time optimal control and that* (5.4.5) *is satisfied. Then there exists a regular functional separating $\bar y - S_\Sigma(T)\zeta$ from $B^\infty_{w,1}(T)$.*

*Proof.* Separating $\bar y - S_\Sigma(T)\zeta$ from the ball $B^\infty_{w,\rho}(T)$ we have the right to require that $\langle\langle \xi, \bar y - S_\Sigma(T)\zeta\rangle\rangle = 1$. The inclusion $S_\Sigma(T)\Sigma(\overline{\Omega}) \subseteq D(A_\Sigma)$ and (5.4.5) imply that $\bar y - S_\Sigma(T)\zeta \in \overline{D(A_\Sigma)}$, so that $\xi$ cannot be a singular functional; $\xi = \xi_z + \xi_s$ with $\xi_z \ne 0$. It then follows from Lemma 5.3.3 that $\xi_z$ does the separation job as well. The same argument works for time optimal controls, since they are norm optimal.

**Theorem 5.4.2.** *Assume* (5.4.5) *holds. Then* (*a*) *If $\bar\mu(\cdot)$ is norm optimal it satisfies* (5.4.4). (*b*) *If $\bar\mu(\cdot)$ is time optimal then* (5.4.4) *holds with $\rho = 1$.*

*Proof.* Using Theorem 5.4.1 we have the separation inequality (5.4.3) with a nonzero regular functional $\xi_z$. This inequality is then equivalent to (5.4.3) with $z \ne 0$, thus, via the argument following (5.4.3) the maximum principle (5.4.4) follows.

**Theorem 5.4.3.** *Assume $\bar\mu(\cdot) \in L^\infty_w(0, T, \Sigma(\overline{\Omega}))$ satisfies the maximum principle* (5.4.4) *with $z \in Z_c(T)$, $z \ne 0$, $\rho$ the maximum norm. Then $\bar\mu(t)$ is norm optimal.*

**Theorem 5.4.4.** *Assume $\bar\mu(\cdot) \in L^\infty_w(0, T, \Sigma(\overline{\Omega}))$ satisfies the maximum principle* (5.4.4) *with $z \in Z(T)$, $z \ne 0$, $\rho = 1$ and that, either*

$$\begin{aligned} &(a)\quad \bar y \in D(A_\Sigma)\,,\ \|A_\Sigma \bar y\|_{\Sigma(\overline{\Omega})} < 1\,, \quad \textit{or} \\ &(b)\quad \zeta \in D(A_\Sigma)\,,\ \|A_\Sigma \zeta\|_{\Sigma(\overline{\Omega})} < 1\,. \end{aligned} \tag{5.4.6}$$

*Then $\bar\mu(t)$ is time optimal.*

The proofs of Theorem 5.4.3 and 5.4.4 are mirror images of those of Theorems 4.7.3 and 4.7.4 via the replacement chart (5.1.17); the space $L^\infty_w(0, T; L^\infty(\Omega))$ becomes $L^\infty_w(0, T; \Sigma(\overline{\Omega}))$, $S_\infty(t)$ becomes $S_\Sigma(t)$, $S_1'(t)$ becomes $S_c'(t)$ and the norm $\|S_1'(t)\|_{L^1(\Omega)}$ becomes $\|S_c'(t)\|_{C(\overline{\Omega})}$. Since there are no surprises, we omit the details.

We note that a key ingredient of the proof of Theorem 5.4.4 is the fact that the semigroup $S_c'(t)$ is analytic, thus the fact that $z \neq 0$ implies that the costate $S_c'(T-\sigma)z$ is nonzero in the whole interval $0 \leq t \leq T$.

It follows from Theorem 3.3.1 that for any multiplier $\psi$ the costate

$$z(t,x) = S_c'(T-t)\psi(x)$$

is infinitely differentiable in $[0,T) \times \Omega$ and satisfies the reverse adjoint equation

$$\begin{aligned} \frac{\partial z(t,x)}{\partial t} &= -A'z(t,x) \quad (0 \leq t < T,\ x \in \Omega) \\ \beta' y(t,x) &= 0 \qquad\qquad\quad (0 \leq t < T,\ x \in \Gamma) \end{aligned} \tag{5.4.7}$$

in the classical sense. We restate the results in this section as theorems on optimal control of the distributed parameter system

$$\begin{aligned} \frac{\partial y(t,x)}{\partial t} &= Ay(t,x) + \mu(t) \quad (0 \leq t \leq T,\ x \in \Omega) \\ y(0,x) &= \zeta(x) \qquad\qquad\quad (x \in \Omega) \\ \beta y(t,x) &= 0 \qquad\qquad\qquad (0 \leq t \leq T,\ x \in \Gamma) \end{aligned} \tag{5.4.8}$$

with controls in $L_w^\infty(0,T;\Sigma(\overline{\Omega}))$. The initial condition $\zeta$ actually belongs to $\Sigma(\overline{\Omega})$, and the equation and initial condition are understood as in Theorem 5.1.3. The maximum principle (5.4.4) can be written in the form

$$\int_{\overline{\Omega}} z(t,x)\bar{\mu}(t,dx) = \max_{\|\mu\|_{\Sigma(\overline{\Omega})} \leq \rho} \int_{\overline{\Omega}} z(t,x)\mu(dx)\,. \tag{5.4.9}$$

To figure out what this means, define

$$\begin{aligned} \mathcal{M}^+(z) &= \left\{x \in \overline{\Omega};\, z(x) = \max_{x \in \overline{\Omega}} |z(x)| = \|z\|_{C(\overline{\Omega})}\right\}, \\ \mathcal{M}^-(z) &= \left\{x \in \overline{\Omega};\, z(x) = -\max_{x \in \overline{\Omega}} |z(x)| = \|z\|_{C(\overline{\Omega})}\right\}. \end{aligned} \tag{5.4.10}$$

**Lemma 5.4.5.** *Let $z(\cdot) \in C(\overline{\Omega})$, $\bar{\mu} \in \Sigma(\overline{\Omega})$ be such that*

$$\langle z, \bar{\mu} \rangle = \max_{\|\mu\|_{\Sigma(\overline{\Omega})} \leq \rho} \langle z, \mu \rangle\,. \tag{5.4.11}$$

*Then $\bar{\mu}$ is supported by the closed set*

$$\mathcal{M}(z) = \left\{x \in \overline{\Omega};\, |z(x)| = \max_{x \in \overline{\Omega}} |z(x)| = \|z\|_{C(\overline{\Omega})}\right\} = \mathcal{M}^+ \cup \mathcal{M}^-\,, \tag{5.4.12}$$

*and*

$$z(x)\bar{\mu}(dx) \geq 0\,, \quad \|\mu\|_{\Sigma(\overline{\Omega})} = \int_{\mathcal{M}(z)} |\bar{\mu}|(dx) = \rho\,. \tag{5.4.13}$$

*Proof.* For any $\mu \in \Sigma(\overline{\Omega})$ with $\|\mu\|_{\Sigma(\overline{\Omega})} \leq \rho$ we have

$$\langle z, \mu \rangle = \int_{\overline{\Omega}} z(x)\mu(dx) \leq \int_{\overline{\Omega}} |z(x)|\,|\mu|(dx) \leq \rho \|z\|_{C(\overline{\Omega})}\,.$$

On the other hand, if (5.4.13) holds then $\bar{\mu}(dx) \geq 0$ in $\mathcal{M}^+(z)$ and $-\bar{\mu}(dx) \geq 0$ in $\mathcal{M}^-(z)$, so that

$$\begin{aligned}
\langle z, \bar{\mu} \rangle &= \int_{\overline{\Omega}} z(x)\bar{\mu}(dx) \\
&= \|z\|_{C(\overline{\Omega})} \int_{\mathcal{M}^+(z)} \bar{\mu}(dx) + \|z\|_{C(\overline{\Omega})} \int_{\mathcal{M}^-(z)} (-\bar{\mu})(dx) \\
&= \|z\|_{C(\overline{\Omega})} \int_{\mathcal{M}(z)} |\bar{\mu}(dx)| = \rho \|z\|_{C(\overline{\Omega})}\,,
\end{aligned}$$

so that $\bar{\mu}$ attains the maximum in (5.4.11).

Conversely, assume $\bar{\mu}$ satisfies (5.4.11). If $\bar{\mu}$ is not supported by $\mathcal{M}(z)$ there exists a Borel set $\mathcal{E}$ with $|\bar{\mu}|(\mathcal{E}) > 0$ such that $|z(x)| < \|z\|_{C(\overline{\Omega})}$ $(x \in \mathcal{E})$. Define

$$\mathcal{E}(\epsilon) = \left\{ x \in \mathcal{E}; |z(x)| \leq \|z\|_{C(\overline{\Omega})} - \epsilon \right\}.$$

Then $\mathcal{E} = \cup_{n \geq 1} \mathcal{E}(1/n)$, so that $|\bar{\mu}|(\mathcal{E}(\epsilon)) = \delta > 0$ for $\epsilon$ small enough. We have

$$\begin{aligned}
\int_{\overline{\Omega}} z(x)\bar{\mu}(dx) &\leq (\|z\|_{C(\overline{\Omega})} - \epsilon) \int_{\mathcal{E}(\epsilon)} |\bar{\mu}|(dx) + \|z\|_{C(\overline{\Omega})} \int_{\overline{\Omega} \setminus \mathcal{E}(\epsilon)} |\bar{\mu}|(dx) \\
&\leq (\|z\|_{C(\overline{\Omega})} - \epsilon)\delta + \|z\|_{C(\overline{\Omega})}(\rho - \delta) < \rho \|z\|_{C(\overline{\Omega})}\,.
\end{aligned}$$

However, by the Hahn - Banach theorem the maximum in (5.4.11) must equal $\rho \|z\|_{C(\overline{\Omega}}$, so we have a contradiction. Granted then that $\bar{\mu}$ is supported by $\mathcal{M}(z)$, if $\bar{\mu}$ attains the maximum we must have

$$\int_{\mathcal{M}(z)} z(x)\bar{\mu}(dx) = \rho \|z\|_{C(\overline{\Omega})} = \int_{\overline{\Omega}} \|z\|_{C(\overline{\Omega})} |\bar{\mu}|(dx)\,.$$

Since $z(x)\bar{\mu}(\mathcal{C}) \leq \|z\|_{C(\overline{\Omega})} |\bar{\mu}|(\mathcal{C})$ for every Borel set $\mathcal{C}$, it follows that

$$z(x)\bar{\mu}(dx) = \|z\|_{C(\overline{\Omega})} |\bar{\mu}|(dx)\,,$$

which shows (5.4.13) and ends the proof of Lemma 5.4.5.

After Lemma 5.4.5, the maximum principle (5.4.9) becomes

$$\begin{gathered}\operatorname{supp}\bar{\mu}(t) \subseteq \mathcal{M}(t,z) = \big\{x \in \overline{\Omega}; |z(t,x)| = \max_{x\in\overline{\Omega}} |z(t,x)| = \|z(t,\cdot)\|_{C(\overline{\Omega})}\big\}, \\ z(t,x)\bar{\mu}(t,dx) \geq 0, \qquad \|\bar{\mu}\|_{\Sigma(\overline{\Omega})} = \int_{\mathcal{M}(t,z)} |\bar{\mu}|(t,dx) = \rho \end{gathered} \tag{5.4.14}$$

a. e. in $0 \leq t \leq T$. In terms of $z(t,x)$, condition

$$\int_0^T \|S_c'(T-\sigma)z\|_{C(\overline{\Omega})} = \int_0^T \|S_c'(\sigma)z\|_{C(\overline{\Omega})} < \infty$$

on multipliers in $Z_c(T)$ is

$$\int_0^T \|z(t,\cdot)\|_{C(\overline{\Omega})}dt = \int_0^T \Big(\max_{x\in\overline{\Omega}} |z(t,x)|\Big)dt < \infty. \tag{5.4.15}$$

In "semigroup-free" formulation, Theorems 5.4.2, 5.4.3 and 5.4.4 become Theorems 5.4.6, 5.4.7 and 5.4.8 below. Controls given by (5.4.9) $\Leftrightarrow$ (5.4.14) are said to be *associated* with the costate $z(t,x)$.

**Theorem 5.4.6.** *Assume*

$$\bar{y}(\cdot) \in \overline{D(A_\Sigma)}. \tag{5.4.16}$$

*Then, if $\bar{\mu}(t)$ is norm optimal it satisfies* (5.4.14) *with a nontrivial solution $z(t,x)$ of the reverse adjoint equation* (5.4.7) *satisfying the growth condition* (5.4.15). *Same for a time optimal control, with $\rho = 1$.*

**Theorem 5.4.7.** *Let $z(t,x)$ be a nontrivial solution of the reverse adjoint equation* (5.4.7) *satisfying the growth condition* (5.4.15). *Then any $\bar{\mu}(t)$ satisfying* (5.4.14) *is norm optimal.*

**Theorem 5.4.8.** *Let $z(t,x)$ be a nontrivial solution of the reverse adjoint equation* (5.4.7) *satisfying the growth condition* (5.4.15). *Then any $\bar{\mu}(t)$ satisfying* (5.4.14) *with $\rho = 1$ and*

$$\begin{aligned} &(a) \quad \bar{y}(\cdot) \in D(A_\Sigma), \ \|A_\Sigma \bar{y}(\cdot)\|_{\Sigma(\overline{\Omega})} < 1, \quad \textit{or} \\ &(b) \quad \zeta(\cdot) \in D(A_\Sigma), \ \|A_\Sigma \zeta(\cdot)\|_{\Sigma(\overline{\Omega})} < 1, \end{aligned} \tag{5.4.17}$$

*is time optimal.*

**Theorem 5.4.9.** *There exists a strongly singular norm optimal control, that is, a norm optimal control $\bar{\mu}(\cdot)$ with $\|\bar{\mu}(\cdot)\|_{L^\infty(0,T;\Sigma(\overline{\Omega}))} = 1$ such that* (5.4.14) *does not hold, even if the growth condition* (5.4.15) *is given up.*

The proof is exactly the same as that of Theorem 4.7.9, with $S_\infty(t)$ replaced by $S_\Sigma(t)$ and $S_c(t)$ replaced by $S_1(t)$.

**Problem 5.4.10.** *Are there any weakly singular (time, norm) optimal controls, that is, any optimal controls obeying* (5.4.14) *but not with any solution* $z(t,x)$ *of the reverse adjoint equation restricted by* (5.4.15)?

In Theorems 5.4.6, 5.4.7 and 5.4.8 "nontrivial" means "not identically zero". The following result, in combination with (5.4.14) gives a *concentration principle;* except in certain specific situations, both time and norm optimal controls have support in small sets. We recall from **3.3** that if $\beta$ is the Dirichlet boundary condition then $\beta' = \beta$; on the other hand, if $\beta$ is a variational boundary condition

$$\beta y(x) = \partial^\nu y(x) - \gamma(x)y(x) = 0 \quad (x \in \Gamma) \tag{5.4.18}$$

($\partial^\nu$ the conormal derivative on the boundary $\Gamma$), we have

$$\beta' y(x) = \partial^\nu y(x) - (\gamma(x) + b(x))y(x) = 0 \quad (x \in \Gamma) \tag{5.4.19}$$

with $b(x) = \sum_{j=1}^m b_j(x)\nu_j$, $(\nu_1, \nu_2, \dots, \nu_m)$ the outer normal vector at $\Gamma$. It follows that $\beta'$ is the Neumann boundary condition if and only if

$$\gamma(x) + \sum_{j=1}^m b_j(x)\nu_j = 0\,. \tag{5.4.20}$$

In the result below, one of the assumptions is that $\beta'$ is *not* the Neumann boundary condition; as a condition on $\beta$, this means that $\beta$ is Dirichlet or is variational with $\gamma(x) + \sum_{j=1}^m b_j(x)\nu_j$ not identically zero on $\Gamma$.

**Theorem 5.4.11.** *Let* $z(t,x)$ *be a nontrivial solution of* (5.4.7). *Assume that the coefficients of* $A$ *and* $\beta$ *are analytic, and that* $\beta'$ *is not the Neumann boundary condition. Then the set* $\mathcal{M}(t,z)$ *in* (5.4.14) *has Lebesgue measure zero for all* $t$. *In space dimension* $m = 1$, $\mathcal{M}(t,z)$ *is finite.*

*Proof.* We set $\theta(t,x) = z(T-t,x)$ so that $\theta(t,x)$ satisfies the forward equation

$$\frac{\partial\theta(t,x)}{\partial t} = A'\theta(t,x) \quad (0 \le t < T,\ x \in \Omega) \tag{5.4.21}$$

$$\beta'\theta(t,x) = 0 \qquad (0 \le t < T,\ x \in \Gamma)\,. \tag{5.4.22}$$

As a particular case of Gevrey's results in [1913], [1914], [1918], solutions of the parabolic equation (5.4.21) in an open cylinder in $I\!R \times I\!R^m$ are analytic in $x$ for each $t$ fixed. If $\theta(\bar t, \bar x) = \pm\|\theta(t,\cdot)\|_{C(\overline{\Omega})}$ for some $(\bar t, \bar x) \in [0,T] \times \overline{\Omega}$ then $x \to \theta(\bar t, x)$ has a maximum or a minimum at $\bar x$ and, if $\bar x \in \Omega$, $\nabla\theta(t,\bar x) = 0$. It follows that

$$\mathcal{M}(t,z) \cap \Omega \subseteq \mathcal{N}\Big(t, \frac{\partial\theta}{\partial x_1}\Big) \cap \ldots \cap \mathcal{N}\Big(t, \frac{\partial\theta}{\partial x_n}\Big) = \mathcal{N}(t, \nabla\theta)\,, \tag{5.4.23}$$

where $\mathcal{N}(t,z) \subset \Omega$ indicates nodal set (= nullset) at time $t$ :

$$\mathcal{N}(t,z) = \{x \in \Omega;\ z(t,x) = 0\}\,.$$

To show that $|\mathcal{M}(t,z)| = 0$ it is enough to show that $|\mathcal{M}(t,z)\cap\Omega| = 0$. If for some $\bar{t}$ all nodal sets in the intersection (5.4.23) have positive measure then, by analyticity of the derivatives,

$$\frac{\partial\theta(\bar{t},x)}{\partial x_1} = 0\,, \ldots, \frac{\partial\theta(\bar{t},x)}{\partial x_n} = 0 \quad (x\in\Omega)\,,$$

so that $\theta(\bar{t},x)$ is constant in $x$. If $\beta$ (hence $\beta'$) is the Dirichlet boundary condition, $\theta(\bar{t},x) = 0$ on the boundary $\Gamma$, hence $\theta(\bar{t},x) = 0$ $(x\in\Omega)$. On the other hand, assume $\beta$ is not Dirichlet. Since $\beta'$ is not the Neumann boundary condition, it is of the form $\partial^\nu\theta(t,x) = \delta(x)\theta(t,x)$ with $\delta(x)$ not identically zero. At any point $\bar{x}\in\Gamma$ where $\delta(\bar{x})\neq 0$ we must have $\theta(t,\bar{x}) = 0$ thus, since $\theta(\bar{t},x)$ is constant we again have $\theta(\bar{t},x) = 0$ $(x\in\Omega)$. Now, if $0 < t \leq \bar{t}$,

$$\theta(\bar{t},\cdot) = S_c(\bar{t}-t)\theta(t,\cdot)\,,$$

and the semigroup $S_c(t)$, being analytic, is one-to-one, so that $\theta(t,\cdot) = 0$ for every $0 < t \leq \bar{t}$; for $t > \bar{t}$ we have $\theta(t,\cdot) = S_c(t-\bar{t})\theta(\bar{t},\cdot) = 0$. This shows that $\theta(t,x)$ is a trivial solution against the hypotheses.

In dimension 1, $\mathcal{N}(t, d\theta/dx)$ must be finite unless $\theta(t,x)$ is constant in $x$, again ruled out by the boundary condition $\beta$. This ends the proof of Theorem 5.4.11.

Of course, even in dimension $> 1$ what is proved is a lot more than $\mathcal{M}(t,z)$ just being a null set. For instance, let $f(x_1,x_2)$ be a nonzero real analytic function of two variables in $\Omega\subseteq\mathbb{R}^2$. Denote by $L(a)$ the line $x_1 = a$ in $\mathbb{R}^2$, and let

$$\mathcal{N}(a) = \{(x_1,x_2)\in L(a)\cap\Omega; f(x_1,x_2) = 0\}\,.$$

Then each $\mathcal{N}(a)$ is finite except for a finite number of values of $a$.

The boundary condition excluded in Theorem 5.4.11 leads to this result.

**Theorem 5.4.12.** *Let $\beta'$ be the Neumann boundary condition* (*that is, let $\beta$ be a variational boundary condition* (5.4.18) *satisfying* (5.4.20)) *and assume that*

$$\sum_{j=1}^{m}\frac{\partial}{\partial x_j}b_j(x) = c(x)\,. \tag{5.4.24}$$

*Then every $\bar{\mu}(\cdot)\in L^\infty_w(0,T;\Sigma(\overline{\Omega}))$ with*

$$\|\bar{\mu}(t)\|_{\Sigma(\overline{\Omega})} = \rho \quad a.\ e.\ in\ 0\leq t\leq T \tag{5.4.25}$$

*and such that, either $\bar{\mu}(t)\geq 0$ a. e. or $\bar{\mu}(t)\leq 0$ a. e. is norm optimal. If $\rho = 1$ and conditions* (5.4.17) *are satisfied, $\bar{\mu}(t)$ is time optimal.*

*Proof.* (5.4.24) implies that $A'$ has no term of order zero, thus every constant $z(t,x) = z$ solves the reverse adjoint equation (5.4.7). If $z\neq 0$ we have $\mathcal{M}(t,z) = \Omega$

for all $t$ and any control such that $\bar{\mu}(t) \geq 0$ and $\|\bar{\mu}(t)\| = \rho$ a. e. satisfies (5.4.14), thus is norm optimal and time optimal if $\rho = 1$ and one of conditions (5.4.17) holds. If $z < 0$ we change $\bar{\mu}(t) \geq 0$ by $\bar{\mu}(t) \leq 0$.

Theorem 5.4.12 illustrates the only situation known to us where the bang-bang property (2.1.19) comes very close to being sufficient for optimality (we don't know of any example where it is *actually* sufficient). The controls singled out in Theorem 5.4.12 are not the only optimal controls; see Example 5.6.3.

**Miscellaneous notes.** The results in this section are in the author [2002:1]. It seems not unlikely that results on the sets $\mathcal{M}(t, z)$ may be amenable with much less than analyticity assumptions to the methods used for the study of nodal sets in Han - Lin [1994] and other papers (see **4.7**) but we don't know of any results.

**5.5. Existence of optimal controls; uniqueness and stability of supports.** Up to a certain point, the results in this section are the mirror image of those in **4.8** via the replacements in the chart (5.1.17). The equation is

$$y'(t) = A_1 y(t) + \mu(t), \quad y(0) = \zeta \tag{5.5.1}$$

with $\zeta \in \Sigma(\overline{\Omega})$ and $\mu(\cdot) \in L^\infty_w(0, T; \Sigma(\overline{\Omega}))$. For the norm optimal problem, a minimizing sequence $\{\mu_n(\cdot)\}$ satisfies

$$y(T, \zeta, \mu_n) = y_n \to \bar{y}, \quad \|\mu_n(\cdot)\|_{L^\infty_w(0,T;\Sigma(\overline{\Omega}))} \to \rho, \tag{5.5.2}$$

where $\rho$ is the infimum norm. For the time optimal problem, a minimizing sequence satisfies

$$y(t_n, \zeta, \mu_n) = y_n \to \bar{y}, \quad \|\mu_n(\cdot)\|_{L^\infty_w(0,t_n;\Sigma(\overline{\Omega}))} \leq 1, \quad t_n \to T \tag{5.5.3}$$

where $T$ is the infimum of all driving times from $\zeta$ to $\bar{y}$. Under the assumption $y(t_n, \zeta, \mu_n) = \bar{y}$ we may additionally require for the norm optimal problem that the norms $\|\mu_n(\cdot)\|_{L^\infty_w(0,T;L^\infty(\Omega))}$ be decreasing, and for the time optimal problem that the control intervals $[0, t_n]$ be decreasing. It is obvious from the definitions that, if there exists a control $\mu(\cdot) \in L^\infty_w(0, T; \Sigma(\overline{\Omega}))$ driving $\zeta$ to $\bar{y}$ in time $T$ there exists a minimizing sequence for the norm optimal problem in $0 \leq t \leq T$; if there exists a control $\mu(\cdot) \in L^\infty_w(0, T; \Sigma(\overline{\Omega}))$ with $\|\mu(\cdot)\|_{L^\infty_w(0,T;\Sigma(\overline{\Omega}))} \leq 1$ driving $\zeta$ to $\bar{y}$ in any time $t$ then there exists a minimizing sequence for the time optimal problem.

We have proved in Theorem 5.1.1 that

$$L^\infty(0, T; \Sigma(\overline{\Omega})) = L^1(0, T; C(\overline{\Omega}))^*$$

so that we can take weak limits of (subsequences of) minimizing sequences.[4]

[4] $C(\overline{\Omega})$ is separable, thus $L^1(0, T; C(\overline{\Omega}))$ is also separable and the topology of the unit ball of $L^\infty_w(0, T; \Sigma(\overline{\Omega}))$ is defined by a metric. We only need to use sequences and subsequences, not generalized sequences.

**Theorem 5.5.1.** *Assume a minimizing sequence for the norm optimal problem exists. Then an optimal control exists. Same for the time optimal problem.*

*Proof:* Let $\mu(\cdot) \in L^\infty_w(0, T; \Sigma(\overline{\Omega}))$ drive $\zeta$ to a target $y \in L^1(\Omega)$ in time $t$. Let $u \in L^1(\Omega)^* = L^\infty(\Omega)$. Then we have

$$\begin{aligned}
\langle u, y - S_\Sigma(t)\zeta\rangle &= \left\langle u, \int_0^t S_\Sigma(t-\sigma)\mu(\sigma)d\sigma \right\rangle \\
&= \lim_{h\to 0} \left\langle u,\ S_1(h)\int_0^t S_\Sigma(t-\sigma)\mu(\sigma)d\sigma \right\rangle \\
&= \lim_{h\to 0} \left\langle S'_\infty(h)u\,,\ \int_0^t S_\Sigma(t-\sigma)\mu(\sigma)d\sigma \right\rangle \\
&= \lim_{h\to 0} \int_0^t \langle S'_\infty(h)u\,,\ S_\Sigma(t-\sigma)\mu(\sigma)\rangle d\sigma \\
&= \lim_{h\to 0} \int_0^t \langle S'_c(t-\sigma)S'_\infty(h)u\,,\ \mu(\sigma)\rangle d\sigma \\
&= \lim_{h\to 0} \int_0^t \langle S'_\infty(t-\sigma+h)u\,,\ \mu(\sigma)\rangle d\sigma \\
&= \int_0^t \langle S'_\infty(t-\sigma)u\,,\ \mu(\sigma)\rangle d\sigma\,, 
\end{aligned} \tag{5.5.4}$$

where the first two angled brackets in (5.5.4) indicate the duality of $L^1(\overline{\Omega})$ and $L^\infty(\Omega)$, the others the duality of $C(\overline{\Omega})$ and $\Sigma(\overline{\Omega})$. The limit in the last line is taken using continuity of $S'_\infty(t)$ in $t > 0$ and the dominated convergence theorem.

If $\{\mu_n(\cdot)\} \subset L^\infty_w(0, T; \Sigma(\overline{\Omega}))$ is a minimizing sequence for the norm optimal problem we have

$$y_n - S(T)\zeta = \int_0^T S_\Sigma(T-\sigma)\mu_n(\sigma)d\sigma$$

and $\|\mu_n(\cdot)\|_{L^\infty_w(0,T;\Sigma(\overline{\Omega}))} \to \rho =$ minimum norm. Selecting a subsequence we may assume that $\mu_n(\cdot) \to \bar\mu(\cdot) \in L^\infty_w(0, T; \Sigma(\overline{\Omega}))$ $L^1(0, T; C(\overline{\Omega}))$-weakly, where the norm of $\bar\mu(\cdot)$ equals $\rho$. We then write (5.5.4) for $u \in L^\infty(\Omega)$, $t = T$ and $\mu(\cdot) = \mu_n(\cdot)$,

$$\langle u, y_n - S_\Sigma(T)\zeta\rangle = \int_0^T \langle S'_\infty(T-\sigma)u\,,\ \mu_n(\sigma)\rangle d\sigma\,,$$

note that $S'_\infty(T - \cdot)\mu \in L^1(0, T; C(\overline{\Omega}))$ and take limits. Using (5.5.4) the result is

$$\langle u,\, \bar y - S(T)\zeta\rangle = \int_0^T \langle S_\Sigma(T-\sigma)u\,,\ \bar\mu(\sigma)\rangle d\sigma = \left\langle u\,,\ \int_0^t S_\Sigma(t-\sigma)\mu(\sigma)d\sigma \right\rangle.$$

Since $u \in L^\infty(\Omega)$ is arbitrary, we obtain

$$\bar{y} - S(T)\zeta = \int_0^T S_\Sigma(T-\sigma)\bar{\mu}(\sigma)d\sigma\,, \tag{5.5.5}$$

which shows that $\bar{\mu}(\cdot)$ drives $\zeta$ to $\bar{y}$ in time $T$ and completes the proof of the part of Theorem 5.5.1 dealing with the norm optimal problem.

If $\{\mu_n(\cdot)\}$, $\mu_n(\cdot) \in L_w^\infty(0, t_n; \Sigma(\overline{\Omega}))$ is a minimizing sequence for the time optimal problem we have

$$y_n - S_\Sigma(t_n)\zeta = \int_0^{t_n} S_\Sigma(t_n-\sigma)\mu_n(\sigma)d\sigma\,, \tag{5.5.6}$$

and $t_n \to T =$ optimal time, $\|\mu_n(\cdot)\|_{L^\infty(0,T;\Sigma(\overline{\Omega}))} \leq 1$. Pick $t > t_n$ and extend $\mu_n(\cdot)$ to $t_n \leq \sigma \leq t$ setting $\mu_n(t) = 0$ there. Selecting a subsequence we can assume that $\mu_n(\cdot) \to \bar{\mu}(\cdot) \in L_w^\infty(0, t; \Sigma(\overline{\Omega}))$ $L^1(0, t; C(\overline{\Omega}))$-weakly. The passage to the limit in (5.5.6) to obtain (5.5.5) is handled similarly (see Theorem 4.8.1 for the treatment of the variable upper limit of integration) and we omit the details. This ends the proof of Theorem 5.5.1.

The existence result for norm optimal controls gives

**Corollary 5.5.2.** *Let $\bar{y} \in R_w^\infty(T)$. Then there exists $\bar{\mu}(\cdot) \in L_w^\infty(0, T; \Sigma(\overline{\Omega}))$ with*

$$\bar{y} = \int_0^T S_\Sigma(T-\sigma)\bar{\mu}(\sigma)d\sigma\,, \quad \|\bar{y}\|_{R_w^\infty(T)} = \|\bar{\mu}(\cdot)\|_{L_w^\infty(0,T;\Sigma(\overline{\Omega}))}\,. \tag{5.5.7}$$

Mirror similarity with the setup in Chapter 4, especially **4.8**, ends here. As we see below (Theorem 5.5.6) there is in general no uniqueness for optimal controls, even for those that satisfy the maximum principle (5.4.14). All we can show is that if two controls drive the same initial condition to the same target norm or time optimally, they "share the same costate" in the sense that both satisfy (5.4.14) with the same costate. This, however, does not mean that costates are unique for a given optimal control; see Remark 5.5.11.

The argument below requires the condition

$$\bar{y} \in \overline{D(A_\Sigma)}\,. \tag{5.5.8}$$

Assume $\bar{\mu}(t)$ drives $\bar{\zeta}$ to $\bar{y}$ norm or time optimally in $0 \leq t \leq T$. According to Theorem 5.4.6, $\bar{\mu}(t)$ satisfies the maximum principle. In the form (5.4.9), it says

$$\int_{\overline{\Omega}} z(t,x)\bar{\mu}(t,dx) = \max_{\|\mu\|_{\Sigma(\overline{\Omega})} \leq \rho} \int_{\overline{\Omega}} z(t,x)\mu(dx)\,.$$

This is the same as

$$\int_{\overline{\Omega}} z(t,x)(\bar{\mu}(t,dx) - \mu(dx)) \geq 0 \quad (\|\mu\|_{\Sigma(\overline{\Omega})} \leq \rho) \tag{5.5.9}$$

($\rho = 1$ for the time optimal problem). If a control $\mu(\cdot) \in L^\infty_w(0,T;\Sigma(\overline{\Omega}))$ with norm $\|\mu(\cdot)\|_{L^\infty(0,T;\Sigma(\overline{\Omega}))} \le \rho$ drives $\zeta$ to $y$ in the same time interval we have

$$\begin{aligned}
&\int_0^T \int_{\overline{\Omega}} z(t,x)(\bar\mu(t,dx) - \mu(t,dx))dt \\
&= \int_0^T \langle S_c'(T-t)z \,,\, \bar\mu(t) - \mu(t)\rangle dt \\
&= \left\langle\!\!\left\langle \xi_z \,,\, \int_0^T S_\Sigma(T-t)\bar\mu(t)dt - \int_0^T S_\Sigma(T-t)\mu(t)dt \right\rangle\!\!\right\rangle \\
&= \langle\!\langle \xi_z \,,\, (y(T,\bar\zeta,\bar\mu) - S_\Sigma(T)\bar\zeta) - (y(T,\zeta,\mu) - S_\Sigma(T)\zeta)\rangle\!\rangle \,. \qquad (5.5.10)
\end{aligned}$$

If $\zeta = \bar\zeta$ and $y = y(T,\zeta,\mu) = \bar y = y(T,\zeta,\bar\mu)$ then both controls drive $\bar\zeta$ to $\bar y$ optimally and the right hand side of (5.5.10) vanishes. Combining with (5.5.9),

$$\int_{\overline{\Omega}} z(t,x)\bar\mu(t,dx) = \int_{\overline{\Omega}} z(t,x)\mu(t,dx) \quad \text{a. e. in } 0 \le t \le T.$$

Accordingly, $\mu(t,dx)$ also satisfies (5.4.9). We conclude that both $\bar\mu(t,dx)$ and $\mu(t,dx)$ satisfy (5.4.14) with the same $z(t,x)$, which proves

**Theorem 5.5.3.** *Assume $\bar\mu(t)$, $\mu(t)$ both drive $\zeta$ to a target $\bar y \in \overline{D(A_\Sigma)}$ time optimally or norm optimally (the latter in the same interval $0 \le t \le T$). Then, if $\bar\mu(t)$ satisfies (5.4.9) $\Leftrightarrow$ (5.4.14) for a costate $z(t,x)$, $\mu(t)$ satisfies (5.4.9) $\Leftrightarrow$ (5.4.14) with the same costate $z(t,x)$. In particular,*

$$\operatorname{supp}\mu(t) \subseteq \mathcal{M}(t,z)\,, \quad \operatorname{supp}\bar\mu(t) \subseteq \mathcal{M}(t,z)\,.$$

As we shall see in Theorem 5.5.5, Theorem 5.5.3 does *not* imply uniqueness of optimal controls, thus the results in **4.9** on strong convergence of sequences of suboptimal controls have no counterpart here. However, (5.5.9) and (5.5.10) yield results on continuous dependence of supports, of which the following is an example.

Assume $\bar\mu(\cdot)$ drives an initial condition $\zeta$ to a target $\bar y \in D(A_\Sigma)$ time or norm optimally in $0 \le t \le T$ (in the latter case we take $\rho = 1$ for simplicity). Let $\{\mu_n(t)\}$ be a sequence of controls with $\|\mu_n(\cdot)\|_{L^\infty_w(0,T;\Sigma(\overline{\Omega}))} \le 1$ driving an initial condition $\zeta_n$ to another target $y_n = y(T,\zeta,\mu_n)$. Using (5.5.10) and (5.5.9) we obtain

$$\begin{aligned}
0 &\le \int_0^T \phi(t)dt = \int_0^T \left(\|z(t,\cdot)\|_{C(\overline{\Omega})} - \int_{\overline{\Omega}} z(t,x)\mu_n(t,dx)\right)dt \\
&= \int_0^T \int_{\overline{\Omega}} z(t,x)(\bar\mu(t,dx) - \mu_n(t,dx))dt \\
&\le \|z\|_{Z(T)}\,\|(\bar y - S_\Sigma(T)\zeta) - (y_n - S_\Sigma(T)\zeta_n)\|_{R^\infty_w(T)} \\
&\le C\|z\|_{Z(T)}\,\|(\bar y - S_\Sigma(T)\zeta) - (y_n - S_\Sigma(T)\zeta_n)\|_{D(A_\Sigma)} = \delta_n\,, \qquad (5.5.11)
\end{aligned}$$

where the constant $C$ comes from the first imbedding (5.1.20). Now, (5.5.11), the fact that $\phi(t) \geq 0$ and and Chebyshev's inequality imply

$$\big|\{t \in [0,T]; \phi(t) \geq \sqrt{\delta_n}\}\big| \leq \frac{1}{\sqrt{\delta_n}}\,\delta_n = \sqrt{\delta_n}\,.$$

Equivalently, there exists a set $e_n \subseteq [0,T]$ whose complement $e_n^c$ is of measure $|e_n^c| \leq \sqrt{\delta_n}$ and such that

$$0 \leq \|z(t,\cdot)\|_{C(\overline{\Omega})} - \int_{\overline{\Omega}} z(t,x)\mu_n(t,dx) \leq \sqrt{\delta_n} \quad (t \in e_n)\,,$$

thus

$$\int_{\overline{\Omega}} z(t,x)\mu_n(t,dx) \geq \|z(t,\cdot)\|_{C(\overline{\Omega})} - \sqrt{\delta_n} \quad (t \in e = \cap e_n)\,. \tag{5.5.12}$$

Assume now that $\delta_n \to 0$ in (5.5.11), which is certainly the case if $\|\zeta_n - \zeta\|_{\Sigma(\overline{\Omega})} \to 0$, $\|y_n - \bar{y}\|_{D(A_\Sigma)} \to 0$. Given $\delta > 0$ select a subsequence of $\{\delta_n\}$ (named in the same way) such that

$$\sum_{n=1}^{\infty} \sqrt{\delta_n} \leq \delta\,,$$

and call $e = e(\delta)$ the set[5] $\cap_n e_n$ in (5.5.12). We have

$$|e(\delta)^c| = \Big|\Big(\bigcap_{n=1}^{\infty} e_n\Big)^c\Big| = \Big|\bigcup_{n=1}^{\infty} e_n^c\Big| \leq \sum_{n=1}^{\infty} \sqrt{\delta_n} \leq \delta\,. \tag{5.5.13}$$

Given $\epsilon > 0$ define

$$\mathcal{M}(t,\epsilon,z) = \big\{x \in \overline{\Omega};\, |z(t,x)| > \|z(t,\cdot)\|_{C(\overline{\Omega})} - \epsilon\big\}\,. \tag{5.5.14}$$

Combining with (5.5.12) we obtain

$$\begin{aligned}
\|z(t,\cdot)\|_{C(\overline{\Omega})} - \sqrt{\delta_n} &\leq \int_{\overline{\Omega}} z(t,x)\mu_n(t,dx) \\
&\leq \int_{\overline{\Omega}\setminus\mathcal{M}(t,\epsilon,z)} |z(t,x)||\mu_n|(t,dx) + \int_{\mathcal{M}(t,\epsilon,z)} |z(t,x)||\mu_n|(t,dx) \\
&\leq \big(\|z(t,\cdot)\|_{C(\overline{\Omega})} - \epsilon\big)|\mu_n|(t,\overline{\Omega}\setminus\mathcal{M}(t,\epsilon,z)) + \|z(t,\cdot)\|_{C(\overline{\Omega})}|\mu_n|(t,\mathcal{M}(t,\epsilon,z)) \\
&= \|z(t,\cdot)\|_{C(\overline{\Omega})}|\mu_n|(t,\overline{\Omega}) - \epsilon|\mu_n|(t,\overline{\Omega}\setminus S_n(t,\epsilon,z)) \\
&\leq \|z(t,\cdot)\|_{C(\overline{\Omega})} - \epsilon|\mu_n|(t,\overline{\Omega}\setminus S_n(t,\epsilon,z))\,,
\end{aligned} \tag{5.5.15}$$

[5] Strictly speaking, the set $e(\delta)$ depends not only on $\delta$ but on the subsequence $\{\delta_n\}$, but this doesn't affect the argument that follows.

since $\|\mu_n(\cdot)\|_{L^\infty_w(0,T;\Sigma(\overline{\Omega}))} \le 1$ implies $\|\mu_n(t)\|_{\Sigma(\overline{\Omega})} = |\mu_n|(t,\overline{\Omega}) \le 1$ a. e. Hence

$$|\mu_n|(t,\overline{\Omega}\setminus \mathcal{M}(t,\epsilon,z)) \le \frac{\sqrt{\delta_n}}{\epsilon} \quad (t \in e(\delta))\,. \tag{5.5.16}$$

**Theorem 5.5.4.** *Assume that* $\delta_n \to 0$ *in* (5.5.11) *Then there exists a set* $e$ *of full measure in* $[0,T]$ *such that*

$$\lim_{n\to\infty} |\mu_n|(t,\overline{\Omega}\setminus \mathcal{M}(t,\epsilon,z)) = 0 \quad (t \in e)\,. \tag{5.5.17}$$

*for all* $\epsilon > 0$.

*Proof.* In case the conclusion of Theorem 5.5.4 fails there exists $\epsilon > 0$, a set $d \subseteq [0,T]$ of positive measure and a subsequence such that

$$|\mu_n|(t,\overline{\Omega}\setminus \mathcal{M}(t,\epsilon,z)) \ge \alpha > 0 \quad (t \in d,\ n = 1,2,\dots)\,. \tag{5.5.18}$$

We take $\delta < |d|$ and construct a subsequence of this subsequence such that (5.5.16) holds. Since, by (5.5.13) we have $|e(\delta)| \ge T-\delta$, it follows that the intersection $e(\delta)\cup d$ has positive measure. In this intersection (5.5.16) and (5.5.18) hold at the same time, a contradiction which completes the proof of the theorem. Note that (5.5.17) says that, for any $\epsilon > 0$, $\mathcal{M}(t,\epsilon,z)$ is an *attractor* for the support of the measure $\mu(t,dx)$; the measure $|\mu_n|$ of the portion of the support not contained in $\mathcal{M}(t,\epsilon,z)$ tends to zero.

**Theorem 5.5.5.** *Given* $T > 0$ *there exists a target* $\bar{y} \in L^1(\Omega)$ *such that* $0$ *can be driven to* $\bar{y}$ *norm optimally in the interval* $0 \le t \le T$ *by different controls.*

The proof is the exact counterpart of Theorem 4.8.5 and we leave it to the reader. The control constructed here does not satisfy (5.5.8). In contrast with Theorem 4.8.5, which gives the only known example of nonuniqueness in the $L^\infty(\Omega)$ setting (and only for the norm optimal problem) the interest of Theorem 5.5.5 is mitigated by Theorem 5.5.6 below, which addresses also the time optimal case and doesn't mind condition (5.5.8) on the target. This result refers to the one dimensional controlled diffusion equation

$$\frac{\partial y(t,x)}{\partial t} = \frac{\partial^2 y(t,x)}{\partial x^2} + \mu(t)\,, \quad y(t,0) = y(t,\pi) = 0\,. \tag{5.5.19}$$

The reverse adjoint equation is

$$\frac{\partial z(t,x)}{\partial t} = -\frac{\partial^2 z(t,x)}{\partial x^2}\,, \quad z(t,0) = z(t,\pi) = 0\,. \tag{5.5.20}$$

**Theorem 5.5.6.** *There exists a target* $\bar{y} \in D(A_\Sigma) \subset L^1(0,\pi)$ *such that* $0$ *can be driven to* $\bar{y}$ *time and norm optimally by different controls.*

*Proof.* One of the controls doing the driving is of the form

$$\mu(t,dx) = \sum_{j=1}^{n} \alpha_j(t)\delta(x-x_j)dx \tag{5.5.21}$$

with $0 < x_1 < \ldots < x_n < \pi$. We write the solution $y(t, x, 0, \mu)$ of (5.5.19) with initial condition $y(0, x, 0, \mu) = 0$ as a Fourier series,

$$y(t, x, 0, \mu) = \sum_{n=1}^{\infty} a_n(t) \sin nx \,. \tag{5.5.22}$$

The Fourier coefficients of (5.5.21) are

$$\begin{aligned} c_n(t) &= \frac{2}{\pi} \int_0^{\pi} \mu(t, dx) \sin nx \\ &= \frac{2}{\pi} \sum_{j=1}^{n} \alpha_j(t) \int_0^{\pi} \delta(x - x_j) \sin nx dx = \frac{2}{\pi} \sum_{j=1}^{n} \alpha_j(t) \sin nx_j \,, \end{aligned}$$

so that the $a_n(t)$ solve

$$a_n'(t) = -n^2 a_n(t) + \frac{2}{\pi} \sum_{j=1}^{n} \alpha_j(t) \sin nx_j \,, \quad a_n(0) = 0 \,.$$

This gives

$$a_n(t) = \frac{2}{\pi} \sum_{j=1}^{n} \sin nx_j \int_0^t e^{-n^2(t-\sigma)} \alpha_j(\sigma) d\sigma \,. \tag{5.5.23}$$

The construction of the nonuniquess example announced in Theorem 5.5.6 is based on two independent facts: $(a)$ if $\nu(t, dx)$ is of the form (5.5.21),

$$\nu(t, dx) = \sum_{j=1}^{n} \beta_j(t) \delta(x - x_j) dx \,,$$

and

$$\int_0^T e^{-n^2(T-\sigma)} \alpha_j(\sigma) d\sigma = \int_0^T e^{-n^2(T-\sigma)} \beta_j(\sigma) d\sigma \,, \tag{5.5.24}$$

then it follows from (5.5.22)-(5.5.23) that $y(T, x, 0, \mu) = y(T, x, 0, \nu)$, $(b)$ if the $x_j$, the $\alpha_j(t)$ and the $\beta_j(t)$ are adequately chosen, both $\mu(t, dx)$ and $\nu(t, dx)$ will be (norm, time) optimal controls. The two controls will be different if $\beta_j(\cdot) \neq \alpha_j(\cdot)$ for some $j$.

In order to achieve (5.5.24), we need an auxiliary result. Let $\Lambda = \{\lambda_1, \lambda_2, \ldots\}$ be an increasing sequence of positive numbers and $\mathcal{L}^2(0, T, \Lambda)$ the closed subspace of $L^2(0, T)$ generated by $\{1, e^{-\lambda_1 t}, e^{-\lambda_2 t}, \ldots\}$. This theorem (a variant of Müntz's theorem, Kaczmarz - Steinhaus [1935, p. 86]) is in Schwartz [1959, p. 54].

**Theorem 5.5.7.** *Assume*

$$\sum_{n=1}^{\infty} \frac{1}{\lambda_n} < \infty \,.$$

*Then* $\mathcal{L}(0, T, \Lambda) \neq L^2(0, T)$.

**Corollary 5.5.8.** *Under the assumptions of Theorem* 5.5.7 *there exists a function* $\eta(\cdot) \in L^\infty(0, T)$, $\eta(\cdot)$ *not identically zero and such that*

$$\int_0^T e^{-\lambda_n t}\eta(t)dt = 0 \quad (n = 1, 2, \ldots).$$

*Proof.* By Theorem 5.5.7 there exists $h(\cdot) \in L^2(0, T)$ orthogonal to every function in the set $\{1, e^{-\lambda_1 t}, e^{-\lambda_2 t}, \ldots\}$. Define

$$\eta(t) = \int_0^t h(\sigma)d\sigma\,,$$

so that $\eta(\cdot) \in H^1(0, T) \subset L^\infty(0, T)$. We have $\eta(0) = 0$ and, by orthogonality to 1, $\eta(T) = 0$. Accordingly,

$$\int_0^T e^{-\lambda_n \sigma}\eta(\sigma)d\sigma = -\frac{e^{-\lambda_n \sigma}}{\lambda_n}\eta(\sigma)\Big|_{\sigma=0}^{\sigma=T} + \frac{1}{\lambda_n}\int_0^T e^{-\lambda_n \sigma}h(\sigma)d\sigma = 0$$

for $n = 1, 2, \ldots$. This ends the proof.

**Lemma 5.5.9.** *Let* $\mu \in \Sigma(\overline{\Omega})$. *Then*

$$\int_0^T S_\Sigma(T - \sigma)\mu d\sigma = \int_0^T S_\Sigma(\sigma)\mu d\sigma \in D(A_\Sigma)\,.$$

*Proof.* Let $y \in D(A'_c)$. We have

$$\Big\langle \int_0^T S_\Sigma(\sigma)\mu d\sigma, A'_c y\Big\rangle = \int_0^T \langle S_\Sigma(\sigma)\mu, A'_c y\rangle d\sigma$$
$$= \int_0^T \langle \mu, S'_c(\sigma)A'_c y\rangle d\sigma = \Big\langle \mu, \int_0^T S'_c(\sigma)A'_c y d\sigma\Big\rangle = \langle \mu, S'_c(T)y - y\rangle\,,$$

thus the first angled bracket is a bounded functional of $y$ in the $C(\overline{\Omega})$ norm. This shows that the integral in the first angled bracket belongs to $D(A_\Sigma)$ and completes the proof of Lemma 5.5.9.

*End of proof of Theorem* 5.5.6. We use Theorem 5.4.8 to calibrate (5.5.21) in such a way that it becomes an optimal control. The function

$$z_n(t, x) = e^{n^2 t}\sin nx \tag{5.5.25}$$

satisfies (5.4.7) for every $n$; the summability condition (5.4.15) is automatic as $z(t, x)$ is everywhere smooth. For $n = 3$ we have

$$\mathcal{M}(t, z) = \Big\{\frac{\pi}{6}, \frac{3\pi}{6}, \frac{5\pi}{6}\Big\}. \tag{5.5.26}$$

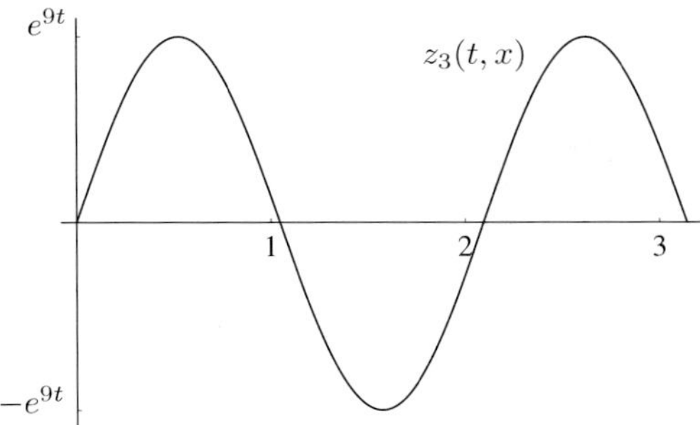


Figure 5.5.1

According to (5.4.14), every control of the form

$$\mu(t,dx) = \alpha_1(t)\delta\Big(x - \frac{\pi}{6}\Big)dx - \alpha_2(t)\delta\Big(x - \frac{3\pi}{6}\Big)dx + \alpha_3(t)\delta\Big(x - \frac{5\pi}{6}\Big)dx \quad (5.5.27)$$

where $\alpha_1(t), \alpha_2(t), \alpha_3(t)$ are arbitrary measurable functions satisfying

$$\alpha_1(t),\, \alpha_2(t),\, \alpha_3(t) \ge 0\,, \quad \alpha_1(t) + \alpha_2(t) + \alpha_3(t) = 1 \quad \text{a. e. in } \ 0 \le t \le T\,,$$

drives 0 to $y(T,x,0,\mu)$ norm and time optimally. Let $\eta(\cdot)$ be the function in Corollary 5.5.8; assume (as we may) that $\|\eta(\cdot)\|_{L^\infty(0,T)} = 1$. Taking

$$\alpha_1(t,\gamma) = \frac{1}{2} + \gamma\eta(t)\,, \quad \alpha_2(t,\gamma) = 0\,, \quad \alpha_3(t,\gamma) = \frac{1}{2} - \gamma\eta(t)\,,$$

we obtain the family of controls

$$\mu_\gamma(t,dx) = \Big(\frac{1}{2} + \gamma\eta(t)\Big)\delta\Big(x - \frac{\pi}{6}\Big)dx + \Big(\frac{1}{2} - \gamma\eta(t)\Big)\delta\Big(x - \frac{5\pi}{6}\Big)dx\,,$$

all of which are time and norm optimal for $-1/2 \le \gamma \le 1/2$. On the other hand,

$$\int_0^T e^{-n^2(T-\sigma)}\alpha_1(\sigma,\gamma)dt = \int_0^T e^{-n^2(T-\sigma)}\alpha_3(\sigma,\gamma)dt = \frac{T}{2}\,,$$

$$\int_0^T e^{-n^2(T-\sigma)}\alpha_2(\sigma,\gamma)dt = 0 \quad \Big(n = 1, 2, \ldots,\ -\frac{1}{2} \le \gamma \le \frac{1}{2}\Big)\,,$$

thus

$$y(T,x,0,\mu_\gamma) = y(T,x,0,\mu_{\gamma'})$$

for $-1/2 \le \gamma, \gamma' \le 1/2$ (see 5.5.24 and preceding comments). We have constructed the nonuniqueness example and the proof of Theorem 5.5.6 is finished.

The following result is the companion of Corollary 4.9.8 and refers to the space $L^1(0, \pi)$ and the semigroup $S_1(t)$ corresponding to the equation (5.5.19). We can't, however, get this result out of the way by stating cavalierly that "the proof is similar to that of Corollary 4.9.8" since the latter result uses uniqueness of optimal controls, which has just been overthrown in Theorem 5.5.6.

**Theorem 5.5.10.** *There exists $y \in R^\infty(T)$ such that*

$$y = \int_0^T S_1(T-\sigma)u(\sigma)d\sigma\,, \quad u(\cdot) \in L^\infty(0, T; L^1(\Omega))$$

*implies*

$$\|u(\cdot)\|_{L^\infty(0,T;L^1(\Omega))} > \|y\|_{R^\infty(T)}.$$

*Proof.* We have proved in Corollary 4.9.8 that if $A = d^2/dx^2$ in $\Omega = (0, \pi)$ with Dirichlet boundary conditions, then the $\mathcal{L}(L^\infty(\Omega), L^\infty(\Omega))$ norm of the semigroup $S_\infty(t) = S'_\infty(t)$ is bounded by 1 for $t \geq 0$. Since $S'_\infty(t)$ is an extension of $S'_c(t)$,

$$\|S'_c(t)\|_{\mathcal{L}(C_0(\overline{\Omega}), C_0(\overline{\Omega}))} \leq 1 \quad (t \geq 0)\,,$$

and taking adjoints,

$$\|S_\Sigma(t)\|_{\mathcal{L}(\Sigma_0(\overline{\Omega}), \Sigma_0(\overline{\Omega}))} \leq 1 \quad (t \geq 0)\,.$$

Theorem 5.1.6 says

$$R^\infty_w(T) = R^\infty(T)\,, \quad \|y\|_{R^\infty_w(T)} = \|y\|_{R^\infty(T)}\,. \tag{5.5.28}$$

Define a control in $L^\infty_w(0, T; \Sigma_0(\overline{\Omega}))$ by $\bar{\mu}(t, dx) = \bar{\mu}(dx) = \delta(x - \pi/2)dx$. This control satisfies (5.4.14) with the costate $z_1(t, x)$ in (5.5.25), thus is norm optimal. In view of the first equality (5.5.28) there exists $u(\cdot) \in L^\infty(0, T; L^1(\Omega))$ such that

$$\begin{aligned} y &= \int_0^T S_\Sigma(T-\sigma)\bar{\mu}(\sigma)d\sigma \\ &= \int_0^T S_1(T-\sigma)u(\sigma)d\sigma = \int_0^T S_\Sigma(T-\sigma)u(\sigma)d\sigma\,. \end{aligned} \tag{5.5.29}$$

Assume we can pick $u(\cdot) \in L^\infty(0, T; L^1(\Omega))$ with $\|u(\cdot)\|_{L^\infty(0,T;L^1(\Omega))} = \|y\|_{R^\infty(T)}$. Then, using (5.5.28) we have

$$\begin{aligned} \|u(\cdot)\|_{L^\infty(0,T;\Sigma_0(\overline{\Omega}))} &= \|u(\cdot)\|_{L^\infty(0,T;L^1(\Omega))} \\ &= \|y\|_{R^\infty(T)} = \|y\|_{R^\infty_w(T)} = \|\bar{\mu}(\cdot)\|_{L^\infty_w(0,T;\Sigma_0(\Omega))}\,, \end{aligned}$$

so that $u(\cdot)$ is norm optimal as well. In view of Lemma 5.5.9, the element $y$ defined in (5.5.29) belongs to $D(A_\Sigma)$, thus Theorem 5.5.3 applies to show that

$\operatorname{supp} u(t) \subseteq \mathcal{M}(t, z_1) = \{\pi/2\}$ which implies that $u(t) = 0$ for all $t$ as an element of $L^1(\Omega)$, absurd. This ends the proof of Theorem 5.5.10.

**Remark 5.5.11.** The family of costates (5.5.25) gives numerous examples of optimal controls can be associated with different costates, that is, that satisfy (5.4.14) for different $z(t, x)$. To this end, note that, since

$$\sin\frac{3\pi}{2} = \sin\frac{7\pi}{2} = \sin\frac{11\pi}{2} = \ldots = -1$$

then[6]

$$\frac{\pi}{2} \in \mathcal{M}^-(t, z_{3+4n}) \quad (0 \le t \le T,\ n = 1, 2, \ldots),$$

which implies that the control

$$\bar{\mu}(t, dx) = -\delta\Big(x - \frac{\pi}{2}\Big)dx \tag{5.5.30}$$

satisfies (5.4.17) for all costates $z_{3+4n}(t, x)$. These are not all; we also have

$$-\sin\frac{\pi}{2} = -\sin\frac{5\pi}{2} = -\sin\frac{9\pi}{2} = -1\,,$$

thus (5.5.30) is also associated with all costates $-z_{1+4n}(t, x)$.

**Remark 5.5.12.** Theorem 5.5.3 can used to construct special controls $\bar{\mu}(t)$ that drive an initial condition $\zeta$ optimally to a target $\bar{y} = y(T, x, \zeta, \bar{\mu})$ and have the uniqueness property; they are the only ones that drive $\zeta$ to $y(T, x, \zeta, \mu)$ optimally (in contrast with the example in Theorem 5.5.6). One such example is the control

$$\bar{\mu}(t, dx) = \delta\Big(x - \frac{\pi}{2}\Big)dx \tag{5.5.31}$$

associated with the costate

$$z_1(t, x) = e^t \sin x\,.$$

In view of Lemma 5.5.9, $y(T, x, 0, \bar{\mu}) \in D(A_\Sigma)$, which makes possible application of Theorem 5.5.3. If $\mu(t)$ is any other control driving 0 to $y(T, x, 0, \bar{\mu})$, it must satisfy (5.4.14) with the same costate $z_1(t, x)$, thus it must coincide with (5.5.31). What makes this argument possible is the existence of a costate $z(t, x)$ associated with $\bar{\mu}(t)$ and such that

$$\mathcal{M}(t, z) = \operatorname{supp} \bar{\mu}(t) \quad (0 \le t \le T). \tag{5.5.32}$$

Surprisingly, uniqueness can also be shown for certain controls associated with a costate $z(t, x)$ such that the inclusion $\mu(t) \subset \mathcal{M}(t, z)$ is strict for all $t$. For an example, consider

$$\bar{\mu}(t, dx) = \delta\Big(x - \frac{\pi}{6}\Big)dx\,, \tag{5.5.33}$$

[6] The $+$ and $-$ signs on $\mathcal{M}(t, z)$ have the same meaning as those on $\mathcal{M}(z)$.

a member of the family (5.5.27) corresponding to $\alpha_1(t) = 1$, $\alpha_2(t) = \alpha_3(t) = 0$ associated with the costate $z_3(t, x)$. Again, we deduce from Lemma 5.5.9 that $y(T, x, 0, \mu) \in D(A_\Sigma)$. By virtue of Theorem 5.5.3, if $\mu(t)$ is another control in $L^\infty_w(0, T; \Sigma(\overline{\Omega}))$ with $\|\mu(t)\|_{L^\infty_w(0,T;\Sigma(\overline{\Omega}))} \leq 1$ driving 0 to $y(T, x, 0, \bar{\mu})$, it has to satisfy (5.4.14) with the same costate $z_3(t, x)$. This means, $\mu(t, dx)$ itself must be a member of the family (5.5.27). The condition

$$y(T, x, 0, \bar{\mu}) = y(T, x, 0, \mu)$$

implies that the Fourier series (5.5.22)-(5.5.23) of $y(T, x, 0, \bar{\mu})$ and of $y(T, x, 0, \mu)$ must be the same, hence

$$\begin{aligned}\int_0^T & e^{-n^2(T-\sigma)}\Big(\sin\frac{n\pi}{6}\,\alpha_1(\sigma) + \sin\frac{3n\pi}{6}\,\alpha_2(\sigma) + \sin\frac{5n\pi}{6}\,\alpha_3(\sigma)\Big)d\sigma \\ &= \int_0^T e^{-n^2(T-\sigma)}\sin\frac{n\pi}{6}d\sigma \quad (n = 1, 2, \ldots)\,. \qquad (5.5.34)\end{aligned}$$

For $n = 2$ we have

$$\sin\frac{2\pi}{6} = \frac{\sqrt{3}}{2}\,, \quad \sin\frac{6\pi}{6} = 0\,, \quad \sin\frac{10\pi}{6} = -\frac{\sqrt{3}}{2}\,.$$

Unless $\alpha_1(t) = 1$, $\alpha_2(t) = \alpha_3(t) = 0$ a. e. we will have

$$\sin\frac{2\pi}{6}\,\alpha_1(\sigma) + \sin\frac{6\pi}{6}\,\alpha_2(\sigma) + \sin\frac{10\pi}{6}\,\alpha_3(\sigma) \leq \sin\frac{2\pi}{6}$$

in a set of positive measure, which preempts (5.5.34) for $n = 2$ and shows the uniqueness of $\bar{\mu}(t)$. Uniqueness could be proved in an easier fashion if we had a costate satisfying (5.5.32), but we don't know if such a costate exists.

**Remark 5.5.13.** In applications of Theorem 5.5.3 and its "approximate" partner, Theorem 5.5.4, it is our best interest to use the "minimal" costate $z(t, x)$, that is, the one with the smallest $\mathcal{M}(t, z)$ in case there is a choice. For instance, the control (5.5.31) is associated with all costates $z_{1+4n}(t, x)$ $(n = 0, 1, \ldots)$ but the uniqueness argument in Remark 5.5.12 does not work except with $z_1(t, x)$. As for Theorem 5.5.4, its conclusion is stronger with $z_1(t, x)$ than, say, with $z_5(t, x)$, since

$$\mathcal{M}(t, \epsilon, z_1) = \{x; |\sin x| > 1 - \epsilon\} = \Big(\frac{\pi}{2} - \gamma_1(\epsilon)\,, \frac{\pi}{2} + \gamma_1(\epsilon)\Big)$$

while $\mathcal{M}(t, \epsilon, z_5)$ is the union of five intervals centered at the elements of

$$\mathcal{M}(t, z_5) = \Big\{\frac{\pi}{10}\,, \frac{3\pi}{10}\,, \frac{5\pi}{10}\,, \frac{7\pi}{10}\,, \frac{9\pi}{10}\Big\}\,.$$

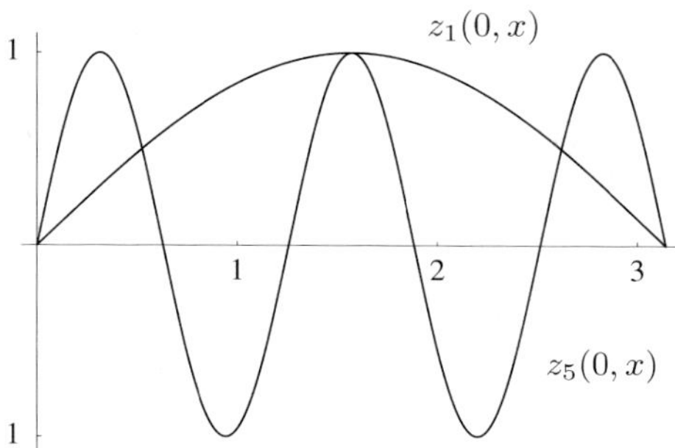


Figure 5.5.2

The optimal costate would be one satisfying (5.5.32), but as pointed out in Remark 5.5.12, we don't know if such a costate exists. For the control (5.5.33), application of Theorem 5.5.4 with the costate $z_3(t, x)$ produces

$$\lim_{n\to\infty} |\mu_n|([0, \pi] \setminus \mathcal{M}(t, \epsilon, z_3)) = 0 \quad (t \in e)\,, \tag{5.5.35}$$

with $e$ a set of full measure in $[0, T]$ and $\mathcal{M}(t, \epsilon, z_3)$ the union of three intervals centered at the three elements of the set (5.5.26). Uniqueness of the control (5.5.33) leads to the suspicion that the limit relation (5.5.35) will remain true if $\mathcal{M}(t, \epsilon, z_3)$ is replaced by the first interval, but we don't have a proof.

**Miscellaneous notes.** The material in this section is mostly taken from the author [2002:1]. Some results (for instance Theorem 5.5.6) are new; at the time [2002:1] was written, uniqueness of optimal controls was in question.

**5.6. Examples and applications.** Example 5.6.1 and the results after it are on the one dimensional controlled diffusion equation

$$\frac{\partial y(t,x)}{\partial t} = \frac{\partial^2 y(t,x)}{\partial x^2} + \mu(t)\,, \quad y(t,0) = y(t,\pi) = 0\,. \tag{5.6.1}$$

The reverse adjoint equation is

$$\frac{\partial z(t,x)}{\partial t} = -\frac{\partial^2 z(t,x)}{\partial x^2}\,, \quad z(t,0) = z(t,\pi) = 0\,, \tag{5.6.2}$$

and we use Theorems 5.4.7 and 5.4.8 to construct (time, norm) optimal controls of the form

$$\begin{aligned} &\operatorname{supp}\bar\mu(t) \subseteq \mathcal{M}(t,z) = \{x \in [0,\pi];\ |z(t,x)| = \|z(t,\cdot)\|_{C_0[0,\pi]}\}\,, \\ &z(t,x)\bar\mu(t,dx) \ge 0\,, \qquad \|\bar\mu\|_{\Sigma_0[0,\pi]} = \int_{\mathcal{M}(t,z)} |\bar\mu|(t,dx) = \rho\,, \end{aligned} \tag{5.6.3}$$

where $z(t,x)$ is a solution of (5.6.2) satisfying

$$\int_0^T \|z(t,\cdot)\|_{C_0[0,\pi]}dt = \int_0^T \Big(\max_{x\in[0,\pi]} |z(t,x)|\Big)dt < \infty\,. \tag{5.6.4}$$

To simplify we take $\rho = 1$. Also, as in **4.9** we agree that "a control $\bar{\mu}(t)$ is norm optimal in the interval $0 \le t \le T$" means that it drives any initial condition $\zeta$ to $y(T,\zeta,\bar{u})$ norm optimally; that "it is time optimal" means that it drives any initial condition $\zeta$ to $y(T,\zeta,\bar{u})$ in optimal time $T$ under the conditions (5.4.17) in Theorem 5.4.8 on the initial condition and the target.

**Example 5.6.1.** Figure 5.6.1 below shows the set $\mathcal{M}(t,z) = \mathcal{M}^+(t,z)$ (thick curve) and the set $\mathcal{N}(t,\nabla z) = \mathcal{N}(t, dz/dx)$ for the costate

$$z(t,x) = e^{t-6}\sin x + e^{9t-30}\sin 3x \tag{5.6.5}$$

in the rectangle $[0,6]\times[0,\pi]$.

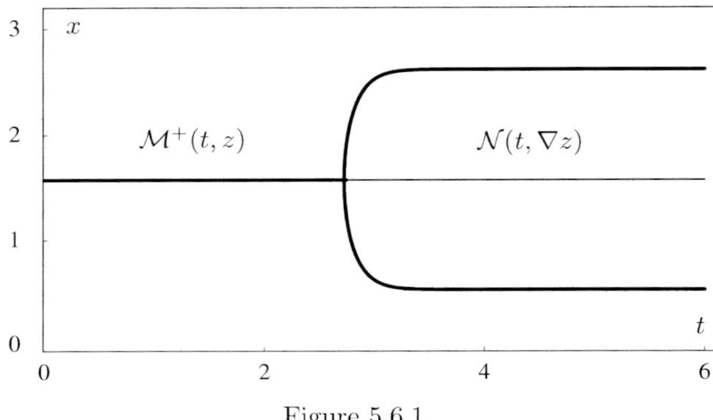


Figure 5.6.1

We have $z(t,x) \ge 0$ in $0 \le t \le 3$. For each $t$, the center line $x = \pi/2$ gives the only maximum of $z(t,x)$ in $0 \le x \le \pi$ until

$$t_0 = 2.72534\ldots$$

where the maximum bifurcates into two equal maxima (the upper and lower curves in Figure 5.6.1). After the bifurcation point the center line is a local minimum which becomes zero at $t = 3$. It becomes negative (and a global minimum) for $t > 3$; however, it plays no role in (5.6.3) as

$$\begin{aligned}\Big|z\Big(t,\frac{\pi}{2}\Big)\Big| &= \Big|e^{t-6}\sin\frac{\pi}{2} + e^{9t-30}\sin\frac{3\pi}{2}\Big| = |e^{t-6} - e^{9t-30}| \\ &= e^{9t-30} - e^{t-6} < e^{9t-30} + \frac{1}{2}e^{t-6} \\ &= e^{t-6}\sin\frac{\pi}{6} + e^{9t-30}\sin\frac{3\pi}{6} \le \|z(t,\cdot)\|_{C_0[0,\pi]} \quad (3 \le t \le 6)\,.\end{aligned}$$

The bifurcation of the maximum and the appearance of the negative minimum are seen in Figure 5.6.2 below:

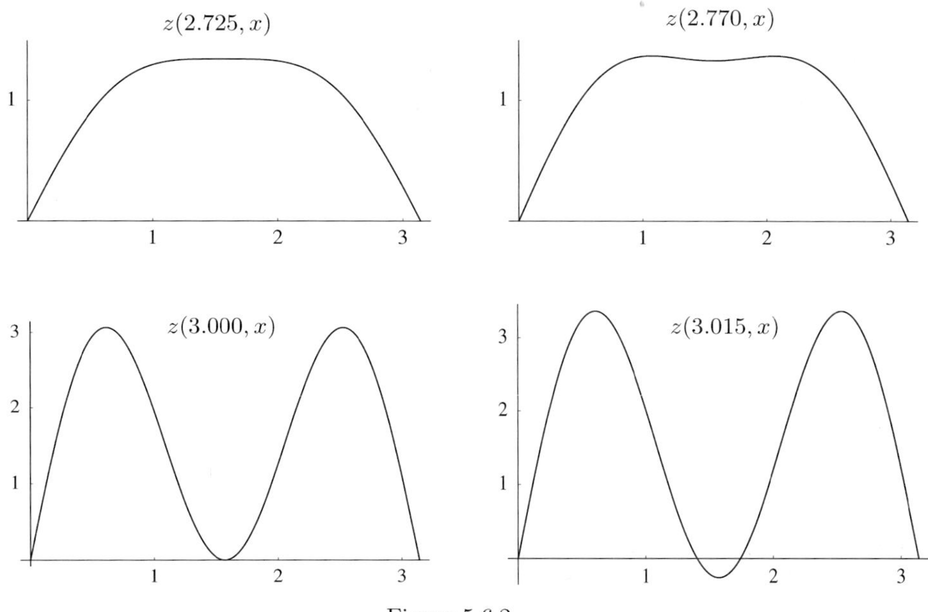


Figure 5.6.2

The optimal controls associated with this costate are

$$\mu(t,dx) = \begin{cases} \delta\Big(x - \dfrac{\pi}{2}\Big)dx & (0 \le t \le t_0) \\ \alpha_1(t)\delta(x - x(t))dx + \alpha_2(t)\delta(x + x(t))dx & (t_0 < t \le 6) \end{cases}$$

where $t \to \pm x(t)$ are the equations of the upper and lower curves and the two measurable functions $\alpha_1(t)$, $\alpha_2(t)$ satisfy

$$\alpha_1(t),\ \alpha_2(t) \ge 0\,, \quad \alpha_1(t) + \alpha_2(t) = 1 \quad \text{a. e. in } t_0 \le t \le 6\,.$$

The following result attempts to emulate (partially) Theorem 4.9.5. We try to characterize optimal controls $\bar{\mu}(t)$ for (5.6.1) whose support (is contained in a finite set that) does not depend on $t$. The controls themselves may depend on $t$.

**Theorem 5.6.2.** *Let*

$$\bar{\mu}(t,dx) = \sum_{j=1}^{n} (-1)^{j-1}\alpha_j(t)\delta\Big(x - \frac{(2j-1)\pi}{2n}\Big)dx \tag{5.6.6}$$

*with the* $\alpha_j(t)$ *measurable and*

$$\alpha_1(t), \ldots, \alpha_n(t) \geq 0\,, \quad \sum_{j=1}^{n} \alpha_j(t) = 1 \quad a.\ e. \tag{5.6.7}$$

*Then* $\bar{\mu}(t)$ *is norm and time optimal. Conversely, let* $\bar{\mu}(t)$ *be an optimal control such that there exists a finite set*

$$\mathcal{M} = \{x_1, x_2, \ldots, x_m\}$$

*with* $supp\,\bar{\mu}(t) = \mathcal{M} \subset [0, \pi]$ *a. e. in* $0 \leq t \leq T$,

$$\big|\{t;\, \mathrm{supp}\,\bar{\mu}(t) \cap \{x_j\} \neq \emptyset\}\big| > 0 \quad (j = 1, 2, \ldots, m)\,, \tag{5.6.8}$$

*and*

$$\int_0^T S_\Sigma(T - \sigma)\bar{\mu}(\sigma)d\sigma \in D(A_\Sigma)\,. \tag{5.6.9}$$

*Then* $\bar{\mu}(t)$ *is of the form* (5.6.6)-(5.6.7).

*Proof.* The function

$$z_n(t, x) = e^{n^2 t} \sin nx \tag{5.6.10}$$

solves the reverse adjoint equation (5.6.2), and the growth condition (5.6.4) is automatic. For every $t$ we have

$$\mathcal{M}(t, z_n) = \left\{\frac{\pi}{2n}\,, \frac{3\pi}{2n}\,, \ldots, \frac{(2n-1)\pi}{2n}\right\}, \tag{5.6.11}$$

or, more precisely,

$$\mathcal{M}^+(t, z_n) = \left\{\frac{\pi}{2n}\,, \frac{5\pi}{2n}\,, \ldots\right\}, \quad \mathcal{M}^-(t, z_n) = \left\{\frac{3\pi}{2n}\,, \frac{7\pi}{2n}\,, \ldots\right\}.$$

Accordingly, any control of the form (5.6.6)-(5.6.7) satisfies the maximum principle (5.6.3) and is thus optimal.

For the converse part, condition (5.6.9) allows us to apply Theorem 5.4.2, thus $\bar{\mu}(t)$ satisfies the maximum principle (5.6.3) for some nontrivial solution $z(t, x)$ of the reverse adjoint equation (5.6.2).[7] Condition (5.6.8) implies

$$|e_j| = \big|\{t;\, (t, x_j) \subseteq \mathcal{M}(t, z)\}\big| > 0 \quad (j = 1, 2, \ldots, m)\,, \tag{5.6.12}$$

and $\mathcal{M}(t, z) \subseteq \mathcal{N}(t, dz/dx)$, thus

$$\left.\frac{\partial z(t, x)}{\partial x}\right|_{x = x_j} = 0 \quad (t \in e_j,\ j = 1, 2, \ldots, m)\,. \tag{5.6.13}$$

[7] Satisfying the growth condition (5.6.4), although this is not germane to the argument.

If $\epsilon > 0$ then $z(t, x)$ is regular in $0 \le t \le T - \epsilon$ so that $z(t, x)$ is given by

$$z(t, x) = \sum_{n=1}^{\infty} e^{-n^2(T-\epsilon-t)} a_n(\epsilon) \sin nx \quad (0 \le t \le T - \epsilon) \tag{5.6.14}$$

where the coefficients $a_n(\epsilon)$ come from the expansion

$$z(T - \epsilon, x) = \sum_{n=1}^{\infty} a_n(\epsilon) \sin nx \,.$$

It follows from (5.6.14) that $z(t, x)$ is analytic in $0 \le t \le T - \epsilon$, thus in $0 \le t < T$. Accordingly, we obtain from (5.6.13) and the fact (stated in (5.6.12)) that the $e_j$ have positive measure that

$$\frac{\partial z(t, x)}{\partial x}\bigg|_{x=x_j} = 0 \quad (0 \le t \le T,\ j = 1, 2, \dots, m)\,. \tag{5.6.15}$$

Differentiating (5.6.14),

$$\frac{\partial z(t, x_j)}{\partial x} = \sum_{n=1}^{\infty} n^2 e^{-n^2(T-\epsilon-t)} a_n(\epsilon) \cos nx_j \tag{5.6.16}$$

in $0 \le t \le T$, and it follows from (5.6.15) and the uniqueness theorem for Dirichlet series (Bernstein [1933 p. 12]) that

$$\cos nx_j = 0 \quad \text{whenever } a_n(\epsilon) \ne 0\,,$$

so that, selecting an arbitrary $a_n(\epsilon) \ne 0$,

$$\mathcal{M} \subseteq \mathcal{M}(t, z_n)$$

with $\mathcal{M}(t, z_n)$ given by (5.6.11), thus we deduce that $\bar{\mu}(t)$ is of the form (5.6.6)-(5.6.7) and complete the proof.

**Remark 5.6.3.** Condition (5.6.8) is not a restriction on the set $\mathcal{M}$; is it is not verified for some $x_j \in \mathcal{M}$, $x_j$ may be discarded. Note also that Theorem 5.6.2 applies in particular to optimal controls with support independent of time.

**Example 5.6.4.** We do the one dimensional controlled diffusion equation with Neumann boundary conditions

$$\frac{\partial y(t, x)}{\partial t} = \frac{\partial^2 y(t, x)}{\partial x^2} + \mu(t)\,, \qquad \frac{\partial y(t, 0)}{\partial x} = \frac{\partial y(t, \pi)}{\partial x} = 0\,. \tag{5.6.17}$$

The reverse adjoint equation is

$$\frac{\partial z(t, x)}{\partial t} = -\frac{\partial^2 z(t, x)}{\partial x^2}\,, \qquad \frac{\partial z(t, 0)}{\partial x} = \frac{\partial z(t, \pi)}{\partial x} = 0\,. \tag{5.6.18}$$

Using the costate $z(t,x) = 1$ and Theorem 5.4.12 we deduce that every control $\bar{\mu}(\cdot) \in L^\infty_w(0,T;\Sigma[0,\pi])$ such that

$$\|\bar{\mu}(t)\|_{\Sigma[0,\pi]} = 1\,, \quad \mu(t) \geq 0 \quad \text{a. e. in } 0 \leq t \leq T \tag{5.6.19}$$

is both norm and time optimal; the costate $z(t,x) = -1$ produces the optimal controls

$$\|\bar{\mu}(t)\|_{\Sigma[0,\pi]} = 1\,, \quad \mu(t) \leq 0 \quad \text{a. e. in } 0 \leq t \leq T\,. \tag{5.6.20}$$

These two families do not exhaust all optimal controls. For instance, for the costate

$$z(t,x) = e^t \cos x$$

we have $\mathcal{M}(t,z) = \mathcal{N}(t,dz/dx) = \{0,\pi\}$ for every $t$, with $z(t,0) = \|z(t,\cdot)\|_{C[0,\pi]}$, $z(t,0) = -\|z(t,\cdot)\|_{C[0,\pi]}$ The associated family of optimal controls is

$$\mu(t,dx) = \alpha_1(t)\delta(x)dx - \alpha_2(t)\delta(x-\pi)dx$$

where the measurable functions $\alpha_1(t), \alpha_2(t) \geq 0$ satisfy $\alpha_1(t) + \alpha_2(t) = 1$ a. e. Unless $\alpha_1(t) = 0$ or $\alpha_2(t) = 0$ none of these controls belong to the families (5.6.19) or (5.6.20). More generally, any member of the family of controls

$$\bar{\mu}(t,dx) = \sum_{j=0}^{n} (-1)^j \alpha_j(t)\delta\Big(x - \frac{j\pi}{n}\Big)dx\,,$$
$$\alpha_1(t), \dots, \alpha_n(t) \geq 0\,, \quad \sum_{j=1}^{n} \alpha_j(t) = 1$$

is optimal; these controls are associated with the costate

$$z_n(t,x) = e^{n^2 t} \cos nx\,,$$

for which

$$\mathcal{M}(t,z_n) = \mathcal{N}(t,dz_n/dx) = \Big\{0, \frac{\pi}{n}, \frac{2\pi}{n}, \dots, \frac{(n-1)\pi}{n}, \pi\Big\}.$$

**Example 5.6.5.** We do the two dimensional controlled diffusion equation in the unit circle $\Omega \subset \mathbb{R}^2$ (boundary $\Gamma$) with Dirichlet boundary condition,

$$\frac{\partial y(t,x)}{\partial t} = \Delta y(t,x) + \mu(t) \;\; (x \in \Omega)\,, \quad y(t,x) = 0 \;\; (x \in \Gamma)\,. \tag{5.6.21}$$

The reverse adjoint equation is

$$\frac{\partial z(t,x)}{\partial t} = -\Delta z(t,x) \;\; (x \in \Omega)\,, \quad z(t,x) = 0 \;\; (x \in \Gamma)\,. \tag{5.6.22}$$

We use Theorems 5.4.7 and 5.4.8 to construct optimal controls of the form

$$\begin{aligned}
&\operatorname{supp} \bar\mu(t) \subseteq \mathcal{M}(t, z) = \big\{x \in \overline\Omega; |z(t, x)| = \|z(t, \cdot)\|_{C_0(\overline\Omega)}\big\}, \\
&z(t, x)\bar\mu(t, dx) \ge 0, \quad \|\bar\mu\|_{\Sigma_0(\overline\Omega)} = \int_{\mathcal{M}(t,z)} |\bar\mu|(t, dx) = \rho,
\end{aligned} \tag{5.6.23}$$

where $z(t, x)$ is a solution of the reverse adjoint equation (5.6.22) satisfying the growth condition

$$\int_0^T \|z(t, \cdot)\|_{C_0(\overline\Omega)} dt = \int_0^T \Big(\max_{x \in \overline\Omega} |z(t, x)|\Big) dt < \infty. \tag{5.6.24}$$

The eigenfunctions of the Laplacian in the unit circle are

$$J_m(\gamma_{mn} r) \sin m\theta, \quad J_m(\gamma_{mn} r) \cos m\theta, \qquad m, n = 0, 1, 2, \ldots$$

corresponding to the eigenvalue $-\gamma_{mn}^2$, where $(r, \theta)$ are the polar coordinates in the plane and, for each $m$, $0 < \gamma_{m1} < \gamma_{m2} < \ldots$ are the zeros of the Bessel function $J_m(r)$. We try as costate the (radially symmetric) function

$$z(t, r) = e^{\gamma_{01}^2 t} J_0(\gamma_{01} r) - 10^{-10} e^{\gamma_{02}^2 t} J_0(\gamma_{02} r) \tag{5.6.25}$$

in the interval $0 \le t \le T = 3$. As always, $\mathcal{M}(t, z) \subseteq \mathcal{N}(t, \nabla z) = \mathcal{N}(t, dz/dr)$ (the inclusion is strict for $t$ in a subinterval, as seen below). Figure 5.6.3 shows a cross section of $\mathcal{N}(t, dz/dr)$. Due to radial symmetry, $0 \in \mathcal{N}(t, dz/dr)$ for all $t$.

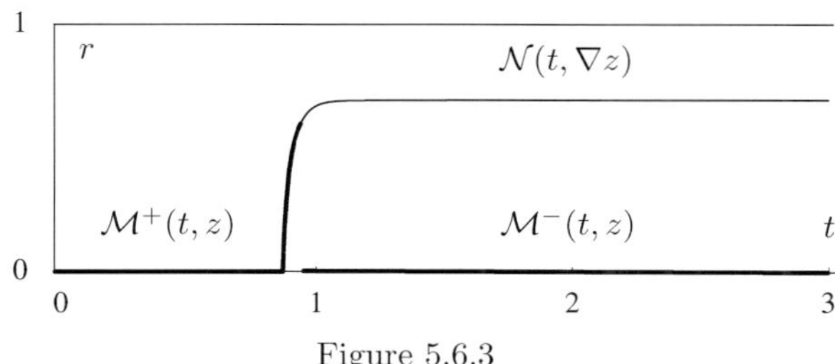


Figure 5.6.3

It remains to identify $\mathcal{M}(t, z)$. For $0 \le t \le t_0$, where

$$t_0 = 0.86535\ldots$$

$z(t, r) \ge 0$ has a global maximum at $r = 0$. At $t_0$, the maximum bifurcates into a circle $r = r(t) > 0$ where $z(t, r)$ attains its maximum and a positive local minimum $r = 0$ (first line of Figure 5.6.4) with $z(t, 0)$ decreasing in $t$.

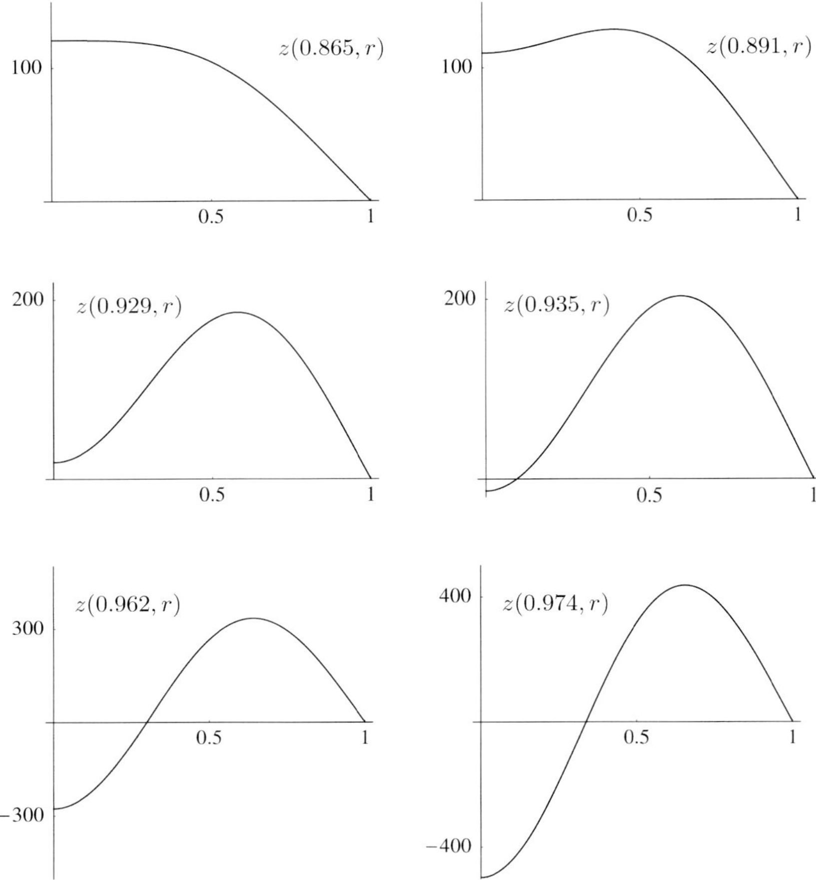


Figure 5.6.4

The local minimum remains positive until

$$t_1 = 0.9325\ldots$$

when it becomes 0, then negative for $t > t_1$ (second line of Figure 5.6.4). We have

$$|z(t,0)| < \|z(t,\cdot)\|_{C_0(\overline{\Omega})} = z(t, r(t)) \tag{5.6.26}$$

until

$$t_2 = 0.96851\ldots.$$

At $t = t_2$ inequality (5.6.26) becomes an equality and then

$$|z(t,0)| = \|z(t,\cdot)\|_{C_0(\overline{\Omega})} > z(t, r(t)) \quad (t_2 < t \leq 3)\,, \tag{5.6.27}$$

If $\mu > c$ and $\alpha \geq 0$ we define

$$H_\alpha = D((\mu I - A)^\alpha)\,.$$

For negative indices $\alpha = -\beta < 0$ the definition is

$$H_\alpha = H_{-\beta} = \mathcal{C}_{-\beta}(H)$$

where $\mathcal{C}_{-\beta}(H)$ is the completion of $H$ in the norm

$$\|y\|_{H_{-\beta}} = \|(\mu I - A)^{-\beta} y\|_H\,.$$

It is plain that the definition is independent of $\mu$. The spaces $\mathcal{H}_\alpha$, $-\infty < \alpha < \infty$, are defined in the same way using the space $\mathcal{H}$ and the operator $\mathcal{A}$. Applying (6.1.7) to the function $F(\lambda) = (\mu + \lambda)^\alpha$ we obtain $(\mu I - A)^\alpha = Q^{-1}(\mu I - \mathcal{A})^\alpha Q$. Equality of the domains yields the "$\alpha \geq 0$" part of the equality

$$H_\alpha = Q^{-1}\mathcal{H}_\alpha Q \quad (\infty < \alpha < \infty)\,. \tag{6.1.8}$$

In the case $\alpha = -\beta < 0$ we have $(\mu I - A)^{-\beta} = Q^{-1}(\mu I - \mathcal{A})^{-\beta} Q$, thus

$$\begin{aligned}\|Qy\|_{\mathcal{H}_{-\beta}} &= \|\{y_n(\cdot)\}\|_{\mathcal{H}_{-\beta}} = \|(\mu I - \mathcal{A})^{-\beta}\{y_n(\cdot)\}\|_{\mathcal{H}} \\ &= \|(\mu I - A)^{-\beta} y\|_H = \|y\|_{H_{-\beta}} \quad (y \in H)\,,\end{aligned}$$

which shows that $Q$ can be extended to an isomorphism $Q : H_{-\beta} \to \mathcal{H}_{-\beta}$ and $H_{-\beta} = Q^{-1}\mathcal{H}_{-\beta}Q$. This completes the proof of (6.1.8) for $\alpha < 0$.

If $\{y_n(\cdot)\} \in \mathcal{H}$ and $\beta > 0$ we have

$$\|\{y_n(\cdot)\}\|_{\mathcal{H}_{-\beta}} = \|(\mu I - \mathcal{A})^{-\beta}\{y_n(\cdot)\}\|_{\mathcal{H}} = \|(\mu + \lambda)^{-\beta}\{y_n(\lambda)\}\|_{\mathcal{H}}\,.$$

On the basis of this equality we can easily show that

$$\mathcal{H}_{-\beta} = \bigoplus\nolimits_n L^2(e_n, \eta_\beta) \tag{6.1.9}$$

where $\eta_\beta(d\lambda) = (\mu + \lambda)^{-\beta}\eta(d\lambda)$. The semigroup $\mathcal{S}(t)$ can be extended to $\mathcal{H}_{-\beta}$ by

$$\begin{aligned}\mathcal{S}(t)\{y_n(\lambda)\} &= \{e^{-\lambda t} y_n(\lambda)\} \\ &= \{(\mu + \lambda)^\beta e^{-\lambda t}(\mu + \lambda)^{-\beta} y_n(\lambda)\} \quad (t > 0)\,,\end{aligned} \tag{6.1.10}$$

with a corresponding extension of $S(t)$ given by

$$S(t) = Q^{-1}\mathcal{S}(t)Q \quad (t > 0)\,. \tag{6.1.11}$$

As a consequence of the definition of $Z(T)$ in **2.3** and **2.4**, $Z(T) \subseteq H_{-1}$. We have

$$\|S(t)z\|_H = \|S(t)Q^{-1}Qz\|_H = \|QS(t)Q^{-1}Qz\|_{\mathcal{H}} = \|\mathcal{S}(t)Qz\|_{\mathcal{H}}\,, \tag{6.1.12}$$

so that condition (2.3.22) characterizing elements of $Z(T)$ is

$$\int_0^T \|S(t)z\|_E dt = \int_0^T \|\mathcal{S}(t)Qz\|_{\mathcal{H}} dt < \infty \tag{6.1.13}$$

(the subindex $w$ in (2.3.22) is not needed since $S(t)^* = S(t)$ is strongly continuous; see (2.4.19) and following comments).

If $\phi(\lambda) = (\mu+\lambda)^\beta e^{-\lambda t}$ we have $d\phi(\lambda)/d\lambda = (\mu+\lambda)^{\beta-1}e^{-\lambda t}(\beta - t(\mu+\lambda)) = 0$ for $\lambda = \beta/t - \mu$, thus

$$(\mu+\lambda)^\beta e^{-\lambda t} \le \left(\frac{\beta}{t}\right)^\beta e^{-\beta+\mu t} \quad (\lambda \ge -\mu)\,,$$

and it follows from (6.1.10) that $\mathcal{S}(t)\mathcal{H}_{-\beta} \subseteq \mathcal{H}$ for $t > 0$ (which implies $\mathcal{H}_{-\beta}$ is a multiplier space) and

$$\|\mathcal{S}(t)\{y_n(\lambda)\}\|_{\mathcal{H}} \le C\,\frac{e^{\mu t}}{t^\beta}\,\|\{y_n(\cdot)\}\|_{\mathcal{H}_{-\beta}} \quad (t > 0)\,. \tag{6.1.14}$$

Let $z \in Z(T)$. Applying $Q$ to both sides of (6.1.4),

$$Q\bar{y} - QS(T)Q^{-1}Q\zeta = \int_0^T \frac{QS(2\sigma)Q^{-1}Qz}{\|S(\sigma)Q^{-1}Qz\|_E} d\sigma\,.$$

We use (6.1.11) in the second term on the left and in the integrand. In the denominator we use (6.1.12), so that formula (6.1.4) is "ported" to the space $\mathcal{H}$ as

$$Q\bar{y} - \mathcal{S}(T)Q\zeta = \int_0^T \frac{\mathcal{S}(2\sigma)Qz}{\|\mathcal{S}(\sigma)Qz\|_{\mathcal{H}}} d\sigma\,. \tag{6.1.15}$$

Now, writing $Q\bar{y} = \{\bar{y}_n(\cdot)\}$, $Q\zeta = \{\zeta_n(\cdot)\}$, $Qz = \{z_n(\cdot)\}$ formula (6.1.15) reduces, coordinate-by-coordinate, to

$$z_n(\lambda)\int_0^T \frac{e^{-2\lambda\sigma}}{\|\mathcal{S}(\sigma)Qz\|} d\sigma = \bar{y}_n(\lambda) - e^{-\lambda T}\zeta_n(\lambda)\,. \tag{6.1.16}$$

Schwartz's inequality gives

$$\begin{aligned}\int_0^T e^{-\lambda\sigma} d\sigma &= \int_0^T \frac{e^{-\lambda\sigma}}{\sqrt{\|\mathcal{S}(\sigma)Qz\|}}\sqrt{\|\mathcal{S}(\sigma)Qz\|}d\sigma \\ &\le \sqrt{\int_0^T \frac{e^{-2\lambda\sigma}}{\|\mathcal{S}(\sigma)Qz\|}d\sigma}\ \sqrt{\int_0^T \|\mathcal{S}(\sigma)Qz\| d\sigma}\,. \end{aligned} \tag{6.1.17}$$

We have $\|\mathcal{S}(T)Qz\| \leq \|\mathcal{S}(T-\sigma)\|\|\mathcal{S}(\sigma)Qz\|$, so that

$$\frac{1}{\|\mathcal{S}(\sigma)Qz\|} \leq \frac{\|\mathcal{S}(T-\sigma)\|}{\|\mathcal{S}(T)Qz\|} \leq C\,, \tag{6.1.18}$$

which makes the first integral on the right of (6.1.17) convergent. For the second integral, $z \in Z(T)$ implies that

$$C(T) = \int_0^T \|\mathcal{S}(\sigma)Qz\|_{\mathcal{H}} d\sigma = \int_0^T \|S(\sigma)z\|_E d\sigma < \infty\,,$$

hence (6.1.17) and (6.1.18) yield

$$\infty > \int_0^T \frac{e^{-2\lambda\sigma}}{\|\mathcal{S}(\sigma)Qz\|} d\sigma \geq \frac{1}{C(T)}\left(\int_0^T e^{-\lambda\sigma} d\sigma\right)^2 = \frac{(1-e^{-\lambda T})^2}{C(T)\,\lambda^2}\,,$$

and it follows from (6.1.16) that

$$|z_n(\lambda)| \leq C(T)\frac{\lambda^2}{(1-e^{-\lambda T})^2}|\bar{y}_n(\lambda) - e^{-\lambda T}\zeta_n(\lambda)|\,. \tag{6.1.19}$$

This inequality means that if $\{\bar{y}_n(\cdot)\} \in \mathcal{H}_\alpha$ then $\{z_n(\cdot)\} \in \mathcal{H}_{\alpha-2}$. A bootstrap argument makes it possible to better this implication considerably. We state the result for the space $H$.

**Theorem 6.1.2.** *Let $\alpha > 1$. Then $\bar{y} \in H_\alpha$ and $z \in Z(T) \Longrightarrow z \in H_{\alpha-1}$.*

*Proof.* Obviously, it is equivalent to prove that $\{\bar{y}_n(\cdot)\} \in \mathcal{H}_\alpha$ and $z \in Z(T)$ $\Longrightarrow \{z_n(\cdot)\} \in \mathcal{H}_{\alpha-1}$. We assume first $\alpha \geq 2$. Inequality (6.1.19) implies that $Qz = \{z_n(\cdot)\} \in \mathcal{H}_{\alpha-2} \subseteq \mathcal{H}$, hence $\|\mathcal{S}(\sigma)Qz\|$ is bounded. Accordingly,

$$\int_0^T \frac{e^{-2\lambda\sigma}}{\|\mathcal{S}(\sigma)Qz\|} d\sigma \geq \frac{1}{C}\int_0^T e^{-2\lambda\sigma} d\sigma = \frac{1}{C}\cdot\frac{1-e^{-2\lambda T}}{2\lambda}\,, \tag{6.1.20}$$

and it follows from (6.1.16) that

$$|z_n(\lambda)| \leq C\frac{\lambda}{1-e^{-\lambda T}}|\bar{y}_n(\lambda) - e^{-\lambda T}\zeta_n(\lambda)|\,, \tag{6.1.21}$$

so that $\{z_n(\cdot)\} \in \mathcal{H}_{\alpha-1}$. In the range $1 < \alpha < 2$, (6.1.19) implies that $\{z_n(\cdot)\} \in \mathcal{H}_{\alpha-2}$, thus it follows from (6.1.14) that

$$\|\mathcal{S}(t)Qz\| = \|\mathcal{S}(t)\{z_n(\cdot)\}\| \leq \frac{C}{t^{2-\alpha}}\,. \tag{6.1.22}$$

We have $\alpha - 2 \geq -1$, so that (6.1.22) gives

$$\int_0^T \frac{e^{-2\lambda\sigma}}{\|\mathcal{S}(\sigma)Qz\|} d\sigma \geq c\int_0^T \sigma^{2-\alpha}e^{-2\lambda\sigma} d\sigma$$

$$= \frac{c}{\lambda^{3-\alpha}}\int_0^T (\lambda\sigma)^{2-\alpha}e^{-2\lambda\sigma} d(\lambda\sigma) = \frac{c}{\lambda^{3-\alpha}}\int_0^{\lambda T} \tau^{2-\alpha}e^{-2\tau} d\tau\,,$$

and it results from (6.1.16) that

$$|z_n(\lambda)| \le C(T, \lambda, \alpha)\lambda^{3-\alpha}|\bar{y}_n(\lambda) - e^{-\lambda T}\zeta_n(\lambda)|\,, \tag{6.1.23}$$

where

$$C(T, \lambda, \alpha) \to C(T, \alpha) = \left(c \int_0^\infty \tau^{2-\alpha} e^{-2\tau} d\tau\right)^{-1} \quad \text{as } \lambda \to \infty\,.$$

Now, $\{\bar{y}_n(\cdot)\} \in \mathcal{H}_\alpha$, hence $\{\bar{y}_n(\cdot) - e^{-\lambda T}\zeta_n(\cdot)\} \in \mathcal{H}_\alpha$, and (6.1.23) implies that

$$\{z_n(\cdot)\} \in \mathcal{H}_{\alpha+\alpha-3} = \mathcal{H}_{2\alpha-3}\,, \quad 2\alpha - 3 > \alpha - 2\,.$$

Choose $\alpha_1$ so that

$$\alpha_1 - 2 = 2\alpha - 3 \quad \text{or} \quad \alpha_1 = 2\alpha - 1 > \alpha\,,$$

and repeat the preceding argument verbatim, obtaining

$$|z_n(\lambda)| \le C(T, \lambda)\lambda^{3-\alpha_1}|\bar{y}_n(\lambda) - e^{-\lambda T}\zeta_n(\lambda)|\,,$$

which implies that

$$\{z_n(\cdot)\} \in \mathcal{H}_{\alpha+\alpha_1-3} = \mathcal{H}_{3\alpha-4}\,, \quad 3\alpha - 4 > 2\alpha - 3\,.$$

Obviously, we can do this as many times as we wish. At the $(n-1)^{\text{th}}$ step we have

$$\{z_n(\cdot)\} \in \mathcal{H}_{n\alpha-(n+1)}\,,$$

and for $n$ large enough we have $n\alpha - (n+1) = n(\alpha - 1) - 1 > 0$. At this point we have shown that $\{z_n(\cdot)\} \in \mathcal{H}$. The argument pertaining to the case $\alpha \ge 2$ (based on inequality (6.1.20)) applies without changes and (6.1.21) is obtained, hence $\{z_n(\cdot)\} \in \mathcal{H}_{\alpha-1}$. This ends the proof.

**Remark 6.1.3.** It is possible that the hypothesis that $z \in Z(T)$ in Theorem 6.1.2 can be weakened, but it cannot be discarded, as Example 3.5.11 shows. In this example, the operator $A$ is self adjoint and $z$ is a multiplier such that $\|S(t)z\| \approx e^{1/t}$ for $t \approx 0$ so that, in view of (6.1.14) $z \notin H_\alpha$ for $-\infty < \alpha < \infty$. However, $\bar{y} \in D(A^\infty)$, thus

$$\bar{y} \in \bigcap_{\alpha=-\infty}^{\infty} \mathcal{H}_\alpha\,.$$

**Theorem 6.1.4.** *Let $-\infty < \alpha < \infty$. Then $z \in \mathcal{H}_\alpha \implies \bar{y} \in \mathcal{H}_{\alpha+1}$.*

*Proof.* We use (6.1.18),

$$\int_0^T \frac{e^{-2\lambda\sigma}}{\|\mathcal{S}(\sigma)Qz\|} d\sigma \le C \int_0^T e^{-2\lambda\sigma} d\sigma = C\frac{1 - e^{-2\lambda T}}{2\lambda}\,,$$

thus obtaining from (6.1.16) that

$$|\bar{y}_n(\lambda) - e^{-\lambda T}\zeta_n(\lambda)| \leq C\frac{1-e^{-2\lambda T}}{2\lambda}|z_n(\lambda)| .$$

This ends the proof of Theorem 6.1.4.

**Miscellaneous notes.** The results in this section are from the author [T.A.:2].

**6.2. Self adjoint semigroups, II.** Theorem 6.1.2 says nothing for $\alpha = 1$. In fact, (6.1.19) only gives $Qz \in \mathcal{H}_{-1}$ or, equivalently, $z \in H_{-1}$, which is redundant since $z \in Z(T) \subseteq H_{-1}$ is a standing assumption; in this case $\sigma^{2-\alpha}$ is not integrable, so we don't get to (6.1.23).

Let $R^\infty(T) \subseteq E$ be the reachable space as defined in **2.1**. The case $\alpha = 0$ of Theorem 6.1.4 gives the inclusion

$$\left\{y \in E;\ y = \int_0^T \frac{S(2\sigma)z}{\|S(\sigma)z\|}d\sigma\ ;\ z \in E\right\} \subseteq B_1^\infty(T) \cap D(A)\,, \tag{6.2.1}$$

where $B_1^\infty(T)$ is the ball of center 0 and radius 1 in $R^\infty(T)$. The missing case $\alpha = 1$ of Theorem 6.1.2 is $y \in D(A)$ and $z \in Z(T) \Longrightarrow z \in H$. If $\subseteq$ can be upgraded to $=$ in (6.2.1), then every $\bar{y} \in B_1^\infty(T) \cap D(A)$ can be reached from $\zeta = 0$ in time $T$ by the control (6.1.3) *with* $z \in E$. It *cannot* be reached in the same time with a control of the form (6.1.3) with $\tilde{z} \in Z(T) \setminus H$. In fact, if both controls

$$\frac{S(T-\sigma)z}{\|S(T-\sigma)z\|}\,, \quad \frac{S(T-\sigma)\tilde{z}}{\|S(T-\sigma)\tilde{z}\|}$$

drive 0 to $\bar{y}$ in time $T$ then both are time optimal, thus by uniqueness of time optimal controls (Corollary 2.1.7) we have

$$\frac{S(T-\sigma)z}{\|S(T-\sigma)z\|} = \frac{S(T-\sigma)\tilde{z}}{\|S(T-\sigma)\tilde{z}\|} \quad \text{a. e.} \tag{6.2.2}$$

If $\mu > c$ then $(\mu I - A)^{-1}$ can be extended to $E_{-1} \supseteq Z(T)$ and $(\mu I - A)^{-1}E_{-1} \subseteq E$. Applying $(\mu I - A)^{-1}$ to both sides of (6.2.2) we obtain

$$\frac{S(T-\sigma)(\mu I - A)^{-1}z}{\|S(T-\sigma)z\|} = \frac{S(T-\sigma)(\mu I - A)^{-1}\tilde{z}}{\|S(T-\sigma)\tilde{z}\|} \quad \text{a. e.}$$

Freezing $\sigma < T$ and setting $c = \|S(T-\sigma)\tilde{z}\|/\|S(T-\sigma)z\|$,

$$\begin{aligned} &S(T-\sigma)(\mu I - A)^{-1}\tilde{z} = S(T-\sigma)c(\mu I - A)^{-1}z \\ &\qquad \Longrightarrow\ (\mu I - A)^{-1}\tilde{z} = (\mu I - A)^{-1}cz\ \Longrightarrow\ \tilde{z} = cz \end{aligned}$$

with $c \neq 0$, which contradicts the fact that $\tilde{z} \notin H$. Accordingly, *Theorem* 6.1.2 *is true for* $\alpha = 1$ *if and only if there is equality in* (6.2.1). Whether or not this is the case seems to be an open problem.

As a particular case of the results in **6.1** we obtain

$$\bar{y} \in D(A^\infty) = \bigcap_{n=1}^{\infty} D(A^n) \iff z \in D(A^\infty) \tag{6.2.3}$$

so that infinite smoothness of the target $\bar{y}$ in (6.1.5) and infinite smoothness of the multiplier $z = \psi$ in (6.1.3) are equivalent. This raises the question of whether similar results exist for semigroups outside the self adjoint class. The answer is in general "no" as the following example shows,

**Example 6.2.1.** We take $E = L^2(0,\infty)$, all functions $y(x)$, $z(x) \dots$ defined in $x \geq 0$ and extended $= 0$ for $x < 0$. The (strongly continuous, isometric) *right translation* semigroup $S(t)$ is

$$S(t)y(x) = \begin{cases} y(x-t) & (x \geq t) \\ 0 & (x < t)\,. \end{cases} \tag{6.2.4}$$

The adjoint semigroup is the *left translation* (and chop-off) semigroup

$$S(t)^* z(x) = \begin{cases} z(x+t) & (x \geq 0) \\ 0 & (x < 0)\,. \end{cases} \tag{6.2.5}$$

These semigroups were introduced in **2.6** and all facts used below are proved there. The infinitesimal generators $A$ of $S(t)$ and $A^*$ of $S(t)^*$ are, respectively,

$$Ay(x) = -y'(x)\,, \quad A^*y(x) = y'(x) \tag{6.2.6}$$

with domain $D(A^*) = \{$all absolutely continuous $y(x)$ with $y'(x)$ in $L^2(0,\infty)\}$; the domain $D(A)$ is the subset of $D(A^*)$ defined by the boundary condition $y(0) = 0$. The space $Z(T)$ consists of all $z(\cdot)$ measurable in $x > 0$ and such that

$$\int_0^T \|S(\sigma)^* z(\cdot)\| d\sigma = \int_0^T \kappa(\sigma, z) d\sigma < \infty\,,$$

where

$$\kappa(\sigma, z) = \|S(\sigma)^* z(\cdot)\| = \sqrt{\int_\sigma^\infty z(x)^2 dx}\,.$$

If the support of $z(x)$ has an intersection of positive measure with $[T - \epsilon, \infty)$ for every $\epsilon > 0$ then $\kappa(\sigma, z) \neq 0$ in $0 \leq \sigma \leq T$ and a control that satisfies formula (6.1.3) is given by

$$\bar{u}(\sigma, x) = \rho \frac{S(T-\sigma)^* z(x)}{\|S(T-\sigma)^* z(\cdot)\|} = \rho \chi(x) \frac{z(x + (T-\sigma))}{\kappa(T-\sigma, z)} \quad (0 < \sigma \leq T)\,,$$

where $\chi(x)$ is the characteristic function of $x \geq 0$. The solution corresponding to this control is

$$y(T, x, \zeta, \bar{u})(x) = \zeta(x - T) + z(x)\omega(T, x, z)\,,$$

where

$$\omega(T, x, z) = \int_0^T \frac{\chi_\sigma(x)}{\kappa(\sigma, z)} d\sigma = \int_0^{\min(x,T)} \frac{d\sigma}{\kappa(\sigma, z)}\,. \tag{6.2.7}$$

If we drive from 0 in time $T$, the target $\bar{y}(x)$ and the costate $z(x)$ are related by

$$\bar{y}(x) = y(T, x, 0, \bar{u}) = z(x)\omega(T, x, z)\,. \tag{6.2.8}$$

Since $A$ is not self adjoint, the pertinent analogue of (6.2.3) is

$$\bar{y} \in D(A^\infty) \iff z \in D((A^*)^\infty)\,. \tag{6.2.9}$$

Both implications in (6.2.9) are false for the semigroup (6.2.4). In fact, assume that $\bar{y}(x)$ is infinitely differentiable in $x \geq 0$ with all derivatives $= 0$ at $x = 0$ (which is equivalent to $\bar{y}(\cdot) \in D(A^\infty)$). If a multiplier $z(\cdot) \in Z(T)$ satisfies (6.2.8) and is not a. e. zero in $[T, \infty)$ then $z(\cdot) \notin D(A^2)$. In fact, $\kappa(T, z) \neq 0$ in $0 \leq t \leq T$, thus $\omega(T, x, z)$, given by (6.2.7), is continuous in $0 \leq x \leq T + \epsilon$ for some $\epsilon > 0$ and continuously differentiable in $0 \leq x < T$ and $T < x < T + \epsilon$ with derivative

$$\omega'(T, x, z) = \begin{cases} \dfrac{1}{\kappa(x, z)} & (x < T)\,, \\ 0 & (x > T)\,. \end{cases} \tag{6.2.10}$$

Accordingly,

$$\Big(\frac{1}{\omega(T, x, z)}\Big)' = \begin{cases} -\dfrac{1}{\kappa(x, z)\omega(T, x, z)^2} & (x < T)\,, \\ 0 & (x \geq T)\,. \end{cases} \tag{6.2.11}$$

We obtain from (6.2.8) that

$$z(x) = \frac{\bar{y}(x)}{\omega(T, x, z)}\,,$$

thus

$$z'(x) = \frac{\bar{y}'(x)}{\omega(T, x, z)} + \bar{y}(x)\Big(\frac{1}{\omega(T, x, z)}\Big)'\,.$$

Accordingly, in view of (6.2.11) and of the fact that $\bar{y}(T) \neq 0$, $z'(x)$ has a jump at $x = T$, thus $z'(\cdot) \notin D(A^*)$. We have then disproved the implication $\Longrightarrow$ in (6.2.9).

To demolish the opposite implication we take. $z(\cdot) \in D((A^*)^\infty)$ (that is, $z(x)$ infinitely differentiable in $x \geq 0$). Assume $z(x) > 0$ for $x > 0$. Then $\kappa(x, z)$ is infinitely differentiable and positive in $x \geq 0$. Differentiating (6.2.8),

$$\bar{y}'(x) = z'(x)\omega(T, x, z) + z(x)\omega'(T, x, z)$$

thus, in view of (6.2.10), the derivative of $\bar{y}(x)$ has a jump at $x = T$, which means $\bar{y}(\cdot) \notin D(A^2)$. It is clear that the argument becomes inoperative unless $z(T) \neq 0$, but it is also possible to construct counterexamples where $z(T) = 0$; we may take $T = 1$ and the costate

$$z(x) = \phi(x)\frac{\sqrt{2\alpha}\, e^{-1/(1-x)^\alpha}}{(1-x)^{(\alpha+1)/2}} \quad (0 \leq x \leq 1)\,, \quad z(x) = 0 \quad (x \geq 1)\,,$$

with $\alpha > 0$, $\phi(x)$ a $C^\infty$ function with $\phi(x) = 0$ in $0 \leq x \leq a < 1/2$, $\phi(x) > 0$ in $x > a$ and $\phi(x) = 1$ in $x \geq 1/2$. We have proved in Theorem 2.6.6 that for this multiplier $\bar{y}(x) = y(1, 0, x, \bar{u}) \in D(A^n)$ if and only if

$$n < 1 + \frac{\alpha}{2}\,,$$

thus $\alpha < 2$ produces a target $\bar{y}(\cdot) \notin D(A^2)$.

Another area where self adjointness seems crucial to the results is that of smoothness of optimal trajectories. To place the problem in perspective, consider an analytic semigroup $S(t)$ with infinitesimal generator $A$ in a Hilbert space $H$. We have proved in Theorem 3.2.3 that every time optimal control satisfies the maximum principle (6.1.2) for some multiplier $\psi$ (not necessarily in $Z(T)$) such that $S(t)^*\psi \neq 0$ for $t > 0$. Accordingly,

$$\bar{u}(t) = \frac{S(T-t)^*\psi}{\|S(T-t)^*\psi\|}\,, \tag{6.2.12}$$

and, since the norm in a Hilbert space is analytic, (6.2.12) says that $\bar{u}(t)$ is "as smooth as $S(t)^*$" in $0 \leq t < T$, although its behavior for $t \approx T$ is open to question. It does not seem possible, however, to obtain smoothness results for trajectories via the variation-of-constants formula

$$y(t, \zeta, \bar{u}) = S(t)\zeta + \int_0^t S(t-\sigma)\bar{u}(\sigma)d\sigma \tag{6.2.13}$$

from smoothness of the controls in $t < T$. Again, something special ensues when $A$ (hence $S(t)$) is self adjoint. Combining (6.2.12) and (6.2.13) we obtain

$$\begin{aligned} y(t, \zeta, \bar{u}) &= S(t)\zeta + \int_0^t S(t-\sigma)\frac{S(T-\sigma)\psi}{\|S(T-\sigma)\psi\|}d\sigma \\ &= S(t)\zeta + \int_0^t \frac{S(T+t-2\sigma)\psi}{\|S(T-\sigma)\psi\|}d\sigma \\ &= S(t)\zeta + S(T-t-\delta)\int_0^t \frac{S(2(t-\sigma)+\delta/2)S(\delta/2)\psi}{\|S(T-\sigma)\psi\|}d\sigma \\ &= S(t)\zeta + y(t, 0, \bar{u})\,. \end{aligned} \tag{6.2.14}$$

This formula shows that

$$y(t, 0, \bar{u}) \in S(T - t - \delta)E \quad (0 < \delta < T - t)\,.$$

On the other hand, it follows from the fact that $S(2(t - \sigma) + \delta/2)$ is analytic in a sector including the interval $0 \le \sigma \le T$ that the integral (6.2.14) is an analytic function of $t$, thus $y(t, \zeta, \bar{u})$ is analytic as well.

**Miscellaneous notes.** Some results in this section are from the author [T.A.:2].

**6.3. Related research.** We include in this final section references to a number of works on the control system

$$y'(t) = Ay(t) + u(t)\,, \quad y(0) = \zeta\,, \tag{6.3.1}$$

on the more general system

$$y'(t) = Ay(t) + Bu(t)\,, \quad y(0) = \zeta\,, \tag{6.3.2}$$

and on nonlinear versions of (6.3.2). For references to some of the early work see the Miscellaneous Notes to **1.3**.

THE TIME OPTIMAL AND THE NORM OPTIMAL PROBLEM. Although the implication time optimality $\Longrightarrow$ norm optimality for the equation (6.3.1) was known at the beginning stages of the theory (the author [1964]), literature on the relation between both optimal control problems has not been plentiful. One item is Krabs - Schmidt [1989], where it is shown that norm optimal controls, being time optimal, satisfy Pontryagin's maximum principle under suitable assumptions. That there exist norm optimal controls that do not satisfy the maximum principle (or even the bang-bang condition $\|u(t)\| =$ minimum norm a. e.) was realized in the author [2000], with further developments in [2001:3]. These counterexamples also led to the discovery that, unlike for time optimal controls, uniqueness of norm optimal controls in smooth spaces is not absolute (it depends on conditions on the semigroup or on the target).

The relation between the norm optimal problem and the time optimal problem attracted more attention for the abstract wave equation $y''(t) = Ay(t) + u(t)$, with $A$ self adjoint and positive definite in a Hilbert space $H$. This relation was studied in several papers, among them Krabs [1985] with $L^2(0, T; H)$ bounds on the controls and Krabs [1989] with $L^\infty(0, T; H)$ bounds. The treatment of abstract wave equations has some points in common with that of reversible systems in **2.2**; in fact, in a suitable direct sum energy space the equation can be reduced to (6.3.2) with $S(t)$ a reversible semigroup and $B$ a projection operator. In common with the treatment in **2.2**, it is not necessary to bring into play multipliers in spaces larger than (the dual of) the state space.

For other works connected with either the time or norm optimal problem see Azé - Cârjă [2000], Cârjă [1984:2], [1984:3], [1993], Egorov [1999], Kluvánek - Knowles [1976].

HYPERSINGULAR EXTREMALS. The singular time optimal controls we have constructed in **2.6** do not satisfy the maximum principle (even in its weak form) in the entire interval $0 \leq t \leq T$. However they do satisfy the strong maximum principle in any interval $0 \leq t \leq T' < T$, where, due to the optimality principle, they are also time optimal. This leads irresistibly to the following definition: the control $\bar{u}(t)$ is *hypersingular* if it is time optimal in an interval $0 \leq t \leq T$ but does not satisfy the weak maximum principle in any subinterval $[a, b]$, where $0 \leq a < b \leq T$. We don't know if hypersingular time optimal controls exist. It is not difficult to show that hypersingular *norm* optimal controls exist using the methods in **2.8**, but this is not very interesting since, unlike time optimality, norm optimality in a subinterval $[a, b]$ is not inherited from the entire interval $[0, T]$.

OTHER COST FUNCTIONALS. Minimization of the functional

$$J(y, u) = \int_0^T f_0(t, y(t), u(t))dt \tag{6.3.3}$$

includes the time optimal problem, where $f_0 \equiv 1$, but not the norm optimal problem. If suitable convexity assumptions are placed on the functional (6.3.3) it is possible to obtain the maximum principle by separation arguments; see the author [1995]. If the convexity assumptions are waived, however, separation is no longer operative, and there is no reason to restrict ourselves to systems with linear dynamics line (6.3.1) and (6.3.2).

SEMILINEAR SYSTEMS. In these, the dynamics is given by a semilinear equation

$$y'(t) = Ay(t) + f(t, y(t), u(t)), \quad y(0) = \zeta \tag{6.3.4}$$

and the cost functional is (6.3.3). One of the objectives is to derive versions of Pontryagin's maximum principle under constraints on the control and on the trajectory. In the absence of convexity, the methods include "local separation" theorems stemming from Ekeland's variational principle and other resources of nonlinear analysis. The theory can be conducted in such a way that the maximum principle appears as a condition of Kuhn - Tucker type for directional derivatives in metric spaces of controls, thus making it a part of nonlinear programming. References are in Barbu [1993], Li - Yong [1995] and the author [1999:2]; for more recent references see the coming monograph by Mordukhovich [T.A.]. It should be pointed out that the linear theory underlies the nonlinear. The role of the reverse adjoint equation is picked up by the (linear) *reverse adjoint variational equation*

$$z'(t) = -(A^* + \partial_y f(t, y(t, \zeta, \bar{u}), \bar{u}(t))^*)z(t) - \partial_y f_0(t, y(t, \zeta, \bar{u}), \bar{u}(t)) \tag{6.3.5}$$

($\partial_y$ indicates Fréchet derivative), where the final condition $z(T)$ is in a multiplier space corresponding to the equation $z'(t) = -A^* z(t)$. The maximum principle for a control set $U$ is

$$\langle z(t), f(t, y(t, \zeta, \bar{u}), \bar{u}(t))\rangle = \max_{u \in U} \langle z(t), f(t, y(t, \zeta, \bar{u}), u)\rangle \quad \text{a. e.} \tag{6.3.6}$$

For some recent papers, see Aizicovici - Motreanu - Pavel [2004], Arada - Raymond [2002], [2003], Mordukhovich - Raymond [2004], Seierstad [1997], Wang - Wang [2001], [2002], You [1999], Zhao - Wang - Zhao [2004]. In the nonlinear realm, the maximum principle is not a sufficient condition for optimality without convexity assumptions; in their absence, optimality depends on second order conditions, descendants of the calculus second derivative test. See Casas - Tröltzsch [2002], Mittelman - Tröltzsch [2001] and Rösch - Tröltzsch [2003].

Existence theory must deal with the fact that weak convergence of controls is not preserved under the nonlinear operator $f(t, y, u)$ and the nonlinear functional $f_0(t, y, u)$, which makes it necessary the introduction of (measure-valued) *relaxed controls*. For two different approaches to relaxed controls, as well as for references see Roubíček [1997] and the author [1999:2, , Part III].

THE LINEAR – QUADRATIC PROBLEM. Among the first finite dimensional control problems to undergo generalization to the infinite dimensional setting is the linear - quadratic problem (linear dynamics, quadratic functional) both in a finite interval (finite horizon) and in the half line (infinite horizon). The first step into infinite dimension was taken by Lukes - Russell [1969]. Without target condition or bounds on the control or the state the problem admits a feedback solution, the feedback operator given by the solution of an operator Riccati equation.

The Lukes - Russell model corresponds to distributed control. When control is exerted on the boundary of the domain the model is (6.3.2) with $B$ unbounded, which introduces numerous complications setting up and interpreting the solutions of the system and of the operator Riccati equation. The linear-quadratic problem is also closely related to stabilization by feedback. For a treatment of these and other problems and for references see Curtain-Zwart [1995] and Lasiecka - Triggiani [2000:1], [2000:2]. Among recent references see Barbu [2000], Barbu - Favini [2002], Barbu - Lasiecka - Triggiani [2000], Curtain [2003], Desch - Milota - Schappacher [2001], Desch - Schappacher - Fašanga - Milota [2002], Favini [2003], Lasiecka-Triggiani [2004], Staffans [1999], Weiss - Zwart [1998], Weiss [2003], Wu [1999].

THE DYNAMIC PROGRAMMING APPROACH. Initiated by Bellman [1957] on a mostly heuristic basis, this approach to optimal problems, although rigorously justified in some situations (Boltyanski [1966]) had to wait until the eighties for the necessary tools (viscosity solutions and nonsmooth analysis among them) to attain a solid mathematical footing. The references below include some recent works on the time optimal problem both with linear and nonlinear dynamics and also other cost functionals. See Albano - Cannarsa [1997], Barbu [1990], [1991],

Barbu - Da Prato [1983], Cannarsa - Cârjă [2004], Cannarsa - Da Prato [1990], Cannarsa - Frankowska [1992], [1996], Cannarsa - Sinestrari [1997], Cârjă [1984:1], [1993], Frankowska [1993], [2002], Frankowska - Plaskacz [2000], Tataru [1992].

THE INPUT - OUTPUT MAP. Given a system described by (6.3.2), it is natural to study the properties of the input - output map $u(\cdot) \to \mathcal{S}u(\cdot) = y(\cdot)$. If $B$ is bounded, $\mathcal{S}$ shares all properties of the input-output map for the equation (6.3.1), in particular the maximal regularity properties used in **3.2** (plus additional properties coming, for instance, from $B$ having finite dimensional domain or, more generally, being compact).

Modeling of partial differential equations with control acting on the boundary leads to the equation (6.3.2) (thus to the study of the input-output map) with $B$ unbounded. Ideas such as observability make it natural to consider the more general map

$$y(t) = \mathcal{F}u(t) = Du(t) + \int_0^t CS(t-\sigma)Bu(\sigma)d\sigma \tag{6.3.7}$$

where $C$ and $D$ are suitable linear operators ($C$ possibly unbounded). The object is to characterize those operators $B, C, D$ that give the map $\mathcal{F}$ suitable properties from the point of view of system theory; see Weiss [1989]. Conversely, one may ask what properties should be asked of a linear map $u(\cdot) \to y(\cdot)$ in order that it will admit the representation (6.3.7). For a treatment of these and related problems and references see Curtain - Zwart [1995] and Lasiecka - Triggiani [2000:1], [2000:2]. For a few more recent works on the map (6.3.7) and related subjects see Idrissi [2003] (bilinear systems), Kowalski - Sadkowski [2003] (systems with integrated semigroups), Staffans - Tucsnak - Weiss [2001], Wu [2003].

# REFERENCES


R. A. Adams

[1975] *Sobolev Spaces,* Academic Press, New York, 1975.

S. Aizicovici, D. Motreanu and N. H. Pavel

[2004] Nonlinear mathematical programming and optimal control, *Dyn. Contin. Discrete Impuls. Syst. Ser. A Math. Anal.* **11** (2004) 503-524.

P. Albano and P. Cannarsa

[1998] Singularities of the minimum time function for semilinear parabolic systems, *ESAIM Proc.* **4**, Soc. Math. Appl. Indust., Paris (1998) 59-72.

D. J. Aldous

[1979] Unconditional bases and martingales in $L^p(F)$, *Math. Proc. Cambridge Phil. Soc.* **85** (1979) 117-123.

H. Amann

[1995] *Linear and Quasilinear Parabolic Problems: Abstract Linear Theory,* Springer, Berlin, 1995.

H. Amann and J. Escher

[1996] Strongly continuous dual semigroups, *Ann. Mat. Pura Appl.* **171** (1996) 41-62.

H. Amann, M. Hieber and G. Simonett

[1994] Bounded $H^\infty$ calculus for elliptic operators, *Differential & Integral Equations* **7** (1994) 1-32.

S. Angenent

[1988] The zero set of a solution of a parabolic equation, *J. Reine Angew. Math.* **390** (1988) 79-96.

H. A. Antosiewicz

[1963] Linear control systems, *Archive Rat. Mech. Analysis* **12** (1963) 313-324.

N. Arada and J.-P. Raymond

[2002] Dirichlet boundary control of semilinear parabolic equations, I, II, *Appl. Math. & Optimization* **45** (2002) 125-143, 145-167.

[2003] Time optimal problems with Dirichlet boundary controls, *Discrete & Continuous Dynamical Systems* **9** (2003) 1549-1570.

D. Azé and O. Cârjă

[2000] Fast controls and minimum time, *Control Cybernetics* **29** (2000) 887-894.

J.-B. Baillon

[1980] Caractère borné de certains générateurs de semi-groupes linéaires dans les espaces de Banach, *C. R. Acad. Sci. Paris Sér A-B* **290** (1980) A757-A760.

A. V. Balakrishnan

[1965] Optimal control problems in Banach spaces, *SIAM J. Control* **3** (1965) 152-180.

S. Banach

[1932] *Théorie des Opérations Linéaires,* Monografje Matematyczne I, Warszawa, 1932.

H. T. Banks and M. Q. Jacobs

[1970] The optimization of trajectories of linear functional differential equations, *SIAM J. Control* **8** (1970) 461-488.

V. Barbu

[1990] The minimal time function for the nonlinear diffusion equation, *Libertas Math.* **10** (1990) 123-130.

[1991] The dynamic programming equation for the time-optimal control problem in infinite dimensions, *SIAM J. Control Optimization* **29** (1991) 445-456.

[1993] *Analysis and Control of Nonlinear Infinite Dimensional Systems,* Academic Press, San Diego 1993.

[2000] State constrained optimal control problems governed by semilinear equations, *Numer. Funct. Anal. Optimization.* **21** (2000) 411-424.

V. Barbu and G. Da Prato

[1983] *Hamilton-Jacobi equations in Hilbert Spaces,* Pitman Research Notes in Mathematics Series, vol. 86, Longman, Harlow, 1983.

V. Barbu and A. Favini

[2002] A degenerate two-point problem, *Progr. Nonlinear Differential Equations Appl.* **50** (2002) 27-37.

V. Barbu, I. Lasiecka and R. Triggiani

[2000] Extended algebraic Riccati equations in the abstract hyperbolic case, *Nonlinear Analysis* **40** (2000) 105-129.

S. Barnett and R. G. Cameron

[1985] *Introduction to Mathematical Control Theory,* Oxford University Press, Oxford, 1985.

R. E. Bellman

[1957] *Dynamic Programming,* Princeton University Press, Princeton, 1957.

R. E. Bellman, I. Glicksberg and O. A. Gross

[1956] On the "bang-bang" control problem, *Quart. Appl. Math.* **14** (1956) 11-18.

A. Bensoussan, G. Da Prato, M. C. Delfour and S. K. Mitter

[1992] *Representation and Control of Infinite Dimensional Systems,* vol. I, Birkhäuser, Basel, 1992.

[1993] *Representation and Control of Infinite Dimensional Systems,* vol. II, Birkhäuser, Basel, 1993.

V. Bernstein

[1933] *Leçons sur les Progrés Récents de la Théorie des Series de Dirichlet,* Gauthier-Villars, Paris, 1933.

V. G. Boltyanski, R. V. Gamkrelidze and L. S. Pontryagin

[1956] On the theory of optimal processes, *Dokl. Akad. Nauk. SSSR* **110** (1956) 7-10.

V. G. Boltyanski

[1966] Sufficient conditions for optimality and the justification of the dynamic programming principle, *SIAM J. Control* **4** (1966) 326-361.

J. Bourgain

[1983] Some remarks on Banach spaces in which martingale difference sequences are unconditional, *Ark. Mat.* **21** (1983) 163-168.

D. L. Burkholder

[1983] A geometric condition that implies the existence of a certain singular integral of Banach space valued functions, *Wadsworth Math. Series,* Wadsworth (1983) 270-286.

A. G. Butkovski

[1965] *Theory of Optimal Control of Distributed Parameter Systems,* Izd. Nauka, Moscow 1965. English translation: Elsevier, New York, 1969.

P. Butzer and H. Berens

[1967] *Semi-Groups of Operators and Approximation,* Springer, Berlin-Heidelberg, 1967.

P. Cannarsa and O. Cârjă

[2004] On the Bellman equation for the minimum time problem in infinite dimensions, *SIAM J. Control Optimization* **43** (2004) 532-548.

P. Cannarsa and G. Da Prato

[1990] Some results on nonlinear optimal control problems and Hamilton-Jacobi equations in infinite dimensions, *J. Functional Analysis* **90** (1990) 27-47.

P. Cannarsa and H. Frankowska

[1992] Value function and optimality conditions for semilinear control problems, *Appl. Math. & Optimization* **26** (1992) 139-169.

[1996] Value function and optimality condition for semilinear control problems, II. Parabolic case, *Appl. Math. & Optimization* **33** (1996) 1-33.

P. Cannarsa and C. Sinestrari

[1997] An infinite dimensional time optimal control problem, *Contemp. Math.* **209** (1997) 29-41.

O. Cârjă

[1984:1] On the minimal time function for distributed control systems in Banach spaces, *J. Optim. Theory Appl.* 44 (1984) 397-406.

[1984:2] On variational perturbations of control problems: minimum-time problem and minimum-effort problem, *J. Optim. Theory Appl.* **44** (1984) 407-433.

[1984:3] The time optimal problem for boundary-distributed control systems, *Boll. Un. Mat. Ital.* **B 3** (1984) 563-581.

[1993] The minimal time function in infinite dimensions, *SIAM J. Control Optimization* **31** (1993) 1103-1114.

E. Casas and F. Tröltzsch

[2002] Second-order necessary and sufficient optimality conditions for optimization problems and applications to control theory, *SIAM J. Control Optimization* **13** (2002) 406-431.

R. Conti

[1968] Time-optimal solution of a linear evolution equation in Banach spaces, *J. Optimization Theory & Appl.* **2** (1968) 277-284.

[1974] *Problemi di controllo e di controllo ottimale,* Unione Tipografica-Editrice Torinese, Torino 1974.

R. F. Curtain

[2003] Riccati equations for stable well-posed linear systems: the generic case, *SIAM J. Control Optimization* **42** (2003) 1681-1702.

R. F. Curtain and H. Zwart

[1995] *An Introduction to Infinite-Dimensional Linear Systems Theory,* Springer, New York, 1995.

W. Desch, J. Milota and W. Schappacher

[2001] Least square control problems in nonreflexive spaces, *Semigroup Forum* **62** (2001) 337-357.

W. Desch, W. Schappacher, E. Fašanga and J. Milota,

[2002] Riccati operators in non-reflexive spaces, *Differential Integral Equations* **15** (2002) 1493-1510.

L. De Simon

[1964] Un'applicazione della teoria degli integrali singolari allo studio delle equazioni differenziali lineari astratte del primo ordine, *Rend. Sem. Mat. Univ. Padova* **34** (1964) 205-223.

G. Dore

[2000] Maximal regularity in $L^p$ spaces for an abstract Cauchy problem, *Advances Diff. Equations* **5** (2000) 293-322.

G. Dore and A. Venni

[1987] On the closedness of the sum of two closed operators, *Math. Zeitschrift* **196** (1987) 189-201.

N. Dunford and J. T. Schwartz

[1958] *Linear Operators,* part I, Interscience, New York, 1958.

A. I. Egorov

[1967] Necessary optimality conditions for distributed parameter systems, *SIAM J. Control* **5** (1967) 352-408.

Yu. B. Egorov

[1962] Certain problems in the theory of optimal control, *Dokl. Akad. Nauk SSSR* **145** (1962) 122-125.

[1963:1] Optimal control in Banach spaces *Dokl. Akad. Nauk SSSR* **150** (1963) 241-244.

[1963:2] Certain problems in the theory of optimal control, *Z. Vycisl. Mat. Fiz.* **5** (1963) 887-904.

[1964] Some necessary conditions for optimality in Banach spaces, *Mat. Sbornik* **64** (1964) 79-101.

[1999] Necessary optimality conditions for control problems, *Russ. J. Math. Phys.* **6** (1999) 174-193.

K. J. Falconer

[1990] *The geometry of fractal sets,* Cambridge University Press, Cambridge, 1990.

H. O. Fattorini

[1964] Time-optimal control of solutions of operational differential equations, *SIAM J. Control* **2** (1964) 54-59.

[1966:1] Control in finite time of differential equations in Banach space, *Comm. Pure Appl. Math.* **19** (1966) 17-34.

[1966:2] Some remarks on complete controllability, *SIAM J. Control* **4** (1966) 686-694.

[1966/7] On Jordan operators and rigidity of linear control systems, *Rev. Un. Mat. Argentina* **23** (1966/7) 67-75.

[1969] Control with bounded inputs, Springer Lecture Notes in Economics and Mathematical Systems, vol. 14 (1969) 92-100.

[1974/5] The time optimal control problem in Banach spaces, *Appl. Math. & Optimization* **1** (1974/5) 163-188.

[1983] *The Cauchy Problem,* Cambridge University Press, Cambridge, 1983.

[1987] Some remarks on convergence of suboptimal controls, A. V. Balakrishnan 60th. Anniversary Volume, Software Optimization (1987) 359-363.

[1995] The maximum principle for linear infinite dimensional control systems with state constraints, *Discrete & Continuous Dynamical Systems* **1** (1995) 77-101.

[1997] Robustness and convergence of suboptimal controls in distributed parameter systems, *Proc. Royal Soc. Edinburgh* **127A** (1997) 1153-1179.

[1999:1] Some remarks on the time optimal control problem in infinite dimension, *Calculus of Variations and Optimal Control,* Chapman & Hall/CRC Research Notes in Mathematics Series Vol. 411, CRC Press, Boca Raton (1999) 77-96.

[1999:2] *Infinite Dimensional Optimization and Control Theory,* Cambridge University Press, Cambridge, 1999.

[2000] The maximum principle in infinite dimension, *Discrete & Continuous Dynamical Systems* **6** (2000) 557-574.

[2001:1] A survey of the time optimal problem and the norm optimal problem in infinite dimension, *Cubo Mat. Educacional* **3** (2001) 147-169.

[2001:2] Time optimality and the maximum principle in infinite dimension, *Optimization* **50** (2001) 361-385.

[2001:3] Existence of singular extremals and singular functionals in reachable spaces, *Jour. Evolution Equations* **1** (2001) 325-347.

[2001:4] The maximum principle for control systems described by linear parabolic equations, *Jour. Math. Anal. Appl.* **259** (2001) 630-651.

[2002:1] Optimal control of diffusions, *Applied Math. & Optimization* **46** (2002) 207-230.

[2002:2] Time and norm optimal controls for linear parabolic equations: necessary and sufficient conditions, *Int. Series Numerical Math.* vol. 143, Birkhäuser, Basel (2002), 151-168.

[2004] Vanishing of the costate in Pontryagin's maximum principle and singular time optimal controls, *Jour. Evolution Equations* **4** (2004) 99-123.

[T.A.:1] (to appear) Sufficiency of the maximum principle for time optimality, *Cubo Mat. Educacional.*

[T.A.:2] (to appear) Smoothness of the costate and the target in the time and norm optimal problems, *Optimization.*

H. O. Fattorini and H. Frankowska

[1990:1] Explicit convergence estimates for suboptimal controls I, *Problems in Control & Information Theory* **19** (1990) 3-29.

[1990:2] Explicit convergence estimates for suboptimal controls II, *Problems in Control & Information Theory* **19** (1990) 69-93.

A. Favini

[2003] The regulator problem for a singular control system, Lecture Notes in Pure and Applied Mathematics vol. 234, Dekker, New York (2003) 191-201.

H. Frankowska

[1993] Set-valued approach to the Hamilton-Jacobi equations, Progress in Systems and Control Theory, vol. 16, Birkhäuser (1993) 105-118.

[2002] Value function in optimal control Parts 1, 2, ICTP Lect. Notes, VIII, Abdus Salam Int. Cent. Theoret. Phys., Trieste (2002) 516-653.

H. Frankowska and Cz. Olech

[1982:1] $R$-convexity of the integral of set-valued functions, *Amer. Jour. Math.* (1982) 117-129.

[1982:2] Boundary solutions of differential inclusions, *Jour. Differential Equations* **44** (1982) 243-260.

H. Frankowska and S. Plaskacz

[2000] Semicontinuous solutions of Hamilton-Jacobi-Bellman equations with degenerate state constraints, *J. Math. Anal. Appl.* 251 (2000) 818-838.

A. Friedman

[1964] Optimal control for hereditary processes, *J. Math. Anal. Appl.* **15** (1964) 396-414.

[1967] Optimal control in Banach spaces, *J. Math. Anal. Appl.* **19** (1967) 35-55.

[1968] Optimal control in Banach spaces with fixed end-points, *J. Math. Anal. Appl.* **24** (1968) 161-181.

L. I. Galchuk

[1968] Optimal control of systems described by parabolic equations, *Vest. Mosk. Univ. Mat. Mekh.* **3** (1968) 21-23. English translation: *SIAM J. Control* **7** (1969) 546-558.

L. Gårding

[1953] On the asymptotic distribution of the eigenvalues of elliptic differential operators, *Math. Scand.* **1** (1953) 237-255.

M. Gevrey

[1913] Sur les équations aux dérivées partielles du type parabolique, *Jour. Math. Pures Appl.* **6** vol. 9 (1913) 305-471.

[1914] Sur les équations aux dérivées partielles du type parabolique, *Jour. Math. Pures Appl.* **6** vol. 10 (1914) 105-148.

[1918] Sur la nature analytique des solutions des équations aux derivées partielles, *Annales École Normale Sup.* **35** (1918) 129-190.

D. Gilbarg and N. S. Trudinger

[1977] *Elliptic Partial Differential Equations of Second Order,* Springer, Berlin, 1977.

J. A. Goldstein

[1985] *Semigroups of Linear Operators with Applications,* Oxford University Press, Oxford, 1985.

I. S. Gradstein and I. M. Rizhyk

[1963] *Tables of integrals, sums, series and products,* Goztejizdat, Moscow, 1963.

A. Halanay

[1968] Optimal control systems for systems with time lag, *SIAM J. Control* **6** (1968) 215-234.

Q. Han and F. H. Lin

[1994] Nodal sets of solutions of parabolic equations II, *Comm. Pure Appl. Math.* **47** (1994) 1219-1238.

E. Hille and R. S. Phillips

[1957] *Functional Analysis and Semi-Groups,* Amer. Math. Soc., Providence, 1957.

L. M. Hocking

[1991] *Optimal Control: an Introduction to the Theory with Applications,* Oxford University Press, Oxford, 1991.

A. Idrissi

[2003] On the unboundedness of control operators for bilinear systems, *Quaest. Math.* **26** (2003) 105-123.

A. Ionescu Tulcea and C. Ionescu Tulcea

[1969] *Topics in the Theory of Lifting,* Springer, Heidelberg, 1969.

R. C. James

[1972] Super-reflexivity in Banach spaces, *Canadian J. Math.* **24** (1972) 896-904.

G. L. Jaratishvili

[1961] The maximum principle in the theory of optimal processes with a delay, *Dokl. Akad. Nauk SSSR* **136** (1961) 39-42.

S. Kaczmarz and H. Steinhaus

[1935] *Theorie der Orthogonalreihen,* Monografje Matematyczne VI, Warszawa, 1935.

R. E. Kalman, Y.-C. Ho and K. S. Narendra

[1963] Controllability of linear dynamical systems, *Contributions to Differential Equations* **1** (1963) 189-213.

L. V. Kantorovich and G. P. Akilov

[1959] *Functional Analysis in Normed Spaces,* Fizmatgiz, Moscow, 1959. English translation [1964] Pergamon Press, New York, 1964.

S. Kaplan

[1985] *The bidual of* $C(X), I$, North-Holland Mathematical Studies vol. 101, Amsterdam, 1985.

I. Kluwánek and G. Knowles

[1976] *Vector Measures and Control Systems,* Notas de Matemática, vol. 58, North-Holland Mathematics Studies, Vol. 20, North-Holland - Elsevier, New York, 1976.

A. N. Kolmogorov

[1931] Über die analytischen Methoden in der Wahrscheinlichkeitsrechnung, *Math. Ann.* **104** (1931) 415-458.

T. Kowalski and W. Sadkowski

[2003] Applications of integrated semigroups for control theory, *Int. J. Differ. Equ. Appl.* **7** (2003) 123-139.

W. Krabs

[1985] On time-minimal distributed control of vibrating systems governed by an abstract wave equation, *Appl. Math. & Optimization* **13** (1985) 137-149.

[1989] On time-minimal distributed control of vibrations, *Appl. Math. & Optimization* **19** (1989) 65-73.

W. Krabs and E. J. P. G. Schmidt

[1980] Time-minimal controllability of linear systems. *Mathematical methods in operations research,* Bulgar. Acad. Sci., Sofia (1981) 65-80.

N. N. Krasovski

[1968] *Theory of Controlled Motion* (Russian), Gostejizdat, Moscow, 1968.

I. Lasiecka and R. Triggiani

[2000:1] *Control Theory for Partial Differential Equations, I. Abstract Parabolic Systems,* Cambridge University Press, Cambridge, 2000.

[2000:2] *Control Theory for Partial Differential Equations, II. Abstract Hyperbolic-like Systems over a Finite Time Horizon,* Cambridge University Press, Cambridge, 2000.

[2004] Optimal control and differential Riccati equations under singular estimates for $e^{At}B$ in the absence of analyticity, *Nonlinear Syst. Aviat. Aerosp. Aeronaut. Astronaut.* **2**, Chapman & Hall/CRC, Boca Raton (2004) 270-307.

X. Li and J. Yong

[1995] *Optimal Control Theory for Infinite Dimensional Systems,* Birkhäuser, Boston, 1995.

F. H. Lin

[1990] A uniqueness theorem for parabolic equations, *Comm. Pure Appl. Math.* **43** (1990) 127-136.

[1991] Nodal sets of solutions of elliptic and parabolic equations, *Comm. Pure Appl. Math.* **44** (1991) 287-308.

J. L. Lions

[1966] Optimisation pour certaines classes d'équations d'évolution non linéaires, *Annali Mat. Pura Appl.* **82** (1966) 275-294.

[1968] *Contrôle Optimal des Systèmes Gouvernés par des Équations aux Derivées Partielles,* Dunod/Gauthier-Villars, Paris, 1968.

D. A. Lukes and D. L. Russell

[1969] Quadratic criterion for distributed systems, *SIAM J. Control* **7** (1969) 101-121.

H. Mittelman and F. Tröltzsch

[2001] Sufficient optimality in a parabolic control problem, *Appl. Optimization* **72** (2001) 305-316.

B. Mordukhovich

[T.A.] (to appear) *Variational Analysis and Generalized Differentiation,* vol I - Basic Theory, vol II - Applications, Springer, New York.

B. Mordukhovich and J.-P. Raymond

[2004] Dirichlet boundary control of hyperbolic equations in the presence of state constraints, *Appl. Math. & Optimization* **49** (2004) 145-157.

B. Mordukhovich and I. Shvartsman

[2004] Optimization and feedback control of constrained parabolic systems under uncertain perturbations, Lecture Notes in Control and Information Sciences, vol. 301, Springer, Berlin (2004) 121-132.

I. P. Natanson

[1957] *Theory of Functions of a Real Variable,* 2nd ed., Gostejizdat, Moscow, 1957. English translation [1955] Vol. I (Chapters 1-9 of 1st ed.) Ungar, New York 1955, [1960] Vol. II (Chapter 10-16, 18 of 2nd ed.), Ungar, New York 1960.

A. Pazy

[1983] *Semigroups of Linear Operators and Applications to Partial Differential Equations,* Springer, New York, 1983.

L. S. Pontryagin, V. G. Boltyanski, R. V. Gamkrelidze and E. F. Mischenko

[1961] *The Mathematical Theory of Optimal Processes,* Gostejizdat, Moscow, 1961. English translation: Wiley, New York, 1962.

A. Rösch and F. Tröltzsch

[2003] Sufficient second-order optimality conditions for a parabolic optimal control problem with pointwise control-state constraints, *SIAM J. Control Optimization* **42** (2003) 138-154.

T. Roubíček

[1997] *Relaxation in Optimization Theory and Variational Calculus*, De Gruyter, Berlin, 1997.

D. L. Russell

[1966] Optimal regulation of linear symmetric hyperbolic systems with finite dimensional controls, *SIAM J. Control* **4** (1966) 276-294.

[1967:1] Nonharmonic Fourier series in the control theory of distributed parameter systems, *J. Math. Anal. Appl.* **18** (1967) 542-560.

[1967:2] On boundary value controllability of linear symmetric hyperbolic systems, *Mathematical Theory of Control,* Academic Press (1967) 312-321.

L. Schwartz

[1959] *Étude des Sommes d'Exponentielles,* Hermann, Paris 1959.

R. Seeley

[1971] Norms and domains of complex powers $A^z$, *Amer. J. Math.* **93** (1971) 299 - 309.

A. Seierstad

[1997] Nonsmooth control problems in Banach state space, *Optimization* **41** (1997) 303-319.

O. J. Staffans

[1999] Quadratic optimal control of well-posed linear systems, *SIAM J. Control Optimization* **37** (1999) 131-164.

O. J. Staffans, M. Tucsnak and G. Weiss

[2001] Well-posed linear systems - a survey with emphasis on conservative systems, *Int. J. Appl. Math. Comput. Sci.* **11** (2001) 7 - 33.

D. W. Stroock and S. R. S. Varadhan,

[1979] *Multidimensional Diffusion Processes,* Springer, New York, 1979.

D. Tataru

[1992] Boundary value problems for first order Hamilton-Jacobi equations, *Nonlinear Analysis* **19** (1992) 1091-1110.

G. Wang and L. Wang

[2001] State-constrained optimal control in Hilbert space, *Numer. Funct. Anal. Optimization* **22** (2001) 255-276.

[2002] State constrained optimal control governed by non-well-posed parabolic differential equations, *SIAM J. Control Optimization* **40** (2002) 1517-1539.

G. Weiss

[1989] Admissibility of unbounded control operators, *SIAM J. Control Optimization* **27** (1989) 527-545.

[2003] Optimal control of systems with a unitary semigroup and with colocated control and observation, *Systems Control Lett.* **48** (2003) 329-340.

G. Weiss and H. Zwart

[1998] An example in linear quadratic optimal control, *Systems Control Lett.* **33** (1998) 339-349.

H. Wu

[1999] Some equivalent conditions for exponential stabilization of linear systems with unbounded control, *Sci. China Ser. E* **42** (1999) 252-259.

[2003] A variation-of-constants formula for a linear abstract evolution equation in Hilbert space, *ANZIAM J.* **44** (2003) 583-590.

K. Yosida

[1978] *Functional Analysis,* $5^{\text{th}}$ ed., Springer, Berlin, 1978.

Y. You

[1999] Optimal control and synthesis of nonlinear infinite dimensional systems, Lecture Notes in Pure and Applied Mathematics, vol. 218, Dekker, New York (2001) 299-336.

C. Zhao, M. Wang and P. Zhao

[2004] Optimal control problems for semilinear non-well-posed parabolic differential equations, *J. Math. Anal. Appl.* **300** (2004) 375-386.

# NOTATION AND SUBJECT INDEX


$A$, $A'$, 135, 174
$\mathcal{A}$, 296
$A_p$, $A'_p$, 135, 136, 175
$A_1$, $A'_1$, 175, 244
$A_\infty$, $A'_\infty$, 187, 245
$A^*_\infty$, 201
$A_c$, $A'_c$, 174, 245
$A_\Sigma$, $A'_\Sigma$, 189, 245
$A^*_\Sigma$, 254
$A_1'^{\bullet}$, 204
$A_c'^{\bullet}$, 257
$[A^*y^*]$, 180

$B^\infty_\rho(T)$, $B^\infty_{w,\rho}(T)$, 2 ($E = \mathbb{R}^m$), 42, 212

$C(\overline{\Omega})$, $C_0(\overline{\Omega})$, 173
$C(\overline{\Omega})^* = \Sigma(\overline{\Omega})$, $C_0(\overline{\Omega})^* = \Sigma_0(\overline{\Omega})$, 183, 184, 244
$C^{(2)}_{\beta'}(\overline{\Omega})$, 174
$C(\overline{\Omega})_{-1}(\lambda)$, $C(\overline{\Omega})_{-1}$, 256
Controls
  admissible, 1
  norm optimal, 1
    bang-bang, 6
    existence of, 10 ($E = \mathbb{R}^m$), 20, 117, 225, 273
    nonuniqueness of, 14, 15, 79, 121, 124, 228, 278
    uniqueness of, 13, 23, 24, 45, 121, 228
    strongly singular, 100, 220, 270
    time independent, 236, 241, 287
    weakly singular, 146, 151, 152
  time optimal, 1
    bang-bang, 6
    existence of, 10 ($E = \mathbb{R}^m$), 20, 117, 225, 273
    nonuniqueness of, 14, 15, 278
    uniqueness of, 13, 23, 24, 35, 228
    optimality principle, 8, 20
    strongly singular, 100
    time independent, 236, 241, 287
    weakly singular, 146, 151, 152
Control set, 17
Control space, 17
Costate, 76
  determinacy of, 235
  indeterminacy of, 235, 283
  minimal, 284
  nonvanishing of, 80
  vanishing of, 77, 87

$D(A)^*$, 52
$D(A_\infty)^*$, 206
$D(A_\Sigma)^*$, 258

$E^\odot$, 56
$E^*_{-1}(\mu)$, $E^*_{-1}$, 48
Extremal, 100
  hypersingular, 306
  regular, 100
  strongly singular, 100
  weakly singular, 100

$F(\mathcal{A})$, 296
Functional
  localized at $T$, 62, 209, 215, 260, 265
  regular, 55, 208, 260
  regular part, 61, 212, 264
  singular, 55, 208, 260
  singular part, 61, 212, 264

$H_\alpha$, $\mathcal{H}_\alpha$, 296
$\mathcal{H}$, 144, 296
$\mathcal{H}^s$, 222
Hausdorff $s$-dimensional outer measure, 222

$\overset{i}{\hookrightarrow}$, 55, 184

$\mathcal{K}$, 96, 125

$\mathcal{L}(X, Y)$, 29
$L^1(0, T; X)^*$, 244
$L^1(\Omega)_{-1}(\lambda)$, $L^1(\Omega)_{-1}$, 203
$L^\infty_w(0, T; L^\infty(\Omega))$, 176
$L^\infty_w(0, T; \Sigma(\overline{\Omega}))$, $L^\infty_w(0, T; \Sigma_0(\overline{\Omega}))$, 245
$L^\infty_w(0, T; X^*)$, 176

$\mathcal{M}$, 96
$\mathcal{M}(z)$, $\mathcal{M}^+(z)$, $\mathcal{M}^-(z)$, 268
$\mathcal{M}(t, z)$, 270
$\mathcal{M}^+(t, z)$, $\mathcal{M}^-(t, z)$, 283
$\mathcal{M}(t, \epsilon, z)$, 277
Maximum principle
  necessity for norm optimality, 2, 42, 69, 217, 220, 267
  sufficiency for norm optimality, 2, 42, 70, 217, 220, 267
  necessity for time optimality, 2, 43, 69, 217, 220, 267
  insufficiency for time optimality, 8, 73
  sufficiency for time optimality, 2, 43, 71, 217, 220, 267
  strong, 125
  weak, 125
Minimizing sequences, 117, 225, 273
  convergence of, 163, 229
  convergence of supports, 276
Multiplier, 1, 76, 125

$\mathcal{N}$, $\mathcal{N}_1(T)$, 98
$\mathcal{N}(z)$, 221
$\mathcal{N}(t, z)$, 271
$\mathcal{N}_A$, 180
$\mathcal{N}_c(\overline{\Omega})$, $\mathcal{N}_{A_\infty}$, 201
$\mathcal{N}_1(\overline{\Omega})$, $\mathcal{N}_{A_\Sigma}$, 254
Nodal set $\mathcal{N}(z)$, 221
  in dimension 1, 232
Norm optimal problem, 1

Optimality principle, 8, 20

$\mathcal{P}(a, b, \mu)$ (property), 101, 107, 110
Parabolic equations
  modeling in $L^p(\Omega)$, 134
  modeling in $C(\overline{\Omega})$, 173
  modeling in in $L^1(\Omega)$, 243
Pontryagin's maximum principle, see Maximum principle

$R^\infty(t)$, $R^\infty(T)$, 2 ($E = I\!R^m$), 22, 27, 196, 252
$R^\infty(t; e)$, $R^\infty_w(t; e)$, 34, 250
$R^\infty(T)^\star$, 22, 47, 56, 213, 264
  direct sum decomposition, 56
  $L^1$ decomposition, 62
$R^\infty_w(t)$, $R^\infty_w(T)$, 191, 195, 213
$R^\infty_w(T)^\star$, 197, 207, 213, 252, 258, 264
  $L^1$ decomposition, 209, 261
$R(\mu; A)^\bullet$, 48
$R(\lambda; A'_1)^\bullet$, 204
$R(\lambda, A'_c)^\bullet$, 256
$\mathcal{R}(T)$, 55
$\mathcal{R}_w(T)$, 208, 260
Reverse adjoint equation, 76
Reversible system, 36
Reversible semigroups, 36

$\mathbf{S}(t)$, 29
$S_p(t)$, $S'_p(t)$, 136, 175
$S_c(t)$, $S'_c(t)$, 174, 245
$S_\infty(t)$, 178
$S_\infty(t)^*$, 203
$S_\Sigma(t)$, $S'_\Sigma(t)$, 189, 245
$S_\Sigma(t)^*$, 256
$S(t)^\bullet$, 50
$S'_1(t)^\bullet$, 204
$S'_c(t)^\bullet$, 257
$\mathcal{S}(T)$, 55
$\mathcal{S}_w(T)$, 209, 260
$\Sigma(\overline{\Omega})$, $\Sigma_0(\overline{\Omega})$, 183, 184, 244
$\Sigma(\overline{\Omega})^*$, 254
  noncharacterization of, 254, 262
$\Sigma_w(\Omega) = L^\infty(\Omega)^*$, 199
State space, 17

Target condition, 1
  approximate, 168
Time optimal problem, 1
Trajectories, 20

Variation-of-constants formula, 2 ($E = I\!R^m$), 20, 178, 190, 246

$W^{p,k}(\Omega)$, $W^{p,k}_\beta(\Omega)$, $W^{p,k}_0(\Omega)$, 135

$\mathcal{Z}$, 51
$Z, Z(T)$, 22, 61
$Z_w, Z_w(T)$, 52
$Z^1(T)$, 205
$Z_c(T)$, 257